# The Global Carbon Cycle and Climate Change

# The Global Carbon Cycle and Climate Change

## Scaling Ecological Energetics from Organism to Biosphere

**David E. Reichle**

Associate Director, retired
Oak Ridge National Laboratory

ELSEVIER

Elsevier
Radarweg 29, PO Box 211, 1000 AE Amsterdam, Netherlands
The Boulevard, Langford Lane, Kidlington, Oxford OX5 1GB, United Kingdom
50 Hampshire Street, 5th Floor, Cambridge, MA 02139, United States

**Notices**
Knowledge and best practice in this field are constantly changing. As new research and experience broaden our understanding, changes in research methods, professional practices, or medical treatment may become necessary.

Practitioners and researchers must always rely on their own experience and knowledge in evaluating and using any information, methods, compounds, or experiments described herein. In using such information or methods they should be mindful of their own safety and the safety of others, including parties for whom they have a professional responsibility.

To the fullest extent of the law, neither the Publisher nor the authors, contributors, or editors, assume any liability for any injury and/or damage to persons or property as a matter of products liability, negligence or otherwise, or from any use or operation of any methods, products, instructions, or ideas contained in the material herein.

**Library of Congress Cataloging-in-Publication Data**
A catalog record for this book is available from the Library of Congress

**British Library Cataloguing-in-Publication Data**
A catalogue record for this book is available from the British Library

ISBN: 978-0-12-820244-9

For information on all Elsevier publications visit our website at
https://www.elsevier.com/books-and-journals

*Publisher:* Candice Janco
*Acquisition Editor:* Marisa LaFleur
*Editorial Project Manager:* Emerald Li
*Production Project Manager:* Prem Kumar Kaliamoorthi
*Cover Designer:* Christian Bilbow

Typeset by TNQ Technologies

# Contents

# List of figures

# List of tables

# Author Bio

**David E. Reichle** was the Associate Laboratory Director of the Oak Ridge National Laboratory for Environmental, Life, and Social Sciences, and the former Director of its Environmental Sciences Division. He was also an adjunct Professor of Ecology at the University of Tennessee. He has authored over 100 scientific articles on radionuclides in the environment and the metabolism of ecosystems, edited four books on productivity and carbon metabolism of ecosystems, and led development of several seminal government reports on greenhouse gas reduction technologies and carbon sequestration. He has served  on many scientific advisory boards for the Department of Energy, the National Science Foundation, the Environmental Protection Agency, the National Academy of Sciences, and other academic institutions and business organizations. He is a fellow of the American Association for the Advancement of Science and recipient of the Scientific Achievement Award from the International Union of Forest Research Organizations, a Distinguished Service Award from the U.S. Department of Energy, and the Muskingum University Distinguished Alumni Service Award. He also served on the national board of governors of the Nature Conservancy and as Chairman of TNC's Tennessee state chapter.

# Foreword

Bioenergetics has long been a subject of research in animal husbandry and ecological research, where it served as an organizing principle in early ecosystem research (Odum, 1959). The metabolism of ecosystems and ecological energetics are subject areas that I always found to be fascinating, and ones that were researched intensely at the Oak Ridge National Laboratory (ORNL), in Oak Ridge Tennessee. I was recruited to ORNL as a new PhD from Northwestern University in 1964 to study the behavior of radionuclides in food chains—uptake, bioaccumulation, and potential pathways leading to human exposure. My postdoctoral fellowship was sponsored by what was then the US Atomic Energy Commission (now the US Department of Energy). One of the little-known facts in the history of American science is that the AEC was the first significant sponsor of modern ecological research in the United States, this role only several decades later being assumed by the National Science Foundation (Egerton, 2017). The Manhattan Project's 1943 Clinton Laboratories, managed by the University of Chicago's Metallurgical Laboratory, was the predecessor of ORNL; Union Carbide Corporation assumed responsibilities in 1947. By the time that I had arrived, WWII was over and research had shifted to the peaceful uses of atomic energy. My job title was "biophysicist" in the Radiation Ecology Section of the Health Physics Division; the Section was later to become the internationally renowned Ecological Sciences Division at ORNL (Auerbach, 1993). Our research team's scope quickly grew from examining the fate and effects of radionuclides in food chains leading to humans to studying the natural biogeochemical cycles that governed the movement of radionuclides in the environment—all of which were ultimately regulated by the metabolism of ecosystems.

Few in the scientific community, much less in the general public, knew what ecology was when the US Atomic Energy Commission began its ecological research programs in the early 1950s (Reichle and Auerbach, 2003). These programs, which antedated major support for ecosystem research by the National Science Foundation by several decades, were the foundation for modern ecosystem research in the United States (Coleman, 2010). Since ecologists at ORNL had been researching ecosystem carbon metabolism, we became the US R&D center for forest ecology and ecosystem modeling when US participation (1964–74) began in the International Biological Program (Smith, 1968; NAS, 2019). International collaboration continued for many years thereafter, and results of research on the deciduous forest biome culminated with publication of *Dynamic Properties of Forest Ecosystems* (Reichle, 1981). This research experience was an important reason why the AEC's successor, the Energy Research and Development Administration, and later the US Department of Energy (DOE), became a leading US agency studying the global carbon cycle.

The mission of DOE and its national laboratories was to promote the safe development of all energy technologies. Both the scientific experience gained from studying the carbon metabolism of ecosystems (Reichle and Auerbach, 1972) and the development of climate models to follow global fallout from weapons testing and the concern about a "nuclear winter" from nuclear weapons deployment, the national laboratories became early leaders in climate change research. The scientific experience gained in early environmental studies of the nuclear industry came full circle in the 1980s to examine the environmental consequences of a fossil fuel−based energy economy.

Ecological energetics is the study of the metabolism of plants, animals, microbes, and ecosystems. Knowledge about the functioning of ecological systems is necessary for our understanding of the metabolism of the biosphere, essential in adressing human-induced climate change, and quite possibly critical to protecting our global environment. This book is the product of a course in ecological energetics that I offered in the early 1970s in the then Graduate Program in Ecology at the University of Tennessee. I had intended the syllabus to be the basis for a textbook in bioenergetics, but somehow never found the time to write the book. Now 45 years later in retirement, I have the time, the field of ecology has matured, and bioenergetics, while an interesting chapter in basic ecology texts of the 1950−70s, has now assumed new societal relevance. Ecological energetics is the foundation for both understanding the metabolism of the biosphere and also the basis for addressing the potential future environmental impacts of climate change.

This book is a journey in time, scale, and complexity. It will be a journey following the flux of solar energy from the sun, and carbon from the atmosphere, through the living systems on earth. It will be a journey in scale—from milligrams to gigatons, from seconds to years, from square centimeters to hectares, and from the cell to the biosphere. This journey has rules which will govern our passage—the principles of thermodynamics, biochemistry, physiology, and ecology. Let us begin.

## SUGGESTED READING

Coleman, D.C., 2010. Big Ecology: The Emergence of Ecosystem Science. Univ. Calif. Press, Berkeley-Los Angeles-London, p. 236. https://epdf.pub/big-ecology-the-emergence-of-ecosystem-science.html.

Egerton, F.N., 2017. History of Ecological Sciences, Part 59: Niches, Biomes, Ecosystems, and Systems. https://www.researchgate.net/publication/320227603_History_of_Ecological_Sciences_Part_59_Niches_Biomes_Ecosystems_and_Systems/.

Odum, E.P., 1959. Fundamentals of Ecology, 2nd Ed. W. B. Sanders Co., Philadelphia and London, p. 546.

Reichle, D.E., Auerbach, S.I., 2003. U.S. RadioecologicL Research Programs of the Atomic Energy Commission in the 1950s. ORNL/TM-2003/280. Oak Ridge National Laboratory, Oak Ridge, TN. http://www.osti.gov/bridge/.

Smith, F.E., 1968. The international biological program and the science of ecology. Proc. Nat'l Acad. Sci. USA 60 (1), 5–11. https://www.ncbi.nlm.nih.gov/pmc/articles/PMC539127/.

# Acknowledgments

My career in ecology has been stimulated by a large number of individuals: my graduate school professor, Orlando Park of Northwestern University, one of the authors of "The Great Apes," Alee, Emerson Park, Park, and Schmidt's *Principles of Animal Ecology (1949),* one of the first and perhaps best ecology texts from the "Chicago School" of ecology; Stanley Auerbach, founder of the Ecological Sciences Division (ESD) at the Oak Ridge National Laboratory (ORNL), and mentors and colleagues in ESD: Dac Crossley, Jerry Olson, George Van Dyne, Frank Harris; early leaders in ecological energetics: Howard Odum, Gene Odum, George Woodwell, Bob Whittaker, Dick Wiegert, David Gates, Jerry Franklin; European ecologists John Phillipson, Amian Macfadyen, John Satchel, Kasimierz Petrusewicz, Lech Ryszkowski, Paul Duvigneaud, and Helmut Lieth influenced me profoundly, both personally and through their seminal publications. The book has its origin in a course in ecological energetics that I offered in the Graduate Program in Ecology at the University of Tennessee, decades of research at ORNL, and has been nurtured through the encouragement and patience of my wife, Donna. Brenda Wyatt provided invaluable technical records assistance, and I am grateful to ORNL for providing access to IT library resources.

# Chapter 1

# An introduction to ecological energetics and the global carbon cycle

**1.1 Recommended Reading**     **3**

Energy is essential for life on Earth. An organism with a positive energy balance is generally a successful organism in nature. Organisms and ecosystems have, consequently, evolved as highly efficient thermodynamic systems. Bioenergetics deals with the energy requirements and the processing of energy by organisms. The term is most often used in reference to animals, but also applies to plants. Plants have evolved the unique photosynthetic process, using sunlight to split water molecules and manufacture organic carbon molecules from atmospheric $CO_2$, thus converting radiant energy into chemical energy to support their metabolic requirements. In animal systems bioenergetics encompasses the procurement of the chemical energy in food, the digestion of food, subsequent metabolism, and the eventual energy expenditures required for living and reproducing. Bioenergetics involves, therefore, many aspects of the organism's physiology, thermal relationships, and behavior, and becomes very complex and complicated to quantify. Bioenergetics has become a very sophisticated tool in animal husbandry, for it deals with the efficiency by which animal protein can be produced economically. By the 1960s, bioenergetics as applied to free-living animals had bifurcated into two fields of study, one approach emphasizing behavioral biology where the animal's activity patterns were studied in relation to its energy balance with its environment, and another physiological approach dealing with the metabolism of the free-living organism. In actuality both these approaches are necessary to understand the thermodynamics of organisms in nature (Reichle et al., 1975).

By the 1970s the growing field of ecology began to utilize bioenergetics to understand the functioning of entire ecosystems. Thus, the study of ecological, or ecosystem, energetics developed. Ecosystem energetics addresses the energy balance of the entire ecosystem and all its trophic levels. It consists of the ecosystem's metabolism—its primary productivity, trophic level exchanges,

The Global Carbon Cycle and Climate Change. https://doi.org/10.1016/B978-0-12-820244-9.00001-9
**1**

turnover and decomposition of detritus, growth, and reproduction. Since biologically utilized energy is the energy stored in carbon molecules, ecosystem metabolism necessarily deals with the carbon balance of the entire ecosystem (Lindeman, 1942; Odum, 1957; Smalley, 1960; Teal, 1962; Macfadyen, 1964; Phillipson, 1966; Woodwell and Botkin, 1970; Reichle et al., 1973). Besides plant photosynthesis and trophic level energetics, understanding the carbon metabolism of the entire ecosystem, above and below ground, includes death and decomposition to complete the ecosystem's carbon balance (net ecosystem production) with the environment. This academically intriguing subject suddenly took on tremendous societal relevance beginning in the 1980s, with the growing concern over the combustion of fossil fuels and the resulting $CO_2$ emissions to the atmosphere, leading to the greenhouse effect and global warming.

What did we know about the global cycle of carbon? And, when did we know it (Rich, 2018)? Ecosystem carbon balances for different types of ecosystems, when used with the geographic distribution of ecosystem types, or biomes, formed the basis for early global carbon balance calculations (Craig, 1957; Revelle and Suess, 1957; Bolin, 1970). As the modeling of ecosystem bioenergetics advanced, it became possible to construct dynamic global carbon models of the biosphere, which were functionally based and could, consequently, permit questions to be asked about the biosphere's response to changing atmospheric $CO_2$ levels or rising temperatures or changing land use cover or feedback loops such as oxidation of Arctic tundra, glacial melting, ocean outgassing, etc (Trabalka, 1985; Trabalka and Reichle, 1986). These questions remain very pertinent and central to the debate today on the consequences of climate change.

This text begins with an introduction to ecological energetics in Chapter 1. Chapter 2 defines energy terms, introduces the physical laws of energy, and discusses how the basic principles of thermodynamics govern biological as well as physical systems. Chapter 3 is a primer on energy relationships between organisms and the environment. Chapter 4 covers the biological energy transformations of photosynthesis and energy conversion efficiencies. In Chapter 5 the energy processing by animals, their metabolism, and energy budgets are examined. Chapter 6 examines how species adapt thermally to their environments. Chapter 7 addresses the energy exchange between plants and animals, ecological energetics, food chains, and the trophic level concept. Then in Chapter 8, the complexities of energy flow in ecosystems are covered. Subsequently, Chapter 9 examines the concept of ecosystem productivity; and then in Chapter 10 the global carbon cycle and the biosphere are reviewed. Chapter 11 examines how the anthropogenic emissions of $CO_2$ and land use change have altered the natural global carbon cycle and have influenced climate change. Ultimately, humankind will have policy decisions to make about fossil energy use to avoid the negative climatic consequences of having changed the biosphere's carbon cycle (Chapter 12).

Before continuing, I need to make several brief comments about references. Much of the ecological energetics and productivity data from the decades of the mid-1960s to the mid-1980s, particularly during the International Biological Program, are contained in the proceedings of international conferences, which are now nearly inaccessible to many. I have identified these publications and extracted pertinent information from these sources so that it can remain in the mainstream of scientific literature. Secondly, I have sought permission to reproduce select graphs and illustrations contained in benchmark publications under copyright by publishers of books and scientific journals, so that they are available to those without the privileges of access to these sources through institutional library IT agreements. And lastly, when possible, I have referenced key data whenever they were published in government-sponsored symposia and reports; and since they are in the public domain, I have provided their urls for convenient, direct IT access by the reader.

## 1.1 Recommended Reading

Bolin, B., Degens, E.T., Kempe, S., Ketner, P. (Eds.), 1979. The Global Carbon Cycle. SCOPE, 13. John Wiley and Sons, New York, p. 491. https://www.researchgate.net/publication/40170880_The_Global_Carbon_Cycle_SCOPE_Report_13/.

Odum, H.T., 1957. Trophic structure and productivity of Silver Springs, Florida. Ecol. Monogr. 25, 55−112. https://doi.org/10.2307/1948571.

Phillipson, J., 1966. Ecological energetics. St. Martin's Press, New York, p. 57. https://www.worldcat.org/title/ecological-energetics/oclc/220312888/.

Reichle, D.E., O'Neill, R.V., Harris, W.F., 1975. Principles of Energy and Material Exchange in Ecosystems. In: W. H. van Dobben and R. H. Lowe-McConnell (Ed.), Unifying Concepts in Ecology. W. Junk Pub, The Hague, pp. 27−43, 302 pp. https://link.springer.com/chapter/10.1007/978-94-010-1954-5_3/.

# Chapter 2

# The physical and chemical bases of energy

There is no better way to begin the study of ecological energetics than by starting with an understanding of the pertinent definitions and terminology of physics and physical chemistry. Learn this terminology early, become comfortable with the units of measure, know the basic concepts, and bioenergetics will come a lot easier.

## 2.1 Energy, work, and power

Energy is the capacity to do work. The unit of measure for energy is the erg, which is the work performed when a force of one dyne acts through a distance of one centimeter. The unit of force, the dyne, yields to a mass of one gram the acceleration of one centimeter per second ($cm\ s^{-1}$). Since an erg of energy is such a small quantity, a larger unit, the joule, which is equal to $10^7$ ergs, becomes a more convenient unit of measure. A unit of heat used frequently in

The Global Carbon Cycle and Climate Change. https://doi.org/10.1016/B978-0-12-820244-9.00002-0
**5**

physical chemistry is the calorie (= 4.184 J). The calorie is the heat energy required to raise the temperature of one gram of water from 14.5 °C to 15.5°C. The calorie is defined as being equal to 4.1840 absolute joules. The calorie is a relatively small unit of measure, and for most chemical and biological calculations the kilocalorie ($10^3$ calories) is used. The kilocalorie (kcal) is the unit which is typically used in discussing dietary intake and is often written as Calorie. A Calorie equals $10^3$ calories, or a kcal.

**Calories and Joules.** A calorie is the energy needed to raise the temperature of 1 g of water through 1 °C (also expressed as 4.1868 J, the unit of energy in the International System of Units). A joule is the energy expended when 1 kg is moved 1 m by a force of 1 Newton (N). Use of joules is now recommended by international convention and is the preferred standard unit to measure heat (FAO, 2003). Nutritionists and food scientists concerned with large amounts of energy generally use kiloJoules (kJ = $10^3$ J) or megaJoules (MJ = $10^6$ J). For many decades, food energy has been expressed in calories, and studies in the field of ecological energetics have traditionally used calories as the measure for energy. In order to retain consistency with research reported in the scientific literature, values used for energy in this book are in calories. The conversion factors for joules and calories are: 1 cal = 4.184 J and 1 J = 0.239 cal.

## 2.2 The different forms of energy

Energy can exist in various forms, but those of greatest importance to living organisms are mechanical, chemical, radiant, and heat energy (Table 2.1). Mechanical energy has two forms: kinetic and potential. Kinetic energy, or free energy, can be described as the "useful energy" which a body possesses by

**TABLE 2.1 Units of measure for energy in its various forms and transformations.**

| Energy | Intensity | Capacity |
|---|---|---|
| Mechanical (ergs) | Force (dynes) | Change in distance (cm) |
| Kinetic (ergs) | Velocity (cm s$^{-1}$) | $\frac{1}{2}$ Mass (g) |
| Potential (ergs) | Height x accereraltion (cm s$^{-2}$) | Mass (g) |
| Chemical (calories) | Heat of combustion (cal g$^{-1}$) | Mass (g) |
| Radiant (calories) | Radiation flux (cal cm$^{-2}$) | Surface area (cm$^2$) |
| Heat (calories) | Difference in temperature (° C) | Heat capacity (cal per °C) |
| Vol. expansion (ergs) | Pressure (dynes cm$^{-2}$) | Change in volume (cm$^3$) |
| Electrical (joules) | Difference in potential (volts) | Coulombs (amps x sec.) |
| Surface (ergs) | Surface tension (ergs cm$^{-1}$) | Change in area (cm$^2$) |

value of its motion, and is measured by the amount of work which is done in bringing that body to rest. Examples would be a moving ball or the Brownian movement of molecules. Potential energy is stored energy, which is only potentially useful until its conversion into the kinetic or free energy where it becomes available to accomplish work. Energy may be stored in a system by virtue of position, as for example, a stone above the Earth's surface, a steel spring under compression, or by virtue of chemical properties due to the arrangement of atoms and electrons within a molecule. Conversion of energy from the potential form to the kinetic form involves movement, i.e., motion.

**Chemical energy.** All organisms must work to live, and they require a source of potential energy which can be utilized in order to perform the life processes. This energy can be found in the form of the chemical energy of biomass used as food. Energy can also be in the form of the chemical energy of inorganic molecules utilized as an alternative energy source to radiant energy by chemotrophs. Assemblies of atoms in matter can be rearranged into different groups; thus, by the movement of atoms and the creation of different atomic bonds, chemical energy is liberated. The combustion (oxidation) of coal in a furnace or food by the respiratory processes in a cell releases energy which can be used to accomplish work. Both of these processes illustrate the conversion of chemical to mechanical energy. Life processes on this Earth have evolved around carbon chemistry, and most chemical energy sources are derived from organic compounds. However, as we shall see shortly, there are some notable exceptions.

**Radiant energy.** The sun, a vast incandescent sphere of gas, releases energy by the nuclear transmutation of hydrogen to helium, and it is upon this energy source that life on Earth depends. Radiant energy is the energy of electromagnetic radiation. Because electromagnetic radiation can be conceptualized as a stream of photons, radiant energy can be viewed as photon energy. Alternatively, EM radiation can be viewed as an electromagnetic wave, which carries energy in its oscillating electric and magnetic fields. These two views are completely equivalent and are reconciled to one another in quantum field theory. Solar radiation is energy in the form of electromagnetic waves involving a rhythmic exchange between potential and kinetic energy. Electromagnetic radiation can have frequencies, or wave lengths, of different energy content and interactions (e.g., absorptivities) with matter.

**Heat energy.** This is a very special form of energy resulting from the random movements of molecules, which by virtue of their motion, possess kinetic energy. Heat is evolved when all other forms of energy are transformed and work is performed. All work, including the growth and reproduction of living organisms, represent the transformation of energy and ultimately results in the production of heat. For example, when an animal during respiration releases the potential energy of glucose, approximately two-thirds of it is converted into mechanical energy to be used for work (activity and growth) and one-thirds is given off as heat.

There are instances of work where heat is absorbed (endodermic processes): the cooling unit of a refrigerator or the fixation of atmospheric nitrogen by certain bacteria are examples; but, these processes are not self-supporting energetically. Nitrogen fixation is always accompanied by the exothermic breakdown of organic substrates. Heat energy released by an exothermic process is never used with complete efficiency by the endergonic process, and so whenever work is done the trend is always toward heat production. In natural processes, changes from one form of energy to another (except to heat) are normally incomplete, because the movement, already shown to be necessary for energy conversion, involves either friction or heat production.

**Temperature** is the relative measure which is used to characterize the amount of heat present in a system. Several common systems are used, only two of which concern us: the Celsius and the Kelvin scales. The Celsius (centigrade) scale ($^\circ$C) establishes zero ($0^\circ$) as the freezing point of water and 100 degrees as the boiling point of water. The Kelvin scale is an absolute measure which establishes zero at the temperature ($-273^\circ$C) at which all molecular motion ceases. Because freezing and boiling points limit life processes, we will conventionally utilize the Celsius scale. Heat may be characterized by the properties of two phases: sensible heat and latent heat. The significance of each of these phases will shortly become apparent. Suffice at present to distinguish between the two phases as follows: sensible heat is that which can be measured by an increase in temperature of a body, for example, the warming action of sunlight irradiating a metal plate. Latent heat is the heat absorption by a body without an equivalent increase in temperature, such as the heats of freezing or vaporization of water (heat of vaporization of water $= 539$ cal at $100^\circ$C). Energy flow is expressed as the product of two factors: (1) an intensity factor (or gradient), and (2) a capacity factor (amount).

Energy, work, and heat are all expressed in the same units: calories, joules, or ergs. It should be evident that the different energies may be compared, but that no relationship exists between the capacity factors alone. For example, electric energy may be converted into heat energy, but the rise in temperature cannot be calculated from the voltage, unless the number of coulombs and the heat capacity of the system are known. It is also clear that the same quantity of work can be accomplished by a small quantity of water passing through a turbine from a great height as by a large quantity of water passing through from a short distance.

Thus, arises one of the fundamental principles of thermodynamics—the interconvertibility of energy forms as well as the "trade-offs" between the intensity and capacity factors. All forms of energy are interconvertible; when conversions do occur, they do so according to rigorous laws of exchange. These are the Laws of Thermodynamics.

## 2.3 The Laws of Thermodynamics

**The First Law of Thermodynamics.** The First Law of Thermodynamics is also known as the Law of Conservation of Energy, since it defines that the sum of all energies in an isolated system is constant. In other words, energy may be transformed from one state to another, but it can neither be created nor destroyed. The total energy in universe remains constant, but it is continuously becoming more diffuse throughout the universe.

Remember that the capacity factor for mechanical and chemical energy is mass (m). Thus, Einstein showed that if there is a change in mass, $\Delta m$,

$$\text{Energy} = \Delta mc^2, \tag{2.1}$$

where:

c is the velocity of light $(3 \times 10^{10} \text{ cm s}^{-1})$.

Therefore, 1 g of water is equivalent to $9 \times 10^{20}$ ergs of energy. It should be evident that the Law of Conservation of Energy and the Law of Conservation of Mass are essentially the same, and that no violation of thermodynamics occurs when energy is converted into mass or mass is converted into energy. However, a change in energy of a system will be brought about if the system does work, or if it absorbs or evolves heat. Thus, when a change of any kind occurs in a closed system (where the amount of matter is fixed but energy is able to enter or leave) an increase or decrease occurs in the internal energy (E) of the system itself; heat (q) is evolved or absorbed, and work (w) is done:

$$\underset{\substack{\text{the decrease in} \\ \text{internal energy} \\ \text{(cal)}}}{\Delta E} = \underset{\substack{\text{the heat (cal)} \\ \text{given off by} \\ \text{the system}}}{q} - \underset{\substack{\text{the work (-cal)} \\ \text{done by the system}}}{w} \tag{2.2}$$

**Work.** The work performed by a system is the energy transferred by the system to its surroundings. The negative value of work indicates that a positive amount of work done by the system has led to energy being lost from the system.

The First Law of Thermodynamics also encompasses the more specific relationship of constant heat sums, which is of considerable importance to biologists interested in energy transformations. It states that the total amount of heat produced, or absorbed, from a chemical reaction which takes place in stages is equal to the total amount of heat evolved, or consumed, when the reaction occurs directly in one step. The evolution of living systems has utilized biochemical mechanisms by which chemical compounds can be reduced in steps, thus enabling more efficient energy capture and utilization to occur. A good biological example is the metabolic oxidation of glucose to carbon dioxide and water:

Direct reaction (combustion)

$$C_6H_{12}O_6 + 6O_2 \rightarrow 6H_2O + 6CO_2 + 673 \text{ kcal of energy}$$

Two-stage reaction (fermentation)

**(a)** $C_6H_{12}O_6 \rightarrow 2C_2H_5OH + 2CO_2 + 18$ kcal of energy
**(b)** $2C_2H_5OH + 6O_2 \rightarrow 6H_2O + 4CO_2 + 655$ kcal of energy
**(a) + (b)** $C_6H_{12}O_6 + 6O_2 \rightarrow 6H_2O + 6CO_2 + 673$ kcal of energy

Thus, no matter which pathway a particular reaction follows, the total amount of heat evolved, or absorbed, is always the same. There will be more discussion about the biological significance of this phenomenon later. Several other energy relationships are also pertinent to bioenergetics.

**Enthalpy.** Enthalpy is the total potential energy of a system. In many cases the only work "w" done on a system results in change in calorific value of the available mass. In other words:

$$w = \Delta E + 0 = \Delta E \qquad (2.3)$$

Therefore, the heat absorbed (q) in a process, measured under conditions of constant volume, is equal to the internal energy increase. According to this equation, if no outside work is done, the energy absorbed by the system is equal to the potential internal increase. A biological example of this would be the reverse of the previous chemical reaction, or photosynthesis:

$6 CO_2 + 6H_2O + 709$ kcal $\rightarrow C_6H_{12}O_6 + 6 O_2$, {caloric difference = change in vol. of $CO_2$ versus $O_2$}

Enthalpy is defined as the heat content of a system. Bond energy is amount of energy required to break a chemical bond. The total bond energy is equivalent to the total potential energy of the system, a quantity known as enthalpy (H). The heat absorbed in a process at constant pressure is equal to the change in enthalpy, $\Delta H$. Since a change in enthalpy can occur through both a change in pressure or volume, as well as internal energy, another term is introduced to describe the heat capacity of a substance:

$$\Delta H = \Delta E + \Delta (pv) \qquad (2.4)$$

The specific heat of a substance is defined as the quantity of heat required to raise the temperature of 1 g of substance by 1° Celsius. This is an extremely important relationship for biological systems. It explains the importance of water as a "thermal buffer," since when compared to other solvents water possesses relatively high heats of vaporization and freezing.

**The Second Law of Thermodynamics.** We are all familiar with the fact that many energetic processes occur spontaneously. For example, water runs downhill; gases expand from regions of high pressure to regions of low pressure; chemical reactions proceed to equilibrium; and heat flows from warm bodies to cooler bodies. The Second Law of Thermodynamics states that processes involving transformations will not occur spontaneously, unless there is a degradation of energy from a nonrandom (ordered) form to a random (disordered) form. In natural systems, spontaneous energy transformations result in the degradation of the energy state of the system from a useful form to a dissipated and less usable form of heat. Obviously, as spontaneous processes occur in a system the system loses the ability to do work.

*Living (biological) systems have evolved to exploit these natural energy trans-formations and to utilize energy as it passes from ordered to random states.*

All systems tend to approach states of equilibrium—in thermodynamic properties; this means complete randomness or energy degradation of the system. As a measure of the extent to which this equilibrium has been reached, another thermodynamic term, entropy is introduced.

**Entropy.** Entropy is a measure of the disorder, or randomness, of a system. Organized, usable energy has low entropy, whereas disorganized entropy such as heat has high entropy. The more the molecules in a system are distributed in a disordered or random manner, the more probable is the arrangement and the greater is the entropy. The First Law of Thermodynamics recognizes the interconvertibility of all forms of energy, but it does not predict how complete the conversions will be. This applies to all energy conversions, except the transformation to heat, which is a property of molecules moving around at random. By contrast, all other forms of energy result from an ordered, nonrandom arrangement of the elementary particles of matter. Heat is the only from of energy due to disorder or random movement, and it is the most likely energy form to occur.

## 2.4 Gaia hypothesis

The Gaia Hypothesis proposed by James Lovelock (1972) suggests that living organisms on the planet interact with their surrounding inorganic environment to form a synergetic and self-regulating system that created, and now maintains, the climate and biochemical conditions that make life on Earth possible. Gaia bases this postulate on the fact that the biosphere, and the evolution or organisms, affects the stability of global temperature, salinity of seawater, and other environmental variables. For instance, even though the luminosity of the sun, the Earth's heat source, has increased about 30% since life began almost four billion years ago, the living system has reacted as a whole to maintain temperatures at a level suitable for life. Cloud formation over the open ocean is almost entirely a function of oceanic algae that emit sulfur molecules as waste metabolites which become condensation nuclei for rain. Clouds, in turn, help regulate surface temperatures.

Lovelock compared the atmospheres of Mars and Earth, and noted that the Earth's high levels of oxygen and nitrogen were abnormal and thermodynamically in disequilibrium. The 21% oxygen content of the atmosphere is an obvious consequence of living organisms, and the levels of other gases, $NH_3$

and $CH_4$, are higher than would be expected for an oxygen-rich atmosphere. Biological activity also explains why the atmosphere is not mainly $CO_2$ and why the oceans are not more saline. Gaia postulates that conditions on Earth are so unusual that they could only result from the activity of the biosphere (Lovelock and Margulis, 1974).

## 2.5 Carbon and energy

After the origin of the universe some 13−18 billion years ago with the Big Bang, a condensing sun began to collapse and increasing pressure allowed helium to "burn" to form carbon.

$$^4He + {}^4He \leftrightarrow {}^8Be \tag{2.5}$$

$$^8Be + {}^4He \rightarrow {}^{12}C \tag{2.6}$$

Hydrogen, H, and helium, He, were the original building blocks of the universe. Approximately 3.8 billion years ago when surface temperatures cooled to 100°C, water condensed out of the atmosphere to form the primitive oceans. Water vapor and carbon dioxide—both degassing from the Earth's crust—served as an early greenhouse. This primordial atmosphere kept early Earth from freezing. Without the presence today of water vapor and $CO_2$ in the atmosphere that creates a significant greenhouse effect, the Earth would be about 33°C cooler and covered by ice (Ramanathan, 1988).

**The forms of carbon.** Carbon occurs in many different materials in many different forms. In bioenergetics, one is interested in the carbon content of organic molecules. Carbon content may be categorized as:

*Total Carbon (TC)*—all the carbon in the sample, including both the inorganic and the organic carbon,

*Total Inorganic Carbon (TIC)*— often referred to as inorganic carbon (IC), carbonate, bicarbonate, and dissolved carbon dioxide ($CO_2$),

*Total Organic Carbon (TOC)*—material derived from decaying vegetation, bacterial growth, and metabolic activities of living organisms or chemicals, and

*Elemental Carbon (EC)*—charcoal, coal, and soot. Resistant to analytical digestion and extraction, EC can be a fraction of either TIC or TOC depending upon the analytical approach.

Carbon mass may be calculated from the proportional composition of carbon in the substrate if its molecule composition is known, i.e.,

$$\text{proportional mass of element} = \frac{\text{mass of element in compound}}{\text{total mass of compound}} \tag{2.5}$$

Therefore, the proportion of carbon times the weight of the compound yields the mass of carbon present. Measurement of $CO_2$, as well as $O_2$ and $CH_4$, gas concentrations are typically made using gas chromatography and thermal conductivity (for example, Haskin, 2013; Emerson).

**Measures of carbon.** In geochemistry, carbon is measured in units of mass, typically grams or kilograms ($10^3$ g). Molecular concentrations can be expressed in terms of mass: milligrams per liter (mg $L^{-1}$); or in terms of volume: microliters per liter (ul/L). These two expressions will be the same only if the density of particles is 1 g $cm^{-3}$. Atmospheric $CO_2$ concentrations are given as ppm volume. Atmospheric concentrations of carbon in the form of carbon dioxide ($CO_2$) are expressed as parts per million by volume (ppm volume) and can be compared relative to other molecular gases in the atmosphere. To convert from ppm by mass to ppm by volume, divide by the density of the molecules.

Quantities may be expressed either as mass or as moles (mols). A mole is defined as the amount of a chemical substance that contains as many representative particles, i.e., atoms, as there are atoms in 12 g of $^{12}C$, which is Avogadro's number $= 6.0221,409 \times 10^{23}$. A mole of $CO_2$ is one gram atomic weight, or the mass in grams of one mole of atoms; a mole of carbon dioxide is equivalent to [12 (C) + 16 (O) x 2] = 44g; while a mole of $^{12}C$ is thus 12g C.

**Carbon chemistry.** Organic molecules contain both carbon and hydrogen. Although many organic chemicals contain other elements, it is the carbon-hydrogen bond that defines them as organic. Organic chemistry defines life. Just as there are millions of different types of living organisms on this planet, there are also millions of different kinds of organic molecules, each with different chemical and physical properties. Carbon (C) appears in the second row of the periodic table and has four bonding electrons in its valence shell. Similar to other nonmetals, carbon needs eight electrons to satisfy its valence shell. Carbon, therefore, forms four bonds with other atoms, each bond consisting of one of carbon's electrons and one of the bonding atom's electrons. Organic chemicals get their diversity from the many different ways carbon can bond to other atoms. The simplest organic chemicals, called hydrocarbons, contain only carbon and hydrogen atoms; the simplest hydrocarbon, methane ($CH_4$), contains a single carbon atom bonded to four hydrogen atoms. To add to the complexity of organic chemistry, neighboring carbon atoms can form double and triple bonds in addition to single carbon-carbon bonds and join with other atoms, such as phosphorous and sulfur, forming complex chains and rings (Carpi, 2013; Hamilton, 2017).

Stable chemical bonds release energy as they form, and bond formation thermodynamically happens spontaneously. However, formation reactions often require energy of activation to rearrange bonds and to get reactions over activation barriers, which are usually an exothermic breaking of bonds and the formation of new ones. Chemical bonds "contain" energy, but energy must be added to yield energy. In biochemical reactions, the energy for breaking bonds comes from the formation of stronger bonds. In photosynthesis, energy from the sun breaks the $CO_2$ and $H_2O$ bonds, and a fairly strong $O_2$ bond is formed. Cellular respiration provides energy by forming the strong oxygen bonds in carbon dioxide and water, breaking the weaker bonds in carbohydrates and sugars. The greater the difference between the bond energies of the formed products ($CO_2$ and $H_2O$), and the reactants, the more energy becomes available. More energy is "available" when the weakest bonds are broken in favor of the stronger bonds being formed. For example, ATP, the coenzyme in cellular respiration, provides energy to chemical reactions in the metabolic process when it transfers phosphate moieties to more strongly bonded glucose or fructose phosphates (see Section 4.3 in Chapter 4).

## 2.6 Recommended reading

Lovelock, J.E., Margulis, L., 1974. Atmospheric homeostasis by and for the biosphere: the Gaia hypothesis. Tellus 26 (1−2), 2−10. https://doi.org/10.3402/tellusa.v26i1-2.9731.

Morowitz, H.J., 1970. Entrophy for Biologists. Academic Press, 195 pp. https://books.google.com/books?hl=en&lr=&id=JSrLBAAAQBAJ&oi=fnd&pg=PP1&ots=QaAM4BSjNz&sig=jFAYu5hagX6HTST-xSCq1094Mbk#v=onepage&q&f=false.

Wikipedia, 2019. Laws of Thermodynamics. https://en.wikipedia.org/w/index.php?title=Laws_of_thermodynamics&oldid=888134088/.

# Chapter 3

# Energy relationships between organisms and their environment

The Earth's environment is very complex. The climate component is one of the most important factors determining the energy balance of free-living organisms, and climate is a complex of variables which determine energy gains and

**The Global Carbon Cycle and Climate Change. https://doi.org/10.1016/B978-0-12-820244-9.00003-2**

losses from organisms. Delicate balance between life and the physical environment is delimited by climatic constraints and mediated by the evolutionary adaptations by plants and animals to the basic physical energy properties that climate imposes upon them.

## 3.1 Energy balance

The climate surrounding us is a ubiquitous milieu made up of radiation and fluid. The fluid may be either liquid or gas, but all plants and animals are immersed in it. Climate is the result of the fluid properties on the radiation balance of the system—actually the average of values of meteorological factors. These factors affect the flow of energy between an organism and its environment. Energy is the ability to do work, and work is force acting through distance. All life processes involve work, and the expenditure of energy, for without energy no living process is viable. The environment, through climate, transfers energy to and from all living systems.

**Energy flow**. The flow of energy is by one or more of the following processes:

- radiation (electromagnetic flux),
- convection (or conduction) of heat, or
- mass transport (of water or other fluids)

Radiation flux is the total amount of radiant energy reaching a unit surface area per unit of time, expressed in units of $cal\ cm^{-2}\ min^{-1}$. Latent heat is that gained or lost by a substance or system without an accompanying rise in temperature (the freezing of water into ice at $0°C$ is an example). Sensible heat is that measurable as a temperature gain or loss in the substance or system. The various energy forms have their means of transfer which affect their respective balances. Fig. 3.1 shows the average values for various components of energy for the Earth's Northern Hemisphere.

These processes differentially affect organisms in different habitats. Soil organisms buried in the soil are not exposed to radiation, are nearly without convection, and may have energy exchange through conduction of heat and mass transport of water vapor and other gases. Aquatic organisms immersed in the waters of a lake or stream may have limited radiation, but strong convection and conduction, no transport of water vapor, but mass transport of carbon dioxide and oxygen. Surface dwelling organisms have energy transferred by all possible processes and are, therefore, subjected to the greatest extremes of energy exchange.

## 3.2 Functional interrelationships affecting leaf temperature

**Wind**. Wind speeds of 1 mph ($50\ cm\ s^{-1}$) are the most effective in their relative effect on leaf temperature. Changes of $1°C$ can be brought about by variations of only $5\ cm\ s^{-1}$ in wind speed. Seldom are conditions in nature without air

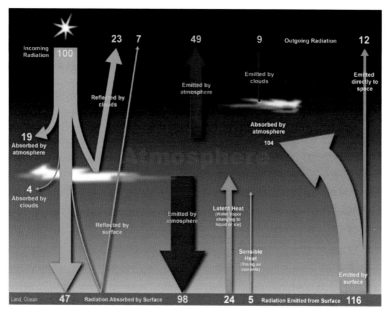

**FIGURE 3.1**   Energy exchange of the Earth and atmosphere for the northern hemisphere (100 units = 0.485 cal cm$^{-2}$ min$^{-1}$) based upon a solar constant value of 1.94 cal cm$^{-2}$ min$^{-1}$ *Source: NOAA, 2019.*

movement of at least 25 cm s$^{-1}$ and this amount of air is significant as it affects leaf temperature. Low wind speeds also significantly affect the rates of water loss. An increase in wind speed can have a strong influence on water loss if the air is relatively dry, while only a moderate effect if the air is moist.

**Temperature**. The temperature of the leaf affects the vapor density of the leaf and, hence, the internal diffusion resistance of the leaf. Transpiration rate and relative humidity also affect the internal diffusion resistance. Significant leaf temperature differences can occur at high radiation intensities, and very striking changes in transpiration may occur. Agricultural crops have leaves with low internal resistances, since they must take in $CO_2$ easily for high productivity. Xeric desert plants typically have high internal resistances to minimize water losses. In general, as temperature increases, transpiration rates will increase unless there are some plant mechanisms, such as leaf shape or waxy cuticle, controlling moisture loss.

**Physiological consequences**. On a sunny day, net photosynthesis in northern latitudes will reach a maximum at 0830 and again a 1600 in the afternoon, with a midday depression which is rather prolonged. If a wind cools off the leaves from near 35°C to as low as 20°C, the noontime depression will nearly disappear. This applies to sunlit leaves; shade leaves may oftentimes photosynthesize more efficiently during the day even though irradiance levels are lower, if temperature does not become a limiting factor.

## 3.3 Solar

Solar radiation is the source of energy which drives the life processes—both the radiant energy which is converted to chemical energy of biomass through photosynthesis, as well as the absorbed and transmitted thermal radiation which determine the temperature regimes of living organisms. The solar constant—the energy flux due to radiation outside the atmosphere on a surface perpendicular to the sun's rays—is approximately 2.0 cal cm$^{-2}$ min$^{-1}$. Radiation flux density (insolation) is the rate of flow of radiation energy through a unit area of material. In most situations the reference material is air and the metric notation for measurement is the gram calorie (252 g cal $=$ 1 Btu and the area unit is the cm$^2$). A convenient notation of insolation is the langley $=$ 1 cal cm$^{-2}$ (41,840 Jm$^{-2}$), and the rate is either in sec$^{-1}$ or min$^{-1}$.

The total solar radiation penetrating the Earth's atmosphere and arriving at the Earth's surface is generally constant with time, although substantial increases may occur in very shortwave length and high frequency ultraviolet and X-ray regions of the solar spectrum due to solar flares. Generally, X-rays are filtered out high in the stratosphere. The solar insolation striking the Earth is depleted in passing through the atmosphere by scattering and absorption by dust, $CO_2$, $O_2$, and $H_2O$ vapor. The daily totals of undepleted solar insolation at any position on the Earth's surface are dependent upon the time of year and latitude. If the axis of rotation of the Earth were perpendicular to the plane of the ecliptic, the pattern of sunlight at a particular point on the Earth's surface would be the same every day of the year. It is the tilting of the Earth's axis that is responsible for the seasons. In summer in the Northern Hemisphere it is tipped toward the sun. In winter the Northern Hemisphere is tipped away and the sun's rays strike less directly. Less solar energy per unit area also is received in winter, because the sun's rays must pass through a greater mass of air before reaching the ground. The pattern of radiation received regionally over the Earth's surface daily and annually is shown in Fig. 3.2.

**Incident radiation**. The solar energy reaching the Earth's surface varies over the year, from an average of less than 0.8 kWh m-$^2$ day$^{-1}$ during winter in northern Europe and Alaska to more than 4 kWh m$^{-2}$ day$^{-1}$ during summer. The seasonal difference decreases closer to the equator. The radiative flux varies with geographical location and is the highest in regions closest to the equator, decreasing toward the poles. The average annual global radiation impinging on a horizontal surface is $\sim$1000 kWh m$^{-2}$ in Central Europe, Central Asia, and Canada reach approx. 1700 kWh m$^{-2}$ in the Mediterranean, rising to $\sim$2200 kWh m$^{-2}$ in most equatorial regions and in desert areas of Africa, Asia, Australia, and the Americas. The average undepleted insolation over a sunlit hemisphere is approximately 0.973 cal cm$^{-2}$ min$^{-1}$. The long-term average value of extraterrestrial insolation over the globe for an entire year is 0.485 cal$^{-2}$ min$^{-1}$.

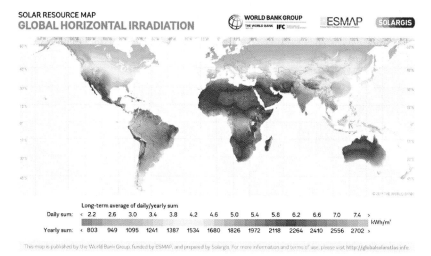

**FIGURE 3.2**   Global map of Global horizontal radiation on the earth's surface, $kWm^{-2}$ *Image credit: SOLARGIS. Source: World Bank, 2017.*

**Transmissivity**. A substantial class of natural materials and structural components of ecosystems transmits a portion of the incident radiation. Therefore, we must recognize transmissivity, which is defined as the fraction of radiation incident upon an object that is passed through that object (Fig. 3.3). Transmissivity in an ecological context becomes most important when considering ecosystems and the amount of incident radiation intercepted by the upper canopy vegetation and that transmitted to lower layers of photosynthetically

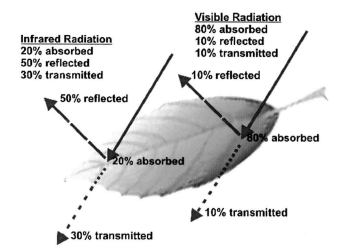

**FIGURE 3.3**   Radiation exchange for a leaf. *Source: Image credit: Shahidan, Salleh and Mustafa, 2007, after Brown and Gillespie, 1995. Courtesy, John Wiley & Sons.*

**TABLE 3.1** Transmission (langleys min$^{-1}$) of direct solar radiation through a canopy of red pine plantation (after Reifsnyder and Lull, 1965).

|  | Clear day | | Cloudy day | |
|---|---|---|---|---|
|  | Long-wave | Short-wave | Long-wave | Short-wave |
| Above canopy | 0.72 | 1.16 | 0.71 | 0.51 |
| Below canopy | 0.66 | 0.16 | 0.67 | 0.14 |
| Below/above (%) | 92 | 14 | 94 | 27 |

active leaves of other plants beneath the canopy. Canopy vegetation strongly reduces short-wave radiation but depletes long-wave radiation very little (Table 3.1). The reduction of short-wave radiation by 73%−86% is perhaps the greatest effect of the forest on any climatic factor. Similar depletion of incident radiation occurs in water bodies due to suspended particles and water molecules themselves, resulting in the photosynthetic zone being limited to the upper meters.

Solar radiation is tracked with a pyrheliometer, an instrument for measurement of direct beam solar irradiance. Sunlight enters the instrument through a window and is directed onto a thermopile which converts heat to an electrical signal that can be recorded. The signal voltage is converted via a formula to measure watts per square meter. It is used with solar tracking systems to keep the instruments aimed at the sun. Solar radiation is measured with actinometers, chemical systems, or physical devices which determine the number of photons in a beam integrally or per unit time. This name is commonly applied to devices used in the ultraviolet and visible wavelength ranges. Solar radiation is measured with a pyranometer, a type of actinometer used for measuring solar irradiance on a planar surface, and it is designed to measure the solar radiation flux density (W m$^{-2}$) from the hemisphere above within a wavelength range 0.3 μm−3 μm. A net radiometer is a type of actinometer used to measure net radiation at the Earth's surface for meteorological applications. The name net radiometer reflects the fact that it measures the difference between downward/ incoming and upward/outgoing radiation from the Earth. It is most commonly used in ecophysiological and energetics studies.

## 3.4 Thermal energy

An organism is coupled to the energy exchange processes by certain specific properties of its own. Consider the radiation incident upon an organism which may be sunlight, skylight, or long-wave thermal radiation from surrounding surfaces. If an organism's surface is highly reflective like a mirror, then it would be relatively uncoupled from the incident flux of radiation due to reflection and the incident radiation would not affect the organism's temperature. If on the other hand, the organism resembles a black body, it absorbs all of the incident radiation and the organism is highly coupled to its environment and its temperature is strongly affected by the incident radiation stream.

Measurements of temperature in environmental studies use one of several measurement systems, depending upon the nature of the temperature phenomenon being monitored. Electrical resistance, and its value, is related to temperature, and this relationship is used in resistance thermometers to measure temperature. Resistance thermometers have a much greater resolution than the liquid-in-glass types, and can measure changes down to fractions of a degree.

Since all objects emit infrared radiation with an intensity approximately proportional to their temperature, temperature can be measured with instruments called radiation thermometers, consisting of a series of optics that focus infrared light onto a special electronic detector. The detector is normally a semiconductor such as silicon, which produces an electric current proportional to the intensity of the infrared radiation. The device calculates the temperature electronically. A key advantage of radiation thermometers is their capability of measuring an object's temperature from a distance.

**Absorptivity**. The coupling factor for thermal energy is the absorptivity of the organism's surface which can be very complicated and is determined by the composition, texture, pigmentation, and geometry of its surface, as well as by the spectral quality and geometry of the incident radiation. A plant leaf has high absorptivity for ultraviolet and visible radiation, a strikingly low absorptivity for near-red radiation, and a high absorptivity for infrared radiation. In fact, there are usually several absorptivity factors simultaneously affecting the thermal balance of an organism at any time (Fig. 3.3). Mean absorptivity, $a$, over the radiant energy spectrum for direct sunlight, $S$, at sea level ($1.34$ cal cm$^{-2}$ min$^{-1}$) is given by:

$$a = \int S_{\text{absorbed}} / S_{\text{incident}} \qquad (3.1)$$

Mean absorptivity is the value of the area under the incident radiation curve divided by the area under the energy absorbed curve. The mean spectral absorptivity for many plant leaves to direct sunlight is between 0.45 and 0.65, but there are exceptions outside this range. The needles of conifers tend to be darker than broadleaved trees and, hence, have absorptivities as high as 0.80. The mean spectral absorptivity for leaves for cloud light is about 15% higher than for direct sunlight.

All bodies that are not at a temperature of absolute zero emit radiation due to thermal excitation of their atoms and molecules. The term thermal radiation refers to the emission of heat, light, or electromagnetic radiation resulting from the temperature of that object. Thermal radiation ranges in wavelength from the longest infrared to the shortest ultraviolet waves. The total emissive power of an object at any given temperature is the rate at which it emits energy of all wavelengths and in all directions per unit area of the radiating surface. According to the Stefan-Boltzman Law, the total radiation emitted ($E$) by a perfect radiator is proportional to the fourth power of the absolute temperature ($^\circ$Kelvin). Thus,

$$E = \sigma\left(T^4 - T_0^4\right) \qquad (3.2)$$

where:

σ is the Stefan-Boltzman constant ($1.32 \times 10^{-12}$ cal cm$^{-2}$ ($^\circ$K$^4$) sec$^{-1}$,
T is the absolute temperature of the object ($^\circ$K), and
T$_0$ is the absolute temperature of the surroundings [at an environmental temperature of 20$^\circ$C (293$^\circ$ K), E is calculated to be $0.73 \times 10^{-3}$ cal cm$^{-2}$ sec$^{-1}$].

**Emissivity.** The hemispherical emissivity of a surface, ε, is defined as https://en.wikipedia.org/wiki/Emissivity - cite_note-ISO_9288-1989-8 defined as the ratio of the radiation emitted by a surface to the radiation emitted by a complete radiator (black body).

$$\varepsilon = M_e / M_e^\circ \qquad (3.3)$$

where:

$M_e$ is the radiant exitance of that surface, and
$M_e^\circ$ is the radiant exitance of a black body at the same temperature as that surface.

For a "gray body," assuming ε is constant with wavelength, the radiation intensity also can be expressed as:

$$\varepsilon = \sigma\left(T^4 - T_0^4\right) / E \qquad (3.4)$$

The emissivity for some representative bodies typical of the natural environment is shown in Table 3.2. Except for luminescent materials, the emissivity cannot exceed unity.

In natural environments, bodies not only emit radiation but they also receive thermal radiation from surrounding surfaces. The net transfer of energy by an object depends upon the temperature of the object and its surroundings which influence the relative rates of emission and absorption. The absorptivity of a surface (ratio of absorbed to incident radiation) and its emissivity are equal. It must, therefore, be apparent that good radiators are good absorbers. The reflectivity (ratio of reflected to incident radiation) of an object is equal to one minus the absorptivity. Therefore, good radiators and good absorbers are poor reflectors. Most surfaces exhibit selective emission, absorption, and reflection. The term albedo usually refers to the reflected portion of a specified spectral band, while reflectivity is usually reserved for reflection of a monochromatic band. Albedo is measured on a scale from 0, corresponding to a black body that absorbs all incident radiation, to 1, corresponding to a body

**TABLE 3.2** Total emissivity, ε, all wavelengths and short-wave absorptivity of common bodies occurring in the natural environment (Handbook of Chemistry and Physics).

| Object | Long-wave emissivity | Short-wave absorptivity |
|---|---|---|
| Black body | 1.00 | 1.00 |
| Dry sand | .90 | .82 |
| Wet soil | .95 | .91 |
| Forest | .90 | .86 |
| Hay field | .90 | .68 |
| Snow | .82 | .13 |
| Aluminum | .05 | .15 |

that reflects all incident radiation. Since most of the solar energy absorbed by matter is converted into thermal energy, the albedo of a surface (whether it be leaf, animal, or ecosystem) is an important parameter determining the energy balance of those objects in nature. Typical values for albedo for various objects in the environment are given in Table 3.3.

**Boundary layer.** Organisms are coupled to the thermal environment of their habitat across a boundary layer of adhering air surrounding the surface of the organism (Fig. 3.4); or for an aquatic organism, the boundary layer would be water. Let us use a terrestrial habitat as a model, The adhering cushion of air acts as a buffer zone between the organism's epithelium and the either warmer or cooler air of its environment. If this cushion of air were a perfect insulator, the organism would be uncoupled from its environment, but then its vital physiological processes such as gas exchange would be impossible. If air were a perfect conductor, the organism would be closely coupled to its environment but subject to thermal shock at every environmental fluctuation. In fact, air is neither an excellent conductor nor insulator for energy; and organisms both benefit from the insulative properties of air and adapt to its conductive properties.

The depth of the boundary layer about an organism depends upon the

- size, shape, and orientation of the surface,
- wind (or current) speed across the surface, and
- temperature gradient across the layer.

The coupling coefficient for the flow of energy along the temperature gradient between the organism and the surrounding air (or water) is called the convection coefficient.

**TABLE 3.3** Typical albedo values for environmental surfaces on earth. *Sources: Eugster et al., 2000; Hollinger et al., 2010; Oke, 1987; Sturman and Tapper, 1986.*

| Surface | Albedo |
|---|---|
| Soils | |
| Wet soil | 0.05 |
| Dry soil | 0.13 |
| Light sand | 0.40 |
| Ecosystems | |
| Evergreen conifers | 0.08–0.11 |
| Deciduous conifers | 0.13–0.15 |
| Evergreen broadleaf | 0.11–0.13 |
| Deciduous broadleaf | 0.14–0.15 |
| Arctic tundra | 0.15–0.20 |
| Grassland | 0.18–0.21 |
| Savannah | 0.18–0.21 |
| Agricultural crops | 0.18–0.19 |
| Desert | 0.20–0.45 |
| Other | |
| Clean snow | 0.75–0.95 |
| Old snow | 0.40–0.70 |
| Sea ice | 0.30–0.40 |
| Clouds | 0.35–0.75 |
| Open ocean | 0.06 |
| Planet Earth | 0.29 |

**Thermal exchange**. Besides radiation, heat can be transferred by two methods: by conduction, where heat must diffuse through solid materials or stagnant fluids (Table 3.4); or by convection, where heat is carried from one point to another by actual movement of the heated material (such as air or water circulation).

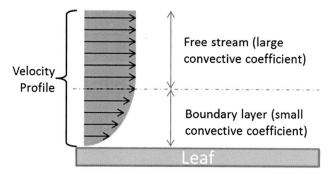

**FIGURE 3.4**  The boundary layer between a leaf and its environment. *Image credit: Dold, Khan, Kohn, and Kaufmann, 2014. Courtesy, The Biomimicry Institute.*

The amount of heat transferred by conduction can be determined using Fourier's Law:

$$q = Q/A = -k(dT/dx) \tag{3.5}$$

where:

q is the heat flux, or heat per unit area per unit time (cal cm$^{-2}$ sec$^{-1}$),
Q is the heat transferred rate, or heat per unit time (cal sec$^{-1}$),
k is the conductivity constant dependent upon the nature of the material and its temperature (cal cm$^{-2}$ sec$^{-1}$ °C),
A is the area perpendicular to the heat flow through which it is passing (cm$^2$), and dT/dx is the thermal gradient in the direction of flow (°C cm)
Table 3.4.

**TABLE 3.4** Typical thermal conductivities of environmental media, biological constituents, and other reference materials at ordinary temperatures.

| Material | Conductivity (kcal/cm$^2$sec$^1$ °C) |
|---|---|
| Air | 5.9971 × 10$^{-7}$ |
| Copper | 9.578 × 10$^{-3}$ |
| Ice | 4.777 × 10$^{-5}$ |
| Mineral oil | 3.296 × 10$^{-6}$ |
| Silver | 1.024 × 10$^{-2}$ |
| Soil (dry) | 3.587 × 10$^{-5}$ |
| Water | 1.433 × 10$^{-5}$ |
| Wood | 9.554 × 10$^{-7}$ |

The amount of heat transferred by free convection in which the temperature difference between the object and surrounding medium (gas or liquid) creates a flow is quantified using a convection coefficient ($h_c$). Since 1 kcal/$(hm^2 °C) = 1.163$ W/$(m^2 - °K)$:

$$h_c = (cal/cm^2 min \ °C) \tag{3.6}$$

for objects of various sizes and orientations. For a flat plate, the convection coefficient in the laminar range is:

$$h_c = 1.31 \times 10^{-4}(\Delta T/L)^{0.25} \tag{3.7}$$

and in the turbulent range is:

$$h_c = 0.465 \times 10^{-4}(\Delta T)^{0.33} \tag{3.8}$$

Convection coefficients for laminar flow are summarized in Table 3.5. The surface of a cold plate facing downward has the same convection coefficient as a warm plate facing upward. A cold plate facing upward has the same convection coefficient as a warm plate facing downward.

Forced convection occurs when there is a wind effect in air, or a current in water, as is the situation in most natural environments. In such cases, it is necessary to take into consideration the velocity of the medium relative to the object. With forced convection in laminar flow over a plane surface it is not important to distinguish between upward and downward facing surfaces:

$$h_c = 5.73 \times 10^{-3} v/L \tag{3.9}$$

where:
v is the velocity of the flow (cm min$^{-1}$), and
L is the characteristic dimension of the object, i.e., the diameter of a sphere or the length of a plate (cm).

**TABLE 3.5** Convection coefficients (cal cm$^{-2}$ min$^{-1}$ °C) for free convection in laminar flow. $\Delta T$ is the temperature difference in °C between the surface of the object and the surrounding air. L is the dimension of the plate in the direction of flow (after Gates, 1962, 1968a).

| | |
|---|---|
| $(\Delta T/D)^{0.25}$ Warm horizontal plate, facing upward | $7.860 \times 10^{-3}$ |
| $(\Delta T/L)^{0.25}$ warm horizontal plate, facing downward | $3.866 \times 10^{-3}$ |
| $(\Delta T/L)^{0.25}$ vertical plate | $6.042 \times 10^{-3}$ |
| $(\Delta T/L)^{0.25}$ horizontal or vertical pipe of diam. D | $6.000 \times 10^{-3}$ |

**TABLE 3.6** Rates of heat transfer (cal·cm$^{-2}$min$^{-1}$) for forced convection across a flat plate as a model for a plant leaf in the environment. Values (cal·cm$^{-2}$min$^{-1}$) are a function of the temperature differential between surface and air, dimension of the surface, and wind speed (after Raschke, 1960; Gates, 1962; Linacres, 1964). *Linacre (1964) lists heat transfer coefficient values reported for leaves for 8 different species of 0.01 − 0.05 cal·cm$^{-2}$min$^{-1}$)°C.*

| | | | Δ temperature | | |
|---|---|---|---|---|---|
| L (cm) | 1°C | 5°C | 10°C | 20°C | 30°C |
| v = 5 mph = 3.72 cm min$^{-1}$ | | | | | |
| 1 cm | 0.085 | 0.425 | 0.850 | 1.600 | 3.200 |
| 5 cm | 0.038 | 0.190 | 0.380 | 0.760 | 1.520 |
| 10 cm | 0.027 | 0.135 | 0.270 | 0.540 | 1.080 |
| v = 10 mph = 7.45 cm min$^{-1}$ | | | | | |
| 1 cm | 0.120 | 0.600 | 1.200 | 2.400 | 4.800 |
| 5 cm | 0.054 | 0.270 | 0.540 | 1.080 | 2.160 |
| 10 cm | 0.038 | 0.19 | 0.380 | 0.760 | 1.520 |
| v = 30 mph = 22.35 cm min$^{-1}$ | | | | | |
| 1 cm | 0.208 | 1.040 | 2.080 | 4.160 | 8.320 |
| 5 cm | 0.093 | 0.465 | 0.930 | 1.860 | 3.720 |
| 10 cm | 0.066 | 0.330 | 0.660 | 1.320 | 2.640 |

Typical values for heat transfer in forced convection across a flat plate as a function of temperature differential, wind speed, and characteristic dimension of the object are illustrated in Table 3.6. Since many objects in nature can be approximated by the shape of a cylinder, heat transfer values under forced convection across a cylinder can be calculated by:

$$h_c = 6.17 \times 10^{-33} \left( v^{-0.3}/D^{0.66} \right) \qquad (3.10)$$

where:

v is the velocity (cm min$^{-1}$), and
D is the diameter of the cylinder (cm).

Although it quickly becomes evident that calculating the conductive heat losses from an organism can be a formidable undertaking, it should also be clear that the organism is closely coupled to the thermal environment of its habitat by a boundary layer of surrounding adhering air or water. This boundary layer acts as a buffer between the organism and its environment. If the covering mantle of air or water were a perfect insulator, the organism would be uncoupled from its environment and the vital physiological processes of photosynthesis and respiration would be impossible. If the organism were completely uncoupled, it would be subject to thermal shock at every environmental fluctuation.

As will be discussed later, organisms both profit from the conductive/convective and insulative properties of their surrounding media, and they display a diverse array of adaptive mechanisms to regulate their heat gains and losses.

**Evapotranspiration.** The process by which water is transferred from the land to the atmosphere by evaporation from the soil and other surfaces and by transpiration from plants is termed evapotranspiration. For the coupling of moisture between the organism and the environment, the coupling factor is the diffusion resistance to moisture flux by the organism's integument and of the boundary layer. The diffusion resistance of the moisture pathway governs the rate at which moisture is lost along a moisture gradient.

The moisture diffusion resistance is determined in plants by the stomatal opening, the permeability of the cuticle, and the complexity of the boundary layer. A thick, waxy cuticle on desert plants effectively decouples them from the moisture gradient. In contrast, the integument of the salamander is very permeable to water, and the animal's skin temperature is strongly influenced by the evaporative cooling produced by the moisture gradient between skin and surrounding air.

Hence, the coupling factors which are related to the properties of the organism affect its energy balance. Let us examine the quantitative relationships between these variables to see how they interact to determine the energy balance of the whole organism.

Relationships between the amount of solar radiation and evapotranspiration permit calculation of either variable if the other is known. For example, since we know that the southwestern region of the United States receives about 650 cal cm$^{-1}$ day$^{-1}$ during the summer, and that the heat of vaporization of water is 580 cal cc$^{-1}$, we can calculate that approximately 0.56 cm day$^{-1}$ of water will be evaporated if half of the radiation is available for evapotranspiration. Thus, one can obtain crude estimates of the potential evapotranspiration in a geographic region (Budyko, 1956):

$$E/r = f(R/L) \qquad (3.11)$$

where:

E is the mean annual evaporation (cm)
r is the mean annual precipitation (cm)

R is the net annual radiation (cal cm$^{-2}$ yr$^{-1}$), and
L is the heat of vaporization of water (580 cal g$^{-1}$).

An excellent reference source for meteorological data affecting evapotranspiration is: http://www.fao.org/docrep/x0490e/x0490e07.htm/. The most widely used method of estimating evapotranspiration from meteorological data is the combined aerodynamic-energy balance equation (Penman, 1963). Under nonlimiting soil-moisture conditions, potential evapotranspiration (PE$_t$) is a constant fraction (f) of the loss from an open-water surface (E$_o$), which is calculated using standard climatological measurements (see Gates, 1968a; 1968b). The pascal (Pa) is the SI derived unit of pressure exerted by a force of magnitude one N perpendicularly upon an area of 1 sq. m. The unit of measurement called standard atmosphere (atm) is defined as 101,325 Pa. Meteorological reports typically state atmospheric pressure in millibars. A millibar $= 0.7500617$ mm of mercury. 1 kPa $= 1000$ Pa which is equal to one centibar. A bar is exactly equal to 100,000 Pa, which is slightly less than the current average atmospheric pressure on the Earth at sea level.

$$E_o = (\Phi Q_n + \mu E_a)/(\Phi + \mu) \tag{3.12}$$

where:

$\Phi$ is the slope of the saturation vapor pressure/temperature curve (Pa C$^{-1}$), Pa $=$ Pascal (kg$\cdot$m$^{-1}$ s$^{-2}$)
$E_a$ is the aerodynamic sink strength term (mm day$^{-1}$) for an open surface,
$\mu$ is the constant of the psychrometer equation (kPa $^{\circ}$C$^{-1}$), and
$Q_n$ is the net radiation balance source term. For an open water surface this can be taken as the:

$$Q_n = = Q_t(1 - \alpha) - L_n \tag{3.13}$$

where:

$Q_t$ is the total short-wave radiation from sun and sky (cal cm$^{-2}$ min$^{-1}$),
$\alpha$ is the short-wave albedo, and
$L_n$ is the net long-wave radiation loss (cal cm$^{-2}$ min$^{-1}$).
$E_a$ is the aerodynamic sink strength term (mm day-1) for an open surface:

$$E_a = 0.35(0.5 + w/161)(e_a - e_d) \tag{3.14}$$

where:

w is the measured wind speed at 2 m (km day$^{-1}$),

$e_a$ is the saturation vapor pressure (mm Hg) of the air at its mean temperature, and

$e_d$ is the mean vapor pressure (mm Hg) of the air.

The mean value of the annual ratios, $f = PE_t/E_o$ is:

$$\frac{\text{potential evapotranspiration forest, } PE_t}{\text{evaporation open water surface, } E_o} \tag{3.15}$$

and for many forests the value for f is $0.85 \pm 0.15$ (Stanhill, 1970).

## 3.5 Energy balance of a leaf

The environmental factors that are most important to the leaf energy balance involve physical relationships for the transfer of energy across the leaf surface. The relevant energy sources are direct and diffuse short-wave radiation, atmospheric long-wave (infrared) radiation, convection, radiant energy losses, transpiration, terrestrial long-wave reradiation, and short-wave reflectance (see Fig. 3.3). To illustrate all these variables for an actual leaf, we will use the example of the leaf of a tulip poplar tree (*Liriodendron tulipifera*) growing in a mesic deciduous forest by McConathy (1976), using the notation of Porter et al. (1973).

**Net environmental heat flow** to the leaf, $Q_e$, is:

$$Q_e = Q_{L,\text{solar}} + Q_{L,\text{IR}} + Q_{L,\text{conv}} - Q_p - Q_{\text{evap}} \tag{3.16}$$

where:

$Q_{L,\text{solar}}$ is the solar energy incident upon the leaf (cal cm$^{-2}$ min$^{-1}$),

$Q_{L,\text{IR}}$ is the net infrared radiation incident upon the leaf (cal cm$^{-2}$ min$^{-1}$),

$Q_{L,\text{conv}}$ is the convective heat transfer between the leaf and the air (cal cm$^{-2}$ min$^{-1}$),

$Q_p$ is the balance between the energy gained through photosynthesis or lost by respiration [a term quite small and assumed here to be zero (Gates, 1965)], and

$Q_{\text{evap}}$ is the heat transferred by transpiration (cal cm$^{-2}$ min$^{-1}$).

**Incident solar radiation**. The solar energy incident upon the leaf, $Q_{L,\text{solar}}$ (cal cm$^{-2}$ min$^{-1}$) is:

$$Q_{L,\text{solar}} = \gamma_1(S + s) + \gamma_2 r(S + s) \tag{3.17}$$

where:

S is direct solar radiation (cal cm$^{-2}$ min$^{-1}$),

s is diffuse skylight radiation (cal cm$^{-2}$ min$^{-1}$),

$\gamma_1$ is the absorbance to direct solar radiation and diffuse sunlight (0.45−0.70)

r is the reflectance of underlying surfaces to direct sunlight and diffuse skylight (dimensionless coefficient), and

$\gamma_2$ is the absorbance to underlying-surface sunlight and diffuse skylight (dimensionless coefficient).

Values for percent reflectance, r, are quite variable, being influenced by the wavelength of the incident light, tree species, leaf age, leaf position, leaf orientation, and season (Reifsnyder and Lull, 1965; Gates and Tautraporn, 1952). McConathy (1976) used an r value of 0.30 and a value of 0.60 for $\gamma_1$ for the *Liriodendron* forest that he studied.

**Net infrared radiation**. The net infrared radiation, $Q_{L,IR}$ (cal cm$^{-2}$ min$^{-1}$), is the sum of the net exchange to the leaf from the atmosphere and the underlying ground surface. Thus, $Q_{L,IR}$, the net infrared balance, is:

$$Q_{L,IR} = \gamma_3 \left[ \sigma T_a^4 \left( 0.44 + 0.08\sqrt{e} \right) + \varepsilon\sigma T_L^4 \right] - \varepsilon\sigma T_a^4 \qquad (3.18)$$

where:

$\gamma_3$ is the absorbance to infrared thermal radiation by the leaf from underlying surfaces (0.97),

$\sigma$ is the Stefan-Boltzman radiation constant ($1.03 \times 10^{-10}$ cal cm$^{-2}$ min$^{-1}$ K$^{-4}$)

$\varepsilon$ is emissivity ($= 0.97$),

$T_L$ is the leaf temperature ($^{\circ}$K),

$T_a$ is the air temperature ($^{\circ}$K) at the leaf, and

e is the vapor pressure of the air at the leaf (g m$^{-2}$ min$^{-1}$).

There are two important input terms in $Q_{L,IR}$, namely:

$IR_a = \sigma T_a^4 (0.44 + 0.8\ \sqrt{e})$ is the atmospheric infrared input (cal cm$^{-2}$ min$^{-1}$), and

$IR_g = \varepsilon\ \sigma\ T_a^4$ is the infrared input from the lower layers (cal cm$^{-2}$ min$^{-1}$).

McConathy reasoned that the leaves in lower layers of a tree would (except for the lowermost leaves) be adjacent to other leaves at a similar temperature and thus justified his formulation for $IR_g$. Although leaf absorption of near infrared is relatively low ($\gamma_3 = 0.20$), the far infrared wavelengths are efficiently absorbed. Values for plant leaves range from 0.94−0.97, and $\gamma_3 = 0.97$ was accepted. The energy reradiated from the leaf was assumed to follow the Stefan-Boltzman Law of radiation with emissivity being assumed equivalent to the absorptivity of leaves for wavelengths greater than 2.5 microns.

**Convective heat transfer**. Determination of $Q_{L,\,conv}$ (cal cm$^{-2}$ min$^{-1}$) is influenced by leaf size, leaf shape, and wind speed:

$$Q_{L,\,conv} = h_c(V/D)^{0.5}(T_L - T_a) \tag{3.19}$$

where:

> $h_c$ is the convection coefficient ($5.7 \times 10^{-3}$ for leaves),
> V is wind speed (cm$^{-2}$ sec$^{-1}$),
> D is mean width of the leaf (cm$^2$),
> $T_a$ is the air temperature ($^\circ$C), and
> $T_L$ is the leaf temperature ($^\circ$C).

**Evaporative heat loss**. Also termed latent heat flux or transpiration, the evaporative heat loss, $Q_{evap}$ (cal cm$^{-2}$ min$^{-1}$), is an important heat loss from the leaf both because of its magnitude and also because it is under control by the plant to a greater extent than other energy transfers. Transpiration is essentially a diffusion phenomenon related to the water vapor concentration gradient between the substomatal cavity and the free air beyond the boundary layer. The tendency for the water vapor in the substomatal cavity, and for the air beyond the leaf, to come to a state of equilibrium (equal water vapor concentrations) is slowed by three components of diffusion resistance:

- *Mesophyll resistance* which is the tendency for mesophyll to slow diffusion,
- *Stomatal resistance* which is the tendency for size of the stomatal aperture to slow diffusion, and
- *Boundary layer resistance* which is the slowing of diffusion by air in the immediate vicinity of the leaf.

In tulip poplar trees, mesophyll resistance is a small and constant component (Richardson et al., 1973). About 90%−95% of the water lost by a leaf is controlled by the stomata, with the remaining 5%−10% being associated with evaporation through the leaf cuticle (Salisbury and Ross, 1969). The stomatal aperture changes as a function of $CO_2$, light intensity, water stress, wind speed, and leaf and air temperatures. At 30°C each gram of water evaporated from the leaf represents a loss of 580 calories. The evaporative heat loss ($Q_{evap}$) (Gates, 1965a) is given by:

$$Q_{evap} = \left[e_{S,L} - RHe_{S,a})/R\right]L \tag{3.20}$$

where:

> $e_{S,L}$ is the saturated water vapor density of free air at temperature $T_a$ (g cm$^{-2}$),
> $e_{S,a}$ is the saturated water vapor density of air in the substomatal cavity (g cm$^{-2}$) at leaf temperature $T_L$ assuming RH to be 1.0,

L is the latent heat of evaporation of water (580 cal g$^{-1}$at 30°C),
RH is the relative humidity of the air (100% = 1), and
R is the total resistance to diffusion (min cm$^{-1}$).

For the tulip poplar leaf, $e_{S,a}$ $_{and}$ $e_{S,a}$ were calculated from the equation (Idle, 1970):

$$E = (217/T)e^{[5427(0.00366-1/T)+1.809]\times10^{-6}} \tag{3.21}$$

where:

e is the air or leaf saturated water vapor density (g cm$^{-3}$), and
T is the air or leaf temperature (K°).

By assuming the mesophyll resistance to be small, the diffusion consists of stomatal resistance, $R_s$ (sec cm$^{-1}$), plus boundary layer resistance ($R_a$). Stomatal resistance can be measured directly using a device called the diffusive resistance parameter (Kanemasu et al., 1969). Then, the formulation to calculate boundary layer resistance (Gates, 1968) is:

$$R_a = k_1\left(D^{0.35}W^{0.20}/V^{0.55}\right) \tag{3.22}$$

where:

$k_1$ is the leaf dimension coefficient (0.042) Gates, 1968,
D is the leaf dimension in the direction of the wind (cm),
W is the leaf dimension at right angles to the wind (cm), and
V is the wind speed (cm s$^{-1}$).

Wind is important in altering transpiration by cooling the leaf and in the $R_a$ equation.

The energy balance of the leaf varies as a function of crown position in the tree canopy. The $Q_{evap}$ heat flux becomes increasingly important with the leaves in the upper crown, and the magnitude of the radiation loss (negative part of the $Q_{IR}$ term) decreases with height of the leaf in the crown. The convective heat loss is relatively uniform regardless of height. The smaller upper-crown leaves lose heat efficiently through evaporation and convection, largely because the boundary layer thickness is a function of leaf size. Also, the smaller upper-crown leaves are more vertically oriented than the lower-crown leaves and thus absorb less solar radiation. Considerable heat is lost from the lower crown by reradiation due in part to the difference in small-scale meteorology in different parts of the tree crown, and also in part to increased stomatal resistance in the leaves of the lower crown.

## 3.6 Radiative energy balance of a forest

The preceding energy relationships and the quantifying equations illustrate the many physical factors influencing the energy balance of just the tree leaf, and how these factors are both measured and calculated. Now to scale up from the leaf to see how the factors contributing the energy balance of the leaf, in summation for the forest stand, and along with all the other structural properties of the forest, contribute to the radiative energy balance of the entire forest (Fig. 3.5). Measured radiative fluxes for the oak forest of Virelles-Blaimont during the International Biological Program (Galoux et al., 1981) illustrate the complexity of the energy balance for an entire forest.

Figure 3.5 is the radiation and heat budget for the forest at Virelles from late May to late October. All the fluxes were measured directly. The direct sun radiation at the extra-atmospheric level was estimated from the tables of Linacre (1969); the albedo taken from Kondratyev (1969); the fraction of terrestrial radiation reaching space was estimated from Kondratyev (1969); heat flow in precipitation was calculated from daily amounts of precipitation and air temperature before and after showers; net ecosystem production (NEP) and heterotrophic respiration were measured through growth analysis of the stand (Duvigneaud et al., 1971); and continuous measurement of $CO_2$ soil flux was determined by titration. The amount of energy for the $Q_i$ (sensible heat in the soil and vegetation layer) and NEP terms in the general energy balance were subtracted from the radiation balance before partitioning latent and turbulent heat fluxes.

**Net radiation**. Net radiation is the energy term for all transfer processes that organize and maintain the ecosystem: convection of latent heat and sensible heat; conduction through the soil; the water, the plant, and air masses; and endergonic photoreaction of chlorophyllous plants. The energy budget equation is expressed as (Galoux, etal., 1981):

$$Q + Q_L + Q_i + K + V + a_pG = 0, \tag{3.23}$$

where: Q is the radiation net balance and $Q_L$ is the advective heat, and the sensible heat flux in all solid, liquid, gaseous masses of soil, interior air masses, biomasses, precipitation water (also with possible fusion or solidification latent heat). $Q_i$ is expressed by:

$$Q_i = \left[ \int_o^z pC_p(\delta T/\delta t)\delta z \right] \tag{3.24}$$

where;

    p is the density
    $C_p$ is the heat capacity
    T is the temperature,
    Z is the height
    V is [$Lk_v(\delta q/\delta z)$], the latent heat of vaporization/condensation,
    q is the specific humidity of air,

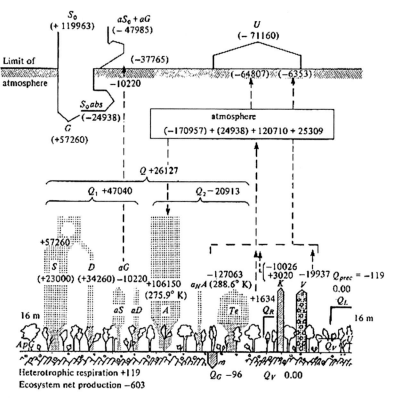

**FIGURE 3.5**  The oak forest of Virelles-Blaimont energy balance from 25 May to 24 October, 1967 (cal cm$^{-2}$). $S_o$, extraatmosphere solar radiation on a horizontal surface (short waves); $aS_o$, extraatmospheric solar radiation reflected by Earth-atmosphere system; $S_o abs$, solar radiation absorbed by atmosphere; $S$, direct solar radiation on a horizontal surface; $U$, extraatmospheric upward radiation (long waves); $D$, diffuse scattered radiation on a horizontal surface (short waves); $G$, global radiation on a horizontal surface $(S + D)$ (short waves); $T_e$, terrestrial radiation (long waves); $A$, atmospheric radiation (long waves); $aS$, reflected solar radiation; $aD$, reflected diffuse radiation, $aG$, reflected global radiation; $a_N A$, reflected atmospheric radiation; $apG$, global radiation utilized in net photosynthesis; $Q_1$, short-wave radiation balance $(G - aG)$; $Q_2$, long-wave radiation balance $(A - T_e)$; $Q$, short- and long-wave radiation balance $(G - aG + A - a_N A - T_e)$; $Q_G$, sensible heat flux in soil; $Q_V$, sensible heat flux in vegetation; $K$, sensible heat turbulent flux; $V$, latent heat in evapotranspiration; $Q_R$, latent heat in water condensation; $Q_h$, advective sensible heat; $Q_{prec}$, sensible heat flux in precipitation water. Parameters of the stand (per ha): biomass, 156 ton; net primary production (ground), 14.6 ton. Exchange aerial surfaces (ha ha$^{-1}$): foliage (2 faces) of trees, 14; bark of trees, 2; herb layer, 2; litter, 1.5; total exchange surfaces (except litter, 18 haha$^{-1}$). Figures in brackets are estimated values (metric ton = $10^6$ g). *Source: Galloux et al., 1981. Courtesy, Cambridge University Press.*

L is the latent heat of vaporization,
$k_V$ is the coefficient of turbulent transfer for vapor,
k is [$pc_p k_K (\delta\theta/\delta z)$], the sensible heat in turbulent air
$\theta$ is the potential temperature of air,

$k_K$ is the coefficient of turbulent transfer for sensible heat,
$a_pG$ is $[\lambda k_C(\delta C/\delta z)]$, the net photosynthesis,
$\lambda$ is the photochemical energy equivalent of carbon dioxide,
$k_c$ is the coefficient of turbulent transfer for carbon dioxide, and
C is the carbon dioxide concentration of the air.

Biomass surfaces are the most active interfaces for transfer of radiant energy. The rates of transfers depend essentially on the masses, the specific heat values, the conductibilities, and the surface-to-mass ratios. In a forest ecosystem, soil and subsoil masses involved in transfers and exchanges can approach 300 g cm$^{-2}$, biomass up to 10 g cm$^{-2}$ (often 2–4 g cm$^{-2}$), water flux through the system ranges from 50–300 g cm$^{-2}$ yr$^{-1}$ (often 80–150 g cm$^{-2}$ yr$^{-1}$), air flow up to 5 metric tons cm$^{-2}$ yr$^{-1}$, and with a chlorophyll mass of only 0.00024 g cm$^{-2}$ yr$^{-1}$. This gives an idea of the distribution of the radiation budget among these masses and the energetic input to the ecosystem functions of warming, convection, evaporation, drainage, and biological production. How the radiative energy balance interacts with the metabolic energy flow for the entire forest canopy will be dealt with in Chapter 8, "Energy Flow in Ecosystems."

## 3.7 Energy exchange of animals

Animals also exchange energy with their environment. But they possess a nervous system and mobility and, consequently, have been able to evolve many complex adaptations to regulate and control that exchange. Among these are:

- Body shape
- Body size
- Epidermal covering
- Sweating
- Control of metabolic rate
- Mobility
- Behavioral adaptations

If an organism is to survive, it can neither gain nor lose net energy over an extended period of time. Short-term transient conditions may exist, such as warming or cooling or drawing upon reserves of fats for metabolic needs, but over the long term the organism must maintain energy equilibrium or die. Energy gained by an organism from its environment is lost through radiation from the organism's surface by reradiation or by heat convection or by conduction or by evaporative cooling and the loss of water or often by all of the processes.

As an example, let us consider the desert iguana, *Dipsosaurus dorsalis*, which lives in an environment where temperature is a very important variable

(Fig. 3.6), using the notation of Porter et al. (1973). The energy balance of the iguana can be expressed as:

$$Q_e = Q_{L,solar} = Q_{U,IR} + Q_{L,conv} \qquad (3.25)$$

where:

$Q_e$ is the net environmental heat flow to the lizard,
$Q_{L, solar}$ is the solar energy incident on the lizard,
$Q_{L,IR}$ is the net infrared radiation exchanged between the sky or ground to the lizard, and
$Q_{L,conv}$ is the convection heat transfer between the lizard and the air.

Since these equations use mechanistic principles from engineering, meteorology, soil physics, and include as well ecological and physiological considerations, it is important to understand the manner in which the terms in the equation were determined and the manner in which they might change as a result of iguana behavior.

**Radiant energy input.** In the open desert, the categories of solar energy incident upon the lizard are direct solar energy, scattered solar energy (not illustrated in Fig. 3.6), and the energy reflected from the ground to the lizard. Total incident radiation ($Q_{L, solar}$) is thus given as:

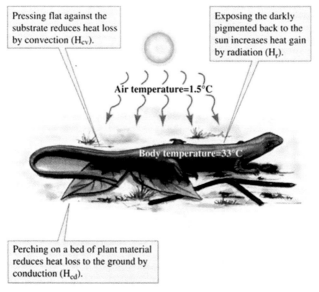

FIGURE 3.6    Energy exchange for a lizard in its natural desert environment, showing the energy flows to the desert surface and to the lizard. *Image credit: Ectotherms Wiki. Courtesy, FANDOM, CC-BY-SA. https://ectotherms.fandom.com/wiki/Leopard_Gecko_-_VK_%26_NM/*

$$Q_{L, \text{solar}} = \gamma_L \left[ A_{p,d} Q_{\text{solar}} + A_{p,r} (1 - \gamma_{so}) Q_{\text{solar}} \right] \qquad (3.26)$$

where:

$Q_{L, \text{solar}}$ is the solar energy incident upon the lizard (cal cm$^{-2}$ min$^{-1}$),
$\gamma_L$ is the maximum absorption of solar energy by the skin of the (cal cm$^{-2}$ min$^{-1}$),
$1 - \gamma_{so}$ is the percentage of the solar energy input that is diffusely reflected from the desert sand (cal cm$^{-2}$ min$^{-1}$),
$A_{p,d}$ is the projected area of the lizard receiving direct plus scattered solar energy (cm$^2$), and
$A_{p,r}$ is the projected area of the lizard for reflected solar energy (cm$^2$).

$Q_{L, \text{solar}}$ varies by the day and season. The maximum absorption by the skin, $\gamma L$, varies according to the body temperature of the lizard. It was assumed that $\gamma L = 0.80$ for all temperatures up to 38°C, and that $\gamma L = 0.60$ for all temperatures above 43°C. Also, assumed was that the value of $\gamma L$ lay along a straight line for all temperatures between 38°C and 43°C. The $A_{p,d}$ parameter was taken as equivalent to the area of the lizard's shadow when under directly-overhead light. $A_{p,r}$ was assumed to be 0.4 $A_{p,d}$.

**Infrared radiation exchange**. The net infrared radiation balance, $Q_{L,IR}$, of the lizard is the sum of the lizard-sky and the lizard-ground exchanges (Fig. 3.6):

$$Q_{L,IR} = A_L F_{L-s} \sigma \left( T_s^4 - T_{sk}^4 \right) + A_L F_{L-sky} \sigma \left( T_{sky}^4 - T_{sk}^4 \right) \qquad (3.27)$$

where:

$A_L$ is the surface area of the lizard (cm$^2$),
$F_{L-s}$ is the shape factor for radiation between lizard and ground,
$F_{L-sky}$ is the shape factor for radiation between lizard and sky,
$\sigma$ is the Stefan-Boltzman constant (0.813 $\times$ 10$^{-10}$ cal cm$^{-2}$ min$^{-1}$ K$^{-4}$),
$T_s$ is the soil temperature ( °C),
$T_{sk}$ is the lizard skin temperature ( °C), and
$T_{sky}$ is the sky temperature ( °C).

Bartlett and Gates (1967) had determined that the radiation shape factor from a lizard to the total environment (the sum of $F_{L-s} + F_{L-sky}$) to be 0.08. By assuming that half the infrared radiation from a lizard goes to the ground and half goes to the sky, Porter et al., 1973 set the same value of $F_{L-s} = F_{L-sky} = 0.40$. As is the situation with the $Q_{L,solar}$ term, the ability of the lizard to alter the $Q_{L,IR}$ term is a result of its ability to change its temperature. In this case, the skin temperature ($T_{sk}$) is the principal controlling variable.

**Convective heat transfer**. The convective heat transfer between the lizard and the air is:

$$Q_{L,conv} = H_L A_L (T_z - T_{sk}) \qquad (3.28)$$

where:

$H_L$ is the heat transfer coefficient,
$T_Z$ is the air temperature at the height of the lizard ( $^\circ$C),
$T_{sk}$ is the lizard's skin temperature ( $^\circ$C), and
$A_L$ is the surface area of the lizard (cm$^2$).

The heat transfer coefficient, $H_L$, was evaluated experimentally using a cast aluminum lizard model. The $T_z$ temperature was estimated using a micro-meteorological model, and the $T_{sk}$ temperature was measured directly. $H_L$ is the variable of wind speed over the lizard, and the lizard can change its position to alter the wind velocity over its surface.

**Energy balance**. Body temperature lizard is a variable under the control of the iguana by its behavior. A temperature model for the iguana was developed that considered the flows of heat between the core of the lizard and an outside layer (skin and fat), and between this outer layer and the environment (Fig. 3.7). The heat stored in the core of the lizard is augmented by energy produced by the metabolism of the animal and reduced by respiratory energy loss. In nature, the iguana tries to optimize an equilibration in temperatures between the core,

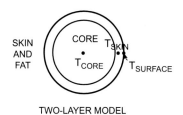

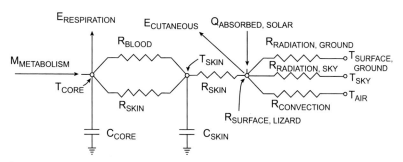

**FIGURE 3.7** Core-shell (two-layer) model for a lizard and a schematic representation of the thermal energy flows with its environment (Porter et al., 1973). *Courtesy, Springer Nature.*

skin surface, and the environment at 38.5°C, with body temperatures between 33.2°C and 41.8°C for over 95% of the time (DeWitt, 1967).

Knowing these thermal relationships, and the temperatures of the iguana's habitat, it was possible for Porter to develop an elaborate model, with the mathematical functions, in Eq. (3.26) and Eq. (3.27), for the thermal fluxes in Fig. 3.7, to predict the behavioral movements of the iguana on a daily basis throughout the year, showing activity as a function of time of day and day of the year (Fig. 3.8). The contour lines show the time-space thermal boundaries of the iguana's environment. Light areas in the figure are the temperature realms above 38°C when the iguana becomes active; the solid bars are field observations of actual iguana activity. In the early spring, the desert iguana is active in the middle of the day. However, with the heat of the summer, this pattern changes to a morning and afternoon pattern. With decreasing temperatures of fall, the animal becomes active again in the middle of the day. The earliest possible yearly emergence predicted by the model is about the first of February, and activity is possible only for a very short period of time, however, as the soil surface temperature does not reach 38°C until about the first of March. Since the desert iguana emerges by lying at the burrow entrance on the sand surface before emerging completely and becoming active, the model predicts that animals would not be active before the first of March. Similarly, after the first of November, the surface temperatures under average conditions drop below 38°C. Thus, the model predicts surface activity would be restricted to between March 1st and November 1st.

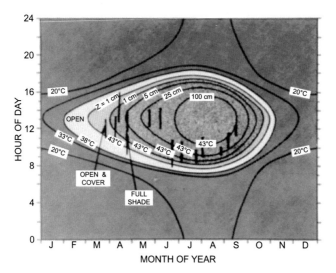

**FIGURE 3.8** Model predicted seasonal behavior patterns for the desert iguana, *Diposaurus dorsalis,* compared to behavioral observations shown as solid bars. *Source: Porter et al., 1973. Courtesy, Springer Nature.*

Clearly the thermal relationships between the desert iguana and its environment are so important to its survival that the animal has evolved mechanisms to optimize its thermal energy balance. We will continue this intriguing subject in Chapter 6 with a more detailed discussion of species adaptations to their energy environments.

## 3.8 Recommended reading

Gates, D.M., 1962. Energy Exchange in the Biosphere. Harper and Row, New York, 151 pp. https://www.worldcat.org/title/energy-exchange-in-the-biosphere/oclc/1180343.

Porter, W.B., Gates, D.M., 1969. Thermodynamic equilibria of animals with environment. Ecological Monographs 39 (3), 227–244. https://doi.org/10.2307/1948545.

Gates, D.M., 1965. Energy, plants and ecology. Ecology 46, 1–13. https://doi.org/10.2307/1935252.

Lovelock, J.E., Margulis, L., 1974. Atmospheric homeostasis by and for the biosphere: the Gaia hypothesis. Tellus Series A Stockholm: International Meteorological Institute 26 (1–2), 2–10. https://doi.org/10.3402/tellusa.v26i1-2.9731.

# Chapter 4

# Biological energy transformations by plants

Biological systems are unique in having evolved to exploit the Laws of Thermodynamics. They neither create nor destroy energy in the Earth system but exploit energy as its entropy increases, i.e., as energy passes from an ordered to unordered state. How living organisms have evolved to capture, and then utilize, energy is the marvel of the evolution of life. Solar energy flux is necessary for the existance, and maintenance, of life on Earth. Albert Szent-Györgi (1961) elegantly described the relationship between energy and life. "It is common knowledge that the ultimate source of all our energy and negative entropy is the radiation of the sun. When a photon interacts with a material particle on our globe it lifts one electron from an electron pair to a higher level. This excited state as a rule has but a short lifetime and the electron drops back within $10^{-7}$ to $10^{-8}$ seconds to the ground state giving off its excess energy in one way or another. Life has learned to catch the electron in the excited state, uncouple it from its partner and let it drop back to the ground state through its biological machinery utilizing its excess energy for life processes."

## 4.1 Solar radiation

The flow of energy reaching the surface of the earth can be summarized (Miller and Urey, 1959) as follows: for sunlight of all wavelengths, 260,000 cal cm$^{-2}$ yr$^{-1}$; wavelengths <2,500 Å, 570 cal cm$^{-2}$ yr$^{-1}$; wavelengths <2,000

The Global Carbon Cycle and Climate Change. https://doi.org/10.1016/B978-0-12-820244-9.00004-4

**43**

Å, 85 cal cm$^{-2}$ yr$^{-1}$; and wavelengths $< 1{,}500$ Å, 3.5 cal cm$^{-2}$ yr$^{-1}$. In addition, electric discharges contribute 4 cal cm$^{-2}$ yr$^{-1}$; cosmic rays, 0.0015 cal cm$^{-2}$ yr$^{-1}$; radioactivity to 2.0 km depth, 0.8 cal cm$^{-2}$ yr$^{-1}$; and volcanoes, 0.13 cal cm$^{-2}$ yr$^{-1}$. The photosynthetic systems of green plants have evolved to exploit the radiant energy in the wavelengths of 380-750 nm, with the action spectra for chlorophyll peaking at 480 and 620 nm. The major source of chemical energy for living systems is derived from the oxidation of organic, carbon-based compounds. The principal method of production of these organic compounds is the utilization of radiant energy by chlorophyll-bearing green plants to synthesize glucose from carbon dioxide and water.

The solar constant—the solar energy flux of radiation outside on a surface perpendicular to the sun's rays—is approximately 2.0 cal cm$^{-2}$ min$^{-1}$. Radiant energy for photosynthesis comes from the visible light spectrum of the sun, within wave lengths ($\lambda$) ranging from 380 to 750 nm (meters x $10^{-9}$). The wavelength distribution of solar energy is shown in Fig. 4.1. About 10% of the energy in the solar stream is in the ultraviolet region from 0.1 to 0.4 micron (micron, $\mu$, $= 10^4$ Ångstroms, Å); the remainder is divided nearly equally between visible light (0.4−0.7 micron) and infrared radiation (0.7−4.0 microns). Because radiation is a flow of energy from one place to another, it can properly be called a flux. The emission of light occurs in the form of discrete packets of waves from the sun termed photons, the energy of which is equal to the product of its frequency, v, and Planck's constant, h ($6.62 \times 10^{-34}$ J s$^{-1}$). Because the energy of electromagnetic radiation is inversely proportional to the wavelength, red light (longest of the visible wavelengths) is the lowest in energy (Table 4.1).

## 4.2 Photosynthesis

There is a direct relationship between the photosynthetic efficiency of different wavelengths and the absorption spectra of photosynthetic pigments. The action

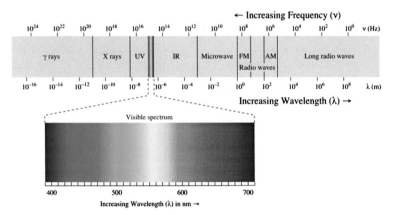

**FIGURE 4.1**   Electromagnetic wavelength distribution of radiant energy. *Image credit: Philip Ronan. Courtesy, Wikimedia Commons, 2019.*

**TABLE 4.1** The energy value of different wavelengths of solar radiation.

| Color | Wavelength, nm | Energy value in kcal mole$^{-1}$ |
|---|---|---|
| Infrared | 620–750 | 41.1150–38.1220 |
| Orange | 592–620 | 48.2960–41.1150 |
| Yellow | 578–592 | 49.4660–48.2960 |
| Green | 500–578 | 57.1830–49.4660 |
| Blue | 464–500 | 61.6190–57.1830 |
| Indigo | 444–464 | 64.3950–61.6190 |
| Violet | 400–444 | 71.4780–64.3950 |
| Ultraviolet | 100–400 | 285.9100–71.4780 |

spectra for chlorophyll *a* and *b* for many different species of crop plants averages peaks at 440 nm (blue-violet) and 620 nm (near red), although other pigments with different absorption spectra may participate; namely, the carotenoids.

Photosynthesis begins with the hydrolysis of water in the chloroplasts of green plants. This is distinguished from chemosynthesis used by certain bacterial species that use the energy-rich sulfur bonds of chemical compounds as an alternative to the radiant energy source used by green plants.

$$\text{Photosynthesis } nCO_2 + 2nH_2O \rightarrow (CH_2O) + nH_2O + nO_2$$

$$\text{Chemosynthesis } nCO_2 + 2nH_2S \rightarrow (CH_2O)n + nH_2O + 2\ nS$$

Photosynthesis occurs in structures called chloroplasts (Fig. 4.2). Chloroplasts are structures (organelles) that occur within the mesophyll cells present in the middle layer of cells in leaves of green plants. Chloroplasts capture radiant energy and transform it into chemical energy. A chloroplast has a double membrane, an outer membrane, and a convoluted folded inner membrane. It is within these inner thylakoid membranes, arranged like stacks of coins, that energy from sunlight is trapped.

Photosynthesis may well have been an evolutionary development that happened only once in ancient cyanobacteria some billion years ago. Molecular data show that Photosystem I likely evolved from the photosystems of green sulfur bacteria (Lockau and Nitschke, 1993). Green plants did not "invent" photosynthesis all over again. Rather, the algal ancestor of green plants merely coopted photosynthetic bacteria within its own cells in a mutualistic relationship (Blankenship, 2010). Both mitochondria and chloroplasts have their own genomes (a closed circle of DNA), and use protein translation machinery (ribosomes) distinct from cellular ribosomes. Both also divide through binary fission, and do not participate in exchange of membrane

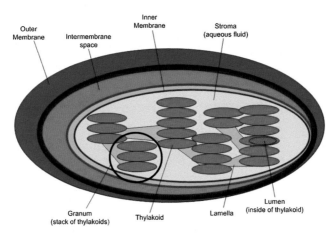

Inner
Membrane

Outer
Membrane

Intermembrane
space

Stroma
(aqueous fluid)

Granum
(stack of thylakoids)

Thylakoid

Lamella

Lumen
(inside of thylakoid)

**FIGURE 4.2** Schematic of a chloroplast from a plant cell. *Image credit: It'sJustMe. Wikimedia Commons, 2014.*

vesicles with other organelles. Closer examination of these (and other) features shows that in all cases these similarities grouped with bacteria rather than with eukaryotes, e.g., bacteria have circular DNA genomes, mitochondria and chloroplast ribosomes are more similar to bacterial ribosomes, and binary fission is a bacterial mode of replication. It is now scientific theory that mitochondria and chloroplasts are the endosymbiotic remnants of once free-living bacteria. Fascinating! It illustrates how unique, and utterly critical for life on Earth has been the evolutionary development of photosynthesis as a mechanism for capturing radiant energy from the sun.

**The Light Reaction in photosynthesis.** The light-driven reaction of photo-synthesis, first propounded by Robert Hill in 1937, is called the light reaction (Hill reaction), and is referred to as the electron transport chain. The electron transport chain of photosynthesis is initiated by absorption of light by Photo-system II ($P_{680}$, or water-plastoquinone oxidoreductase), the first protein com-plex in the light-dependent reactions of photosynthesis. It is located in the thylakoid membrane. Within the photosystem, enzymes capture photons of light to energize electrons that are then transferred through a variety of coenzymes and cofactors to reduce the chemical compound plastoquinone to plastoquinol. The energized electrons are replaced by oxidizing water to form hydrogen ions and molecular oxygen. By replenishing lost electrons with electrons from the splitting of water, photosystem II provides the electrons for all of photosynthesis to occur. The light-dependent splitting of water molecule is called photolysis.

There are two steps in the light reaction. The first is Photosystem II, which initiates the hydrolysis of water (Fig. 4.3). Within the chloroplasts, hydrogen $H^+$ ions are "pumped" across the membrane with energy from ATP and NADPH into the thylakoid space creating an ionic gradient which allows $H^+$

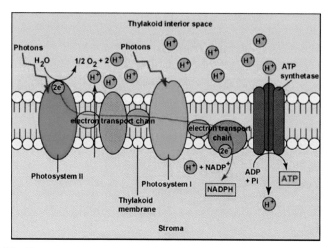

**FIGURE 4.3** Photosystem II, the photolysis of $H_2O$, and Photosystem I, producer of ATP and NADPH, both occurring in the thylakoid membrane of the chloroplast. *Source: Kaiser (2001). Courtesy, Creative Commons.*

ions to diffuse back out to the stoma (the space outside the thylakoid membrane). This allows ATP to be produced (see Fig. 4.4) and become available for the Calvin cycle. Increase in light results in an increase in $H^+$; the whole process is radiant energy dependent, since the pH gradient will not occur if there is no light.

The active hydrogen formed by photochemical hydrolysis does not appear as free hydrogen. Instead, the first compound to be reduced is TPN (triphosphopyridine nucleotide), an organic coenzyme. In this photochemical reduction of TPN, sufficient energy is available to generate an energy-rich phosphate group. In the reduction of TPN (accomplished by acceptance of two electrons and a proton in its pyridine ring), a high-energy phosphoric acid group is formed and ADP is converted to ATP.

The second step is Photosystem I, or plastocyanin-ferredoxin oxidoreductase, and is the second photosystem in the photosynthetic light reactions in the chloroplast (Fig. 4.3). Photosystem I is a membrane protein complex that uses light energy to produce the high energy carriers ATP and NADPH. NADPH, nicotinamide adenine dinucleotide phosphate hydrogen, is a product of Photosystem II and is used to help fuel the reactions that take place in Photosystem I. NADPH is an energy-carrying molecule that also provides energy to fuel the Calvin cycle in the second stage of photosynthesis (dark reaction). Now the energetic basis has been established for the chemical conversion of $CO_2$ into glucose.

An important discovery in photosynthetic mechanisms was made by Frederick Blackman in 1905 when he discovered that under certain conditions photosynthesis could not be accelerated by increasing the illumination level. This result demonstrated a rate limiting nonphotochemical reaction (the dark

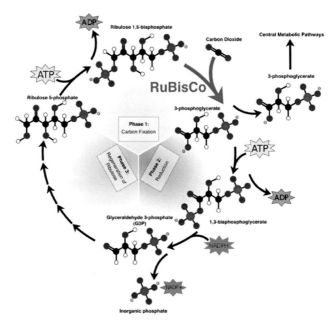

**FIGURE 4.4**   The Calvin Cycle. Atoms are: black - carbon, white - hydrogen, red - oxygen, pink - phosphorus. *Image credit: Mike Jones, Wikimedia contributors, 2019.*

reaction) which involved the actual fixation of $CO_2$. Thus, in principle, all that is required is a hydrogen donor. The light-independent reactions, or dark reactions, of photosynthesis are the chemical reactions that convert carbon dioxide into glucose. These reactions occur in the stroma, the fluid-filled area of a chloroplast outside the thylakoid membranes. These reactions take the products (ATP and NADPH) of light-dependent reactions.

**The Calvin Cycle.** The final chemical reaction which is necessary to be understood is how the chemical energy released from the hydrolysis of water and the reducing $H^+$ ions are captured and utilized to form simple carbohydrates from $CO_2$ (Fig. 4.4). The cycle of reactions in the Calvin cycle, which occur in the second phase of photosynthesis, does not require the presence of light. These biochemical reactions involve the fixation of carbon dioxide and its reduction to carbohydrate and the dissociation of water, using chemical energy stored in ATP.

The Calvin cycle starts (top of cycle in Fig. 4.4) with carbon fixation. The enzyme Rubisco catalyzes the carboxylation of a 5-carbon compound to make a 6-carbon compound that splits in half to form two 3-phosphoglycerate molecules. The enzyme phosphoglycerate kinase catalyzes phosphorylation of 3-phosphoglycerate to form 1,3-biphosphoglycerate. Next the enzyme glyceraldehyde 3-phosphate dehydrogenase catalyzes the reduction of

1,3-biphosphoglycerate by NADPH. Finally, ribulose 1,5-bisphosphate is regenerated. At the end of the regeneration, the net gain of the set of reactions is one glyceraldehyde 3-phosphate molecule per three carbon dioxide molecules.

The overall chemical equation for the Calvin cycle is:

$$3CO_2 + 6NADPH + 5H_2O + 9ATP \rightarrow \text{Glyceraldehyde-3-}$$
$$\text{phosphate} + 2H^+ + 6NADP^+ + 9ADP + 8 \text{ inorganic phosphate}$$

Six runs of the cycle are required to produce one glucose molecule. The surplus glyceraldehyde-3-phosphate that is produced by the reactions can be used to form a variety of carbohydrates, depending on the needs of the plant.

**Energy-rich molecular bonds.** Cells in living organisms require large amounts of free energy to run their biochemical processes. This energy comes from chemical substrates containing energy-rich bonds. Energy-rich molecules are formed by the oxidation of substrates which the cell obtains from the environment. This energy can then be liberated from these molecules by hydrolysis. The energy in chemical bonds is derived from the attractions between atoms or molecules. The strength of chemical bonds may vary considerably, such as the electrostatic force of attraction between oppositely charged ions or through the sharing of electrons as in covalent bonds. Phosphate and mercaptan bonds are high energy bonds in organic molecules. Carbon forms many energy-rich chemical bonds. For example, glycolysis is the metabolic pathway in the cell that takes the glucose, $C_6H_{12}O_6$, formed in photosynthesis and converts it into pyruvate, $CH_3COCOO^- + H^+$. Glycolysis is an oxygen independent metabolic pathway. The free energy released in this process is used to form the high-energy molecule, ATP. Glycolysis is a sequence of 10 enzyme-catalyzed reactions. We briefly covered the high energy molecule NADPH previously in the discussion of Photosystem I. Now we will look in more detail at the high energy molecules ADP (adenosine diphosphate) and ATP (adenosine triphosphate).

Central to the energy metabolism of all cells is the ADP-ATP cycle. The hydrolysis of ATP is highly exergonic and used to drive numerous endergonic biochemical processes of the cell. The phosphorylated coenzyme, adenosine diphosphate or ADP, reacts with energy-rich phosphate groups to form adenosine triphosphate (Fig. 4.5). ATP is universally distributed in all living cells and serves as both the initial receptor of energy liberated in all cellular oxidative reactions, thus driving all the energy-requiring life processes, as well as playing a key role in photosynthesis:

When ATP is hydrolyzed a large amount of heat is liberated (a free energy of hydrolysis of the terminal phosphate bond in ATP is approximately 8000 calories), in contrast to free energy of hydrolysis of an ordinary ester of 2000 to 2000 calories.

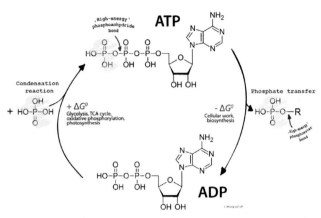

**FIGURE 4.5**   ADP-ATP cycle fueled by the glycolysis of a glucose substrate. *Source: Muessig, 2016. Courtesy, Wikimedia Commons.*

## 4.3 Strategies for coping with environmental constraints

There are many, varied strategies used by plants to either minimize or increase the absorption of the incident radiation flux. A few examples will illustrate the scope of these adaptations.

**Photosynthetic adaptations.** One of the challenges which plants face is that, when opening their stomata to take in $CO_2$ for photosynthesis, they also become vulnerable to increased evapotranspirational water loss. This can be an acute problem for plants growing in hot arid environments. During the Calvin cycle, C3 plants, using the standard photosynthetic pathway, take up $CO_2$ and turn it into sugar energy. C3 includes most plants—those living in temperate, cool, and wet environments; crops such as wheat, rice, barley, and potato, and all woody trees. They utilize a Calvin cycle as described above.

*C4 plants*, typical of warm, sunny environments, such as maize, sugar cane, and millet, avoid excessive water loss. To do this they minimize photorespiration by separating initial $CO_2$ fixation and the Calvin cycle by performing these steps in different cells in the leaf. $CO_2$ is fixed at night into oxaloacetate by PEP carboxylase and then converted into malate which is stored in bundle sheath cells to be made available during daylight hours when $H^+$ ions are being produced. This minimizes opening of stomata during the warmer daytime hours.

*CAM*, for crassulacean acid metabolism, plants adapted to living in hot, dry environments, such as cacti, also minimize photorespiration. But they accomplish this by separating initial $CO_2$ fixation, not in space as do C3 plants, but by separating these steps in time. Malic acid is stored in vacuoles until the next day when the light reactions occur, when stomata can now remain closed.

The difference between C3 and CAM vs. C4 plants is that C4 plants make a four-carbon sugar during the Calvin cycle instead of two three-carbon sugars as in C3 plants. This larger sugar in C4 plants brings more $CO_2$ to the rubisco enzyme, reducing oxygen levels and making the process energy-intensive. More $CO_2$ is brought into the process because of how cells are located. A C4 plant uses this photosynthetic pathway to avoid photorespiration.

Photorespiration is a wasteful reaction that occurs when plants take in oxygen and give out carbon dioxide instead of taking in carbon dioxide and releasing oxygen. Photorespiration (also known as the oxidative photosynthetic carbon cycle, or C2, i.e., C2 photosynthesis) refers to a process in plant metabolism where the enzyme rubisco oxygenates, RuBP, causing some of the energy produced by photosynthesis to be wasted. The desired reaction is the addition of carbon dioxide to carboxylation, a key step in the Calvin cycle; however, approximately 25% of reactions by rubisco instead add oxygen, creating a product that cannot be used within the Calvin cycle. This process reduces the efficiency of photosynthesis, potentially reducing photosynthetic output by 25% in C3 plants (Wikipedia, 2015).

**Modified structures.** The cuticle has been shown to thicken in plants under high radiation stress, which changes the emissivity (Baitrago, 2016). Leaf hairs of a tropical and subtropical coastal dune pioneer plant were shown to increase photosynthesis, and the hair layer increased the leaf's diffusion resistance to water loss, leading to a reduction in transpiration and higher instantaneous water use efficiency (Ripley, 1999). Chemical radio-protectants have been found in plants which scavenge free radicals before significant cellular damage occurs from irradiation (Weiss, 2009). These are just a few examples of the variety of structural and compositional modifications which plants have evolved to deal with radiation stress.

**Altered pigmentation.** Excess light and UV-radiation are hazardous natural stressors for plants, which have evolved a range of avoidance and tolerance strategies. Nonphotosynthetic pigments are used by plants for protection from excess light and UV-radiation. These include secondary metabolites belonging mainly to a variety of flavonoids, $C_6-C_3-C_6$ types, and the closely related anthocyanin flavylium salts, $C_6-C_3-C_6^+$ (Edreva, 2005). Many lichens are able to colonize and dominate extremely dry environments with high exposure to solar radiation, such as deserts or arctic and alpine ecosystems. Lichens are normally exposed to the highest irradiance levels while desiccated. The pigments parietin and melanin occur in two widespread foliose lichens. Parietin is present in the ubiquitous sun-tolerant lichen, *Xanthoria parietina*, while the shade-adapted specimens of the old forest species, *Lobaria pulmonaria*, are melanin-free (Solhaug et al., 2003). Another example is *Chlamydomonas nivalis*, a green alga which lives upon the surface ice of Arctic glaciers. It owes its red color to a bright red carotenoid pigment, which protects the chloroplast from intense visible and also ultraviolet radiation.

## 4.4 Energy conversion efficiencies

Photosynthetic efficiency is the efficiency by which a plant converts the radiative energy that it receives into chemical energy, defined ecologically as the value of the slope of the linear relationship between biomass production and intercepted photosynthetically active radiation. The photosynthetic efficiency of plants depends how light energy is defined − whether it is incident light or only that which is absorbed, and on the kind of light that is used. It takes eight to 10 or more photons to utilize one molecule of $CO_2$. The Gibbs free energy for converting a mole of $CO_2$ to glucose is 114 kcal. Eight moles of photons of wavelength 600 nm contain 381 kcal, which yields a nominal efficiency of 30%. However, photosynthesis can occur with light up to wavelength 720 nm so long as there is also light at wavelengths below 680 nm to keep Photosystem II operating. Using longer wavelengths means less light energy is needed for the same number of photons and, therefore, for the same amount of photosynthesis. For sunlight, where only 45% of the light is in the photosynthetically active wavelength range, the theoretical maximum efficiency of solar energy conversion is approximately 11%. In actuality, however, plants do not absorb all incoming sunlight (due to reflection, respiration requirements of photosynthesis and the need for optimal solar radiation levels) and do not convert all absorbed energy into biomass, which results in a maximum overall photosynthetic efficiency in the range of 3 to 6% of total solar radiation (Wikipedia, 2018).

The maximum efficiency of light utilization by green plants is not a simple value. Algae and higher plants appear to be equal in their inherent capacity to utilize light energy. At very low light intensities, both algae and higher plants can convert up to 20 % of photosynthetically active radiation (PAR), equivalent to the visible light spectrum, into the chemical energy of biomass. However, when growing in full sunlight, photosynthetic conversion in both plants and algae is reduced to around 2−3% of PAR, equivalent to 1−1.5% of total solar radiation. This light saturation effect is explained by the biochemistry of the dark reaction photosystem. For example. in full sunlight one photon is captured per chlorophyll molecule every 0.1 second. Therefore, 200 chlorophyll molecules in a chloroplast will capture 2000 photons/sec (0.5 msec/photon). However, takes 5 msec for enzymatic turnover; thus, the chloroplast can utilize only 1/10th of the photons available.

Not all plants are equal in their photosynthetic efficiency, having sacrificed at times efficiency in their adaptations to the constraints of particular environments in which they have evolved. Photosynthetic efficiency is typically between 1% and 2% (Table 4.2). C4 plants are up to 50% more efficient than C3 plants in converting radiant energy into chemical energy. Highly

**TABLE 4.2** Efficiencies of photosynthetic radiant energy conversion into biomass by plants.

| Plant type | Efficiency% |
|---|---|
| Algae | 1%–3% |
| Natural vegetation | 1%–2% |
| Sugar cane | 5%–7% |
| Corn | 1%–2% |
| Most crop plants | 1%–3% |
| Rain forest | 3% |

productive plants can be useful to humans as efficient food sources, and it is not unexpected that agriculture has both selected for and developed high efficiency plants.

Solar energy striking the Earth's surface is 178,000 TW. With total photosynthetic productivity of Earth being between ~1500 and 2250 TW, or $1.13-1.67 \times 10^{22}$ calories sec$^{-1}$ (47,300–71,000 EJ per year). For actual sunlight, where only 45% of the light is in the photosynthetically active wavelength range, the theoretical maximum efficiency of solar energy conversion is approximately 11%. In actuality, however, plants do not absorb all incoming sunlight (due to reflection and the need for optimal solar radiation levels). Also, plants do not convert all absorbed energy into biomass because of the respiration requirements of photosynthesis, which results in an overall theoretical photosynthetic efficiency of 3%–6% of total solar radiation (FAO, 1997). Theoretical efficiency is far from that actually achieved. With photosynthetic efficiency defined as incident radiation ÷ (energy fixed in photosynthesis, or GPP), the GPP efficiency of the planet is between 0.84% and 1.26% (Wikipedia, 2018). Lieth (1975) calculated the global efficiency of NPP (GPP minus autotrophic respiration) to be 0.13%; aquatic ecosystems (Kozlovsky, 1976) exhibit NPP efficiencies of ~0.4% (range of 0.1–1.0). The NPP efficiency of the rain forest at the Luquillo Experimental Forest, Puerto Rico, was reported to be 3% (Odum, 1970; Lugo, 2004).

## 4.5 Recommended reading

Heldt, H.-W., Piechulla, B., 2011. Plant Biochemistry, fourth ed. Elsevier, 622 pp https://epdf.pub/plant-biochemistry-4th-edition.html.

Osmond, C.B., 1987. Photosynthesis and carbon economy of plants. New Phytologist 106 (1), 161–175. https://doi.org/10.1111/j.1469-8137.1987.tb04688.x.

# Chapter 5

# Energy processing by animals

Plants and bacteria are unique in having developed the ability to transform radiant energy into carbon-rich molecular compounds. Within the plant cell these carbon molecules can be oxidized, releasing their stored chemical energy. The chemical basis of respiration is the same for animals as for plants, so we will cover respiration only once here in animals, where the metabolic processes is more complicated. A most comprehensive treatment of animal energetics. *The Fire of Life* by Max Kleiber (1961), should be a prized reference in one's library.

The Global Carbon Cycle and Climate Change. https://doi.org/10.1016/B978-0-12-820244-9.00005-6

## 5.1 Metabolism

**Metabolism** is the sum of the physical and chemical processes in an organism by which its material substance is produced, maintained, and destroyed, and by which energy is made available. Potential energy is stored energy, whereas kinetic energy is energy of motion. The ultimate source of this "substance" (food energy) is the conversion of radiant energy through the photosynthesis of green plants into chemical energy. This energy may be liberated and made available to do cellular work through the breaking of chemical bonds. When energy-rich carbon molecules pass through the cell membrane, they enter a new environment: the metabolic machinery of the cell. The primary purpose of respiration is to convert nutrients (energy-rich carbon molecules and other sulfur, phosphorus, and nitrogen containing compounds) into useful energy and at the same time to create new carbon compounds for the function and structure of the organisms. For atoms to be joined by specific bonds into molecules, work must be done, i.e., an external source of energy is needed to bring the atoms together to form a stable molecule. Respiration is the oxidation of organic compounds to produce energy for cellular processes. The biochemical processes of living cells are involved in the breakdown (catabolism) and synthesis (anabolism) of protoplasm, and the concomitant energy transformations are collectively termed the metabolism of the cell. Let us briefly return to the Second Law of Thermodynamics and, in this context, examine what metabolic energy constitutes (Fig. 5.1):

**Laws of thermodynamics** govern movement of energy. According to the first law of thermodynamics, energy cannot be created or destroyed, although it can be transferred or changed from one form to another. Total energy in the universe has remained constant, but it is continuously becoming more diffuse throughout the universe. The second law of thermodynamics states that when energy is converted from one form to another, some is usable energy, and some is degraded into a less usable form, i.e., heat.

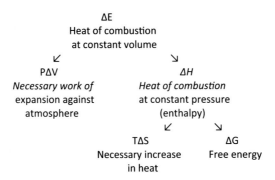

**FIGURE 5.1**   Relationship between enthalpy (H), free energy (G), and entropy (S).

*Entropy* (S) is a measure of disorder or randomness. Organized, usable energy has low entropy, whereas disorganized entropy such as heat has high entropy. All energy conversions have efficiencies <100%. Compare the efficiency of gas-powered internal combustion engines (20%–30%) with aerobic respiration (67%).

*Enthalpy* (H) is the total potential energy of a system. Bond energy is the amount of energy required to break a chemical bond. Total bond energy is equivalent to the total potential energy of the system, a quantity known as enthalpy (H). Free energy is energy that is available to do cellular work. Free energy (G) is the amount of energy available to do work under the conditions of a biochemical reaction. Entropy (S) and free energy (G) are related inversely.

$$G = H - TS \tag{5.1}$$

where:

G = free energy,
H = enthalpy of the system,
T is the absolute temperature expressed in degrees Kelvin, and
S is entropy.

Chemical reactions involve changes in free energy, so that:

$$\Delta G = \Delta H - T\Delta S \tag{5.2}$$

Free energy decreases during an exergonic reaction. When $\Delta G$ is negative, the reaction is exergonic, and releases energy. Exergonic reactions are spontaneous or "downhill."

**The calorigenic effect** is a result of the increased metabolic rate and increased oxygen consumption by cells. When the metabolic rate increases, more heat is generated and body temperature rises. Since energy use is measured in calories, this is known as the calorigenic effect.

## 5.2 Free energy

Free energy or Gibbs free energy G, is the energy available in a system to do useful work and is different from the total energy change of a chemical reaction. Thus,

$\Delta$ total energy $= \Delta$ utilizable energy $+ \Delta$ non $-$ utilizable energy

$\Delta H = \Delta G + T\Delta S$ or $\tag{5.3}$

$\Delta G = \Delta H - T\Delta S$

where:

G is Gibbs free energy (kJ mol$^{-1)}$
H is the heat of combustion, enthalpy (kJ mol$^{-1}$)

T is temperature (°Kelvin), and
S is entropy ($J°K^{-1}$)

To illustrate the concept of free energy, let us return to the familiar example of glucose oxidation. When 1 mol of glucose combines with 6 mol of carbon dioxide and 6 mol of water, the heat of combustion, $\Delta H$, amounts to 673,000 calories:

$$C_6H_{12}O_6 + 6O_2 = 6CO_2 + 6H_2O + 673,000 \text{ cal} \qquad (5.4)$$

Since there is no volume change, the total energy change, $\Delta E$, is equal to the change in enthalpy, $\Delta H$. The heat of formation, H, of glucose from the basic elements of $CO_2$ and $H_2O$ can be calculated by subtracting the heat of combustion of glucose from the heats of formation of 6 mol of $CO_2$ and 6 mol of $H_2O$. The oxidation of 1 g-atomic-weight of solid carbon (graphite) to 1 mol of $CO_2$ gas yields a heat of combustion, $\Delta H$, of −94,240 cal. The heat of formation of 1 mol of water by the combustion of 1 mol of hydrogen gas with $\frac{1}{2}$ mole of oxygen gas amounts to −68,310 cal. The heat of formation, H, of glucose is then calculated as:

6 mol $CO_2$ (g) = 6 × −94,240 cal = −565,440 cal
6 mol $H_2O$ (L) = 6 × −68,310 cal = −409,860 cal
Heats of formation of products of glucose combustion: 975,300 cal
Heat combustion of 1 mol glucose (g) = −673,000 cal
Heat of formation of glucose from $CO_2$ and $H_2O$ = −302,300 cal

The free energy change, $\Delta G$, in the formation of glucose can be calculated from the previous equation:

$$\Delta G = \Delta H - T\Delta S \qquad (5.5)$$

Entropy values for the various atoms can be found in the *Handbook of Physics and Chemistry*. Essentially, the values given are relative heat capacities at very low temperature equivalent to absolute zero (−273°C):

$$6 \times 1.3 + 12 \times 15.62 + 6 \times 24.52 = -342.4 \text{ entropy units}$$
$$\text{(for C)} \quad \text{(for H)} \qquad \text{(for O)} \qquad\qquad\qquad (5.6)$$

One gram of glucose represents −50.7 entropy units.

Entropy for glucose = elementary entropy of = entropy in glucose

$$\Delta S_{298} = -342.1 + 50.7 = -291.4 \text{ entropy units} \qquad (5.7)$$
$$\text{at 25°C} \qquad \text{atmos in glucose}$$

Therefore:

$$\Delta H = -302,300 \text{ cal}$$
$$T\Delta S = 298 \times -291.4 = +86,500 \text{ cal}$$
$$\Delta G298 = -215,800 \text{ cal}$$

This is the free energy change in glucose formation. The free energy involved in the oxidation of glucose is the difference between the free energy of formation of the combustion products and that of glucose, because by definition the free energy of $O_2$ is zero.

G in 6 mols $CO_2$ = 6 × (−94,100) = = −564,600 cal
G in 6 mols $H_2O$ = 6 × (−56.560) = = −339,360 cal
$\sum$G in combustion products = −903,960 cal
G of glucose = −215,800 cal
$\Delta$G in combustion of glucose = −688,166 cal

The change in free energy, $\Delta$G, when glucose is oxidized is thus about 2% greater than the heat of combustion, $\Delta$H, measured in a bomb calorimeter. This is because of the heat capacities of the reaction products and the fact that the reacting systems could also absorb heat from the environment. Although the absolute values of $\Delta$H and $\Delta$G are usually similar, conceptually they are quite different.

At this point you may be confused about why free energy values have been negative. Gibbs free energy is a derived quantity that blends together the two great driving forces in chemical and physical processes, namely enthalpy change and entropy change. Think of entropy as the measure of the random movement of molecules in the system. High entropy means a more random or chaotic state, such as a gas compared to a liquid. Processes in which entropy decreases tend not to occur in nature, unless there is a significant input of energy to cause them to take place.

Consider a reaction run at constant temperature, $\Delta$G = $\Delta$H−T$\Delta$S, where $\Delta$H is the enthalpy change (the heat of the reaction) and $\Delta$S is the change in entropy, If the free energy is negative, there is a change in enthalpy and entropy that favor the process and it occurs spontaneously. When $\Delta$ G is negative, the reaction is exergonic, and releases energy. This includes exothermic reactions in which the entropy increases, or exothermic reactions which have small decreases in entropy (as long as the temperature is relatively high), and endothermic reactions which are accompanied by large increases in entropy (like evaporation of water). When $\Delta$ G is negative, then the reaction is exergonic and releases energy.

## 5.3 Respiration

Respiration is a set of metabolic reactions and processes which occur in the mitochondria of cells, transferring biochemical energy from molecular substrates into the high energy bonds of ATP and some waste byproducts. Respiration in plants and animals is an oxidative metabolic reaction, which is mostly used for the gain of energy in cells. The most commonly used substrate in respiration in plant cells and animal cells are glucose, amino acids, and fatty acids. Molecular oxygen is the common oxidizing agent (electron acceptor).

Depending upon the electron acceptor, respiration is broadly classified as either aerobic respiration where organisms use oxygen as the electron acceptor, or anaerobic respiration in organisms where oxygen is not utilized.

Respiration is one of the key ways a cell releases chemical energy to fuel cellular activity. Although cellular respiration is technically a combustion reaction, it does not resemble one when it occurs in a living cell, because of the slow release of energy from the series of reactions. The chemical energy stored in ATP is used to drive the cellular processes requiring energy, including biosynthesis, locomotion, or transportation of molecules across cell membranes.

**Glycolysis**. Aerobic respiration in plant and animal cells requires $O_2$ in order to create ATP. Under aerobic conditions, one molecule of glucose is converted into two molecules of pyruvate (pyruvic acid), generating energy in the form of two net molecules of ATP (Fig. 5.2).

Pyruvate is oxidized to acetyl-CoA and $CO_2$ by pyruvate dehydrogenase. In the conversion of pyruvate to acetyl-CoA, one molecule of NADH and one molecule of $CO_2$ are formed. The resulting acetyl-CoA enters the Krebs cycle, reacting with $CO_2$ and makes two ATP, NADH, and FADH. The NADH pulls the enzyme's electrons to send them through the electron transport chain. The electron transport chain pulls $H^+$ ions through the chain. From the electron transport chain, the released hydrogen ions make ADP for an end result of 32 ATP. $O_2$ attracts itself to the leftover electron to make water. Lastly, ATP leaves through the ATP channel and out of the mitochondria. The chemical reaction and energetics can be expressed as:

$$C_6H_{12}O_6 + 6O_2 \rightarrow 6CO_2 + 6H_2O + \text{Heat} \tag{5.8}$$

$$\Delta G = -2880 \text{ kJ per mol of } C_6H_{12}O_6 \tag{5.9}$$

The negative $\Delta G$ indicates that the reaction can occur spontaneously.

**Krebs cycle**. This is also called the citric acid cycle or the tricarboxylic acid cycle. When oxygen is present, acetyl-CoA is produced from the pyruvate molecules created in glycolysis. When acetyl-CoA is formed, aerobic or anaerobic respiration can occur. If oxygen is present, aerobic respiration will

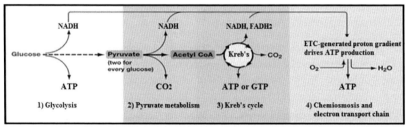

**FIGURE 5.2** Summary of anaerobic respiration: the metabolic pathway of glycolysis. *Image credit: ScienceGal4.0 - Own work, Wikimedia Commons, 2018.*

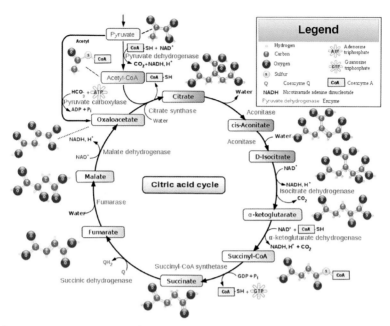

**FIGURE 5.3** The citric acid or Kreb's cycle. *Image credit: Narayanese, WikiUserPedia, Yasi-neMrabet, and TatuBaggins, Wikimedia contributors, 2019. Courtesy Creative Commons.*

lead to the Krebs cycle. If oxygen is not present, fermentation of the pyruvate molecule will occur. In the presence of oxygen, when acetyl-CoA is produced, the molecule enters the Krebs cycle and is oxidized to $CO_2$, while at the same time reducing NAD to NADH. NADH then can be used by the electron transport chain to create further ATP as part of oxidative phosphorylation. To fully oxidize the equivalent of one glucose molecule, two acetyl-CoA must be metabolized by the Krebs cycle. Two waste products, $H_2O$ and $CO_2$, are created during this cycle (Fig. 5.3).

The citric acid cycle is an 8-step process involving 18 different enzymes and coenzymes. During the cycle, acetyl-CoA (2 carbons) + oxaloacetate (4 carbons) yield citrate (6 carbons), which is rearranged to a more reactive form called isocitrate (6 carbons). Isocitrate is modified to become $\alpha$-ketoglutarate (5 carbons), succinyl-CoA, succinate, fumarate, malate, and, finally, oxalo-acetate. The net gain of high-energy compounds from one cycle is three NADH, one $FADH_2$, and one GTP; the GTP may subsequently be used to produce ATP. Thus, the total yield from one glucose molecule (2 pyruvate molecules) is six NADH, two $FADH_2$, and two ATP. Anaerobic respiration may occur in animals under conditions of intense physical activity where lactic acid is formed as the end product:

$$Glucose \rightarrow 2Lactic\ acid\ +2ATP \qquad (5.10)$$

A few microbes like yeast respire in the absence of oxygen in a process called *fermentation* where alcohol is the end product:

$$\text{Glucose} \rightarrow 2\text{Ethanol} + 2CO_2 + 6\text{ATP} \qquad (5.11)$$

Anaerobic respiration is used by few microorganisms which use an inorganic acceptor such as sulfur and does not utilize oxygen or pyruvate.

**Oxidative phosphorylation**. In eukaryotes, oxidative phosphorylation occurs in the mitochondria. It comprises the electron transport chain that establishes a proton gradient (chemiosmotic gradient) across the boundary of the inner membrane by oxidizing the NADH produced from the Krebs cycle. ATP is synthesized by the ATP synthase enzyme when the chemiosmotic gradient is used to drive the phosphorylation of ADP. The electrons are finally transferred to exogenous oxygen and, with the addition of two protons, water is formed.

**Efficiency of ATP production**. Table 5.1 describes the reactions involved when one glucose molecule is fully oxidized into carbon dioxide. It is assumed

**TABLE 5.1** Summary of aerobic respiration: The efficiency of ATP production by glycolysis.

| Step | Coenzyme yield | ATP yield | Source of ATP |
|---|---|---|---|
| Glycolysis preparatory phase | | −2 | Phosphorylation of glucose and fructose 6-phosphate uses two ATP from the cytoplasm. |
| Glycolysis pay-off phase | | 4 | Substrate-level phosphorylation |
| | 2 NADH | 3 or 5 | Oxidative phosphorylation: Each NADH produces net 1.5 ATP (instead of usual 2.5) due to NADH transport over the mitochondrial membrane |
| Oxidative decarboxylation of pyruvate | 2 NADH | 5 | Oxidative phosphorylation |
| Krebs cycle | | 2 | Substrate-level phosphorylation |
| | 6 NADH | 15 | Oxidative phosphorylation |
| | 2 FADH$_2$ | 3 | Oxidative phosphorylation |
| **Total yield** | | **30 or 32 ATP** | From the complete oxidation of one glucose molecule to carbon dioxide and oxidation of all the reduced coenzymes. |

**Source:** Wikipedia contributors: Cellular respiration, 2018.

that all the reduced coenzymes are oxidized by the electron transport chain and used for oxidative phosphorylation.

**The Respiratory quotient**. The ratio of moles of $CO_2$ produced to moles of $O_2$ consumed in any metabolic process is usually referred to in the field of bioenergetics as the respiratory quotient, RQ. Avogadro's Law states that equal volumes of all gases, at the same temperature and pressure, have the same number of molecules. Therefore, equal volumes at standard conditions of all gases contain the same numbers of molecules. Consequently, under the same conditions, the same numbers of moles of different gases occupy the same volumes. The molar ratio may, therefore, be replaced by the ratio of volumes, which is essentially what is measured in a respirometer, i.e., the volume of $O_2$ consumed or the volume of $CO_2$ produced (if the test organism is an animal).

The RQ is of interest bioenergetically because it reveals information about the composition of the substance being metabolized. Catabolism of carbohydrate, fats, and proteins each leads to a characteristic RQ. Therefore, respiratory quotients obtained from respiration trials provide an indirect method to estimate the heats of combustion and the substrates being metabolized by cells. For example:

$$\text{Carbohydrate: } C_6H_{12}O_6 + 6O_2 = 6CO_2 + 6O_2 \tag{5.12}$$
$$RQ = 6/6 = 1.0$$

$$\text{Fat: } 2C_3H_5 \, [CH_3(CH_2)15 \, COO]_3 + 145O_2$$
$$= 102CO_2 + 86H_2O \tag{5.13}$$
$$RQ = 102/145 = 0.703$$

$$\text{Protein: } 2CH_3CH(NH_2)_2 \, COOH + 6O_2$$
$$= CO(NH_2)_2 + 5CO_2 + 5H_2O \tag{5.14}$$
$$RQ = 5/6 = 0.830.$$

The thermal equivalents of $O_2$ and $CO_2$ in respiration are derived from the heats of combustion, $\Delta H$, of the compound and the concomitant moles of gas produced. Typically, caloric equivalents of $O_2$ are used in indirect calorimetry, i.e., converting gas exchange to calories of heat produced. The thermal equivalents of $O_2$ are such that from the RQ alone, it is possible to determine what organic substrate (or mixture thereof) is being metabolized by the organism (Table 5.2).

With all possible intermediates, RQs of higher than 1.0 can be obtained during synthesis of fat from carbohydrate:

$$4C_6H_{12}O_6 + O_2 = C_{16}H_{32}O_2 + 8CO_2 + 8H_2O$$
$$\text{(glucose)} \qquad \text{(palmitic acid)} \tag{5.15}$$
$$RQ = 8/1 = 8.0$$

**TABLE 5.2** Thermal equivalents (kcal $L^{-1}$) for different compounds.

| Compound | RQ | kcal liter$^{-1}$ |
|---|---|---|
| Fat | 0.7 | 4.686 |
| Protein | 0.83 | 4.838 |
| Carbohydrate | 1.0 | 5.047 |

This is typical of well-fed organisms storing food intake as fat. On the other hand, animals under starvation are converting fat to carbohydrate and typically have RQs lower than 0.7:

$$C_{16}H_{32}O_2 + H_2O = 2C_6H_{12}O_6 + 4CO_2 + 4H_2O$$

(palmitic acid)   (glucose)                                        (5.16)

$$RQ = 4/11 = 0.36$$

**Redox potential.** Redox potential is an expression used to describe the "tendency" of an "environment" to supply electrons to a metabolic process. Highly oxygenated environments, oxic environments, have a high redox potential due to the fact that $O_2$ is readily available as an electron acceptor. The redox state of iron, Fe, is widely used as an index of the transition from mildly oxidizing conditions to strongly reducing conditions. Heterotrophic organisms in oxic environments use $O_2$ as a strong electron acceptor. Electrons come from the metabolism of reduced organic compounds originally obtained from the environment and oxidized to $CO_2$. During aerobic respiration by eukaryotic cells, aerobic respiration occurs in the mitochondria. Four electrons flowing across the mitochondrial membrane combine with one $O_2$ and four $H^+$ ions to form two molecules of $H_2O$ in a remarkably efficient biochemical capture of energy.

In the environment, oxidation proceeds most readily in, and at, lower redox potentials in neutral or alkaline conditions, while various kinds of anaerobic metabolism, such as denitrification, occur more readily under acidic conditions. Normally, organic matter contributes to strongly reducing, lower redox potential in water-saturated soils and sediments. Redox potential drops in soils as the heterotrophic respiration of organic carbon depletes available $O_2$. After $O_2$ is depleted by aerobic respiration, denitrification begins when redox reaches 747 mV at pH 7.0; then denitrifying bacteria use nitrate as an alternative electron acceptor during the oxidation of organic matter. When nitrate is depleted, reduction of $Mn^{+4}$ begins at redox 526 mV, then reduction of $Fe^{+3}$ at 47 mV. Below the zone of Mn reduction, most redox reactions are performed by obligate anaerobes. The environments of flooded soils and sediments are in dynamic equilibrium maintained by the availability of $O_2$ at the surface interface with the atmosphere and buried organic carbon as a source of reducing power at depths. This explains why carbon substrates are either oxidized to $CO_2$ or reduced to $CH_4$.

## 5.4 Energy value of foods

To state that the carbon-based chemistry of living systems is complex is a huge understatement. Carbon molecules and carbon compounds are the means by which energy is captured, stored, and transferred in the biosphere. Carbon compounds are used by living systems for various purposes, most importantly for:

- energy to run the metabolic processes,
- structural components for support and protection,
- reproductive gemmules, and
- energy storage.

As energy is transformed and transferred within and between organisms, the energetic values, and costs of production, of organic compounds become important.

Heats of combustion and caloric values are ways of expressing the energy content of compounds. The heat evolved in the complete oxidation of a substance is known as the heat of combustion. Its value is expressed in terms of heats of combustion per gram or heats of combustion per mol. Since only rapid, complete chemical reactions are easily amenable to thermochemical measurement, heats of combustion have been far the most common data in thermochemistry. In developing standardized values for different molecules (Table 5.3), it is essential to burn everything to its highest state of oxidation to ensure that a definite reaction is being measured which will give reproducible results. Samples are ignited electrically in a steel bomb containing oxygen under a pressure of 25 atm. All carbon is burned to $CO_2$ and all hydrogen to $H_2O$.

The overall heat of a chemical reaction at constant pressure is the same, regardless of the intermediate steps involved, Hess' Law, states that the heat evolved or absorbed in a chemical process is the same whether the process takes place in one or in several steps. This is also known as the law of constant heat summation. These are both corollaries of the first law of thermodynamics, the law of conservation of energy. They follow from principles covered in Chapter 2, since the enthalpy change $\Delta H$ for a reaction depends only on the initial and final states and is not affected by the path of the reaction. Energy transfers occurring via the biochemical processes in cells are stepwise resulting in incremental change in enthalpy spreading out heat formation which could damage cellular structure. This is demonstrated by the following examples:

**TABLE 5.3** Heats of combustion to $H_2O$ (L) and $CO_2$ (g) at 25°C and constant pressure.[a]

| Substance | $\Delta H$ in kcal mol$^{-1}$ |
|---|---|
| Hydrogen, $H_2$ (g) | −68.3174 |
| Graphite, C (g) | −94.0518 |
| Carbon monoxide, CO (g) | −67.6361 |
| Methane, $CH_4$ (g) | −212.798 |
| Methanol, $CH_3OH$ (L) | −173.698 |
| Formaldehyde, $CH_2O$ (L) | −134.698 |
| Formic acid, $CH_2O_2$ (L) | −64.598 |
| Glucose, $C_6H_2O_6$ (g) | −673.0 |
| Lactic acid, $CH_3*CHOH*COOH$ (L) | −326.0 |
| Glycogen, $C_6H_{10}O_5$ (g) | −679.0 |
| Ethyl alcohol, $C_2H_5OH$ (L) | −327.6 |

[a]Since the above are exothermic reactions and heat is evolved (lost) from the system, the sign of $\Delta H$ is negative.

**Example 1.**

$$CH_4 \text{ (g)} + 1/2O_2 \text{ (g)} \rightarrow CH_3OH \text{ (L)} \quad \Delta H = -39.1$$

$$CH_3OH \text{ (L)} + 1/2O_2 \text{ (g)} \rightarrow CH_2O \text{ (L)} + H_2O \text{ (L)} \quad \Delta H = -39.0$$

$$CH_2O \text{ (ll)} + 1/2O_2 \text{ (g)} \rightarrow CH_2O_2 \text{ (L)} \quad \Delta H = -70.1$$

$$CH_2O_2 \text{ (L)} + 1/2O_2 \text{ (g)} \rightarrow CO_2 \text{ (g)} + H_2O \text{ (L)} \quad \Delta H = -64.6$$

If these four reactions and their $\Delta H$'s are added, the intermediate compounds cancel out, since they occur in equal quantities on each side of the arrow. The overall reaction and overall $\Delta H$ can then be obtained:

$$CH_4 \text{ (g)} + 2O_2 \text{ (g)} \rightarrow CO_2 \text{ (g)} + 2H_2O \text{ (L)} \quad \Delta H = -212.8$$

The total amount of heat evolved in the oxidation of methane to $CO_2$ and $H_2O$ is the same whether the combustion is carried out in a single step or in a stepwise fashion.

**Example two.**

Reactions in biological systems are not spontaneous in the sense that this expression is used in relation to inanimate systems. Biological reactions are mostly coupled, such as the following example of conversion of sugar to alcohol and carbon dioxide. The fermentation, or glycolysis reactions, also may be considered as examples of internal oxidation-reduction not involving molecular oxygen.

Anaerobic reaction $C_6H_{12}O_6 \rightarrow 2C_2H_5OH + 2CO_2$ (glucose) (ethyl alcohol) $\Delta H = -17.8$

Aerobic reaction. $C_2H_5OH = 3O_2 \rightarrow 2CO_2 + 3H_2O$ $\Delta H = -327.6$

Anaerobic reactions. $C_6H_{12}O_6 \rightarrow 2CH3 * CHOH*COOH$ $\Delta H = -21.$

$C_6H_{10}O_5 \rightarrow 2CH*CHOH*COOH$ $\Delta H = -27.$

Aerobic reaction. $CH_3*CHOH*COO + 41/2O_2 \rightarrow 3CO_2 + 3H_2O$ $\Delta H = -326.$

**Calorific value** is the term describing the total energy released as heat when a substance undergoes complete combustion with oxygen under standard conditions. The chemical reaction is typically a hydrocarbon or other organic molecule reacting with oxygen to form carbon dioxide and water and release heat. The caloric content of food is determined from its enthalpy of combustion ($\Delta H_{comb}$) per gram, as measured in a bomb calorimeter, using the general reaction:

$$\text{Food} + \text{Excess } O_2(g) \rightarrow CO_2(g) + H_2O(l) + N_2(g) \qquad (5.17)$$

The nutritional Calorie (with a capital C) that you see on food labels is equal to 1 kcal (kilocalorie).

## 5.5 Digestion and assimilation

The caloric value of ecological material varies considerably (Table 5.4), depending substantially upon the quantity of the indigestible structural components, e.g., lignins, scleroproteins, calcareous shells, etc., transferred along food chains and between trophic levels. Typical values for food digestibility are given in Table 5.5.

Of any given quantity of food consumed by an animal, only a portion is fully digestible and assimilated by the organism. Undigested residues pass from the gut in the form of feces, composed primarily of lignins, celluloses, bone, scleroproteins, and other compounds difficult to chemically decompose without special enzymes. The digestibility factor varies not only among organisms, but it is also strikingly affected by the food consumed. Meat eaters (predators) digest more of their food and have the highest food assimilation:intake quotient, while herbivores (particularly grazers) have considerably lower food energy assimilation. Ultimately, as we shall see in later chapters, the assimilation factor significantly affects the efficiency of energy.

**TABLE 5.4** Energy values for plant parts and animal taxa.

| Material | Energy value (kcal $g^{-1}$ dry weight) |
|---|---|
| Leaves[a] | 4.229 |
| Stems[a] | 4.267 |
| Roots[a] | 4.740 |
| Litter[a] | 4.298 |
| Seeds[a] | 5.065 |
| Molluscs[a] | 4.600 |
| Annelids[a] | 4.617 |
| Beetle[b] | 6.314 |
| Fish[b] | 5.823 |
| Mammals[a] | 5.163 |

[a]after Golley (1961).
[b]after Slobdkin and Richman (1961).

There are three basic methods for the measurement of assimilation (A):

**Indicator method**—by marking food with a visible, inert, non-assimilable marker; the passage of the food bolus can be detected visibly and assimilation calculated from the increase in concentration of the indicator compound.

$$\%A = 100 \times (1 - C_{food} / C_{feces}) \tag{5.18}$$

where:

A is assimilation,
$C_{food}$ is the concentration of the indicator in the food, and
$C_{feces}$ is the concentration in the feces.

**Gravimetric method**—by weighing the amount of food eaten, passed through the animal in feces, and/or the gain in weight by the animal.

$$\%A = 100 \times (W_c - W_f)/W_c \tag{5.19}$$

where:

$W_c$, is the weight of food eaten and $W_f$ is the weight of feces produced

**Radiotracer method**—by measuring the radioactivity in feces when an animal feeds on radioactive food and when the animal is transferred to nonradioactive food; the whole-body decrease in radioactivity can be used to estimate assimilation as well as the turnover rate (Fig. 5.4). At first the

**TABLE 5.5** Food assimilation for different foods and by different trophic level consumers reported in the scientific literature.

| Trophic level | Organism | Food | Assimilation % |
|---|---|---|---|
| Predator[a] | Lizard | Insects | 83%–88% |
| Predator[b] | Menhaden | Zooplankton | 87%–91% |
| Predator[c] | Harp seal | Cod | 93% |
| Predator[v] | Elephant shrew | Insects | 31%–71% |
| Predator[d] | Turtle | Fish | 91% |
| Predator[j] | Fish | Animals | 89%–97% |
| Herbivore[d] | Turtle | Plants | 49% |
| Herbivore | Caterpillar | Leaves | 10%–40% |
| Herbivore[e] | Elephant | Hay | 44% |
| Herbivore[f] | Mussel | Phytoplankton | 78%–85% |
| Herbivore[g] | Tilapia | Algae | 30%–60% |
| Herbivore (bird)[h] | Plant cutter | Leaves | 48%–70% |
| Fruitlvore[h] | Bird | Fruit | 20%–80% |
| Saprovore[i] | Isopod | Litter | 3%–8% |
| Decomposer[j] | Bivalve mollusc | Suspended detritus | 45% |
| Omnivore[j] | Welk | Algae, sm inverts | 40%–67% |

[a]Xiang (1993).
[b]Durban and Durban (1983).
[c]Lawson at al. (1997).
[d]Woodall and Currie (1989).
[e]Spinage (1994); Spencer (1998).
[f]Hayden (2008).
[g]Talapia Territory.
[h]PoultryHub.
[i]Reichle (1967).
[j]Crisp (1975).

radioactive loss curve ($T_{b1}$) is steep inthe wood roach, *Parcoblatta,* reflecting the passage of undigested food through the gut and its loss as feces. Later the curve becomes less steep ($T_{b2}$) and represents the elimination (turnover rate) of the assimilated isotope. The roach is a detrital feeder and has undigestible components in its ingested food. The snail-predator, *Sphaeroderus,* feeds by biting the snail, inserting its mouthparts into the shell of the snail, and secreting enzyme(s) to digest the snail. *Sphaeroderus* was fed on tagged fluid, had ~ 100% assimilation, and did not exhibit a two-component loss curve. The ordinate axis intercept of the radioactive loss curve can be used to estimate the

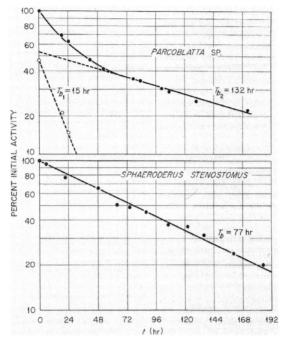

**FIGURE 5.4**   Radioactive elimination curve for two cryptozoan species (*Parcoblatta* sp., the wood roach, and *Sphaeroderus stenostomus*, a snail-feeding carabid ground beetle) fed with $^{134}$Cs isotope-tagged food. *Source: Reichle and Crossley (1966). Courtesy, Health Physics Society.*

percentage isotope assimilated (and presumably the food which it labeled). This is quite a useful technique for minute animals, or to use on free-ranging animals in their natural habitat (Reichle and Crossley, 1966; Reichle, 1969). If the radioactive tag is a nutrient element, then additional useful information is obtained at the same time (see Eqs. 7.11−7.17).

## 5.6 Respiration rates

In 1883, Max Rubner proposed that log of the metabolic rate would vary with log of body mass in just the same way that the log of surface area does, because heat is lost through the external surface of the animal. The rates of many phenomena of metabolism, such as oxygen consumption as a measure of respiration, follow a simple relationship proportional to the surface area of the organism. Temperature influences on animals also result in varied forms of expression demonstrated in animal sizes, appendages, and geographic distributions.

**Body size/surface area**. This relationship can be a means of quantitatively expressing the often-observed characteristic that while the total respiration of large animals is greater than that of small animals, their rates are not. Small animals typically have higher respiration rates (for example, shrews and mice) than do larger animals (for example, buffaloes and elephants). This is why smaller animals are more physically active and why they must spend a greater portion of their lives in securing food.

Application of this surface area relationship to physiological processes has been reviewed my many authors (e.g., Reichle, 1968). Of the theories advanced for interpretation of the surface law of animal metabolism, the most valuable relates metabolism to the rate of heat transfer (Kleiber, 1961). For a given temperature difference between an animal and its environment, the rate of heat transfer per unit area of surface is the same for bodies of all sizes. Different-sized bodies of similar shape have surface areas proportional to the square of their linear dimensions and to the two-thirds power of their volumes. If the dimensions of the bodies are the same, then their surface areas are proportional to two-thirds the power of their weights. This relationship is described by the power function:

$$Y = aX^b \qquad (5.20)$$

with b = 0.67 relative to absolute weight, and

$$Y/X = aX^{b-1} \qquad (5.21)$$

with (b - 1) = −0.33 relative to unit weight.

This is the simplest hypothesis, i.e., the power function b is a constant. Metabolic processes in animals are influenced by many variables, both physiological and environmental, and exact correlation of the two-thirds power of body weight model seldom occurs. Basal energy metabolism does not vary directly with either simple body weight or with anatomical surface area, since animals have evolved numerous control mechanisms which affect their energy balance with the environment. Basal metabolism does vary in a systematic fashion for closely related animals with what has been termed the metabolically effective body weight, $X^b$. Empirical measurements must determine the value of b for different animal types. There appears to be remarkable similarity between the beta coefficients for different animal types, although absolute metabolic rates may differ somewhat (Table 5.6).

**Bergmann's Rule and Allen's Rule**. In endothermic animals, within broadly distributed and closely related taxonomic groups (genera and families), populations and species of larger size are found in colder environments, and species of smaller size are found in warmer regions and more southern latitudes (Allee et al., 1949). This phenomenon was first described by German biologist Carl Bergmann in 1847 who noticed this geographical distribution pattern in many birds and mammals, e.g., foxes, bears, and penguins. Generally larger-bodied animals, tend to conform more closely to Bergmann's Rule than do smaller-bodied animals. This phenomenon is generally related to body surface-volume relationships and body heat conservation in northern regions and heat radiation in southern latitudes. A similar thermal influence on the anatomy of animals first noted in 1876, Allen's Rule (Encyclopedia.com,

**TABLE 5.6** Values for the body weight exponential function, b, for different animal types.

| Animal class | b | Reference |
|---|---|---|
| Annelida | 0.67 | Reichle (1970) |
| Arthropoda | 0.725 | Reichle (1970) |
| Pisces | 0.82 | Urbina and Glover (2013) |
| Reptilia | 0.76 | Nagy (2005) |
| Aves | 0.68 | Glazier (2008) |
| Mammalia | 0.734 | Brody (1945) |

2019), results in the tendency toward lessening of extremities in colder climates, notable examples being tails, ears, and limbs in species as diverse as birds and rabbits. Bergmann's and Allen's rules reflect the animals' responses to the physical, convective property of temperature. Another animal response, the Converse Bergmann Rule operates through the organisms' metabolic, chemical energy balances. The converse Bergmann rule pertains to ectoderms, notably the insecta and reptiles, where animals become smaller in more northern latitudes. This phenomenon is generally related to length of the seasons; more southern habitats allow for longer life cycles and longer feeding periods which permit larger food energy inputs and larger bodies.

## 5.7 Energy costs of digestion

**Calorigenic effect**. Some energy is required to execute the digestive process and to accomplish the metabolic conversion of food nutrient. The heat increment of the catabolism of protein is the main energetic cost; this cost is referred to as the calorigenic effect or specific dynamic action (SDA). Earlier belief that this energy cost was due to the work of digestion (visceral muscular activity) has generally been disproven. The SDA is an energy waste incident to food utilization. While the metabolizable energy (or physiological food value) seems to correspond to the thermodynamic heat of combustion, $\Delta H$, and the net energy available for growth, maintenance and reproduction seem to correspond to the thermodynamic free energy, F, the "food utilization tax," or the calorigenic effect appears to be analogous to the entropy term, $T\Delta S$. Therefore:

$$\Delta F = \Delta H - T\Delta S \quad \text{of thermodynamics} \quad (5.22)$$

$$\text{Net} = \text{metabolizable} - \text{calorigenic} \quad \text{of bioenergetics}$$
$$\text{energy} \quad \text{energy} \quad \text{effect} \quad (5.23)$$

If the postabsorptive energy expense of maintenance of a mature, nongrowing animal, such as a dog, at thermal neutrality is 100 kcal per day, then if the animal consumes 100 kcal in the form of meat, its heat production will increase to 131 kcal per day; the extra 31 kcal is the SDA of 100 kcal of meat (Table 5.7). If the dog is fed 131 kcal, its heat production will increase to 137 kcal, etc.

If one continues to adjust the caloric intake until it equals the caloric expenditure in heat production by the animal (basal metabolism + calorigenic effect), the relationships between the energy in the food and heat production are obtained: for the animal. From Table 5.7 the SDA can be calculated to be 40% of the intake energy for lean meat; for fat about 14.5%; and for sucrose about 6.5%. If the dog ingests 140 kcal of metabolizable energy in the form of protein, about 40 kcal is the entropy tax and 100 kcal is the physiologically free energy available to the animal for maintenance and other activities.

**Environmental temperature and metabolic rates**. Temperature has a pronounced effect on metabolic rates, and this relationship is described by the Q10 approximation. Similarly, we can also expect a temperature effect on food utilization as well as an optimum temperature for food utilization. Between a high environmental temperature, at which an animal eats less than it requires for maintenance, and a low environmental temperature, at which the animal

**TABLE 5.7** The relationship between food energy and heat production, the calorigenic effect or specific dynamic action (SDA), in a dog fed 100 kcal day$^{-1}$ of lean meat (protein) [columns 1-4], compared with the food energy and heat production equivalents to be obtained from a pure fat [columns 5-6] or carbohydrate [columns 7-8] diet.*

| Protein Diet | | | | Fat Diet | | Carbohydrate Diet | |
|---|---|---|---|---|---|---|---|
| Food Energy | heat production | ΔHeat | SDA | food energy | heat production | food energy | heat production |
| 0 | 100 | 0 | 0 | 0 | 100 | 0 | 100 |
| 100 | 130.9 | 30.9 | 31 | 100 | 112.7 | 100 | 106 |
| 130.9 | 137.3 | 6.9 | 6.4 | 112.7 | 114.3 | 106 | 106.4 |
| 137.3 | 139.3 | 2.0 | 2.0 | 114.3 | 114.5 | 106.4 | 106.42 |
| 139.3 | 139.9 | 0.6 | 0.6 | 114.5 | 114.55 | | |
| 139.9 | 140.1 | 0.2 | 0.2 | | | | |
| 140.1 | 140.2 | 0.1 | 0.1 | | | | |

* After Kleiber, 1961.

eats to capacity but cannot take in enough food to maintain a constant body temperature, the temperature relations for food utilization are complex.

Most animals need to maintain their core body temperature within a relatively narrow range. Endotherms use internally generated metabolic heat to maintain steady body temperatures when the environment is cool, and evaporative cooling to offset warm environmental temperatures. Their body temperature tends to stay steady regardless of the environment. Endotherms depend mainly on external heat sources, and their body temperature changes with the temperature of the environment.

The ability of endotherms to maintain stable body temperatures is called homeothermy. The advantage of homeothermy is that it enables the animal to sustain high levels of activity regardless of environmental temperatures. Homeothermy frees the animal from dependence upon sunlight to maintain body temperatures, and it enables animals to occupy thermally inhospitable environments such as those in high altitudes and at the poles. There are costs, however, for homeothermy, including increased metabolic energy requirements that require heavier demands upon and a steadier supply of food resources. Homeotherms have higher risk of starvation, and are susceptible to dehydration from evaporative water loss from higher respiratory rates.

Animals exchange heat with their environment through radiation, conduction—sometimes aided by convection—and evaporation. For both endotherms and ectotherms, body temperature depends on the balance between heat generated by the organism and heat exchanged with (lost to or gained from) the environment. When temperature is so warm that the animal becomes inactive and ingests less food than is necessary for maintenance, there is no gain in weight. When the environmental temperature decreases, the heat requirement of the animal increases and the animal increases its food ingestion rate. Food intake cannot indefinitely follow the increase in heat requirement, because the food capacity of the animal is limited. When environmental temperature becomes so low that the heat requirement exceeds the feeding capacity, the animal eats all it can and, even from an abundant food supply, it starves. This situation is similar to that for hummingbirds when confronted with cool nighttime temperatures. Rather than starving, the hummingbirds abandon homeothermy, enter torpor, sleep, and cool off—similar to hibernating animals. They wake the next morning, resume homeothermy, and restore body temperature after they can meet their heat requirement through feeding.

The metabolic response of the mouse in response to a range of environmental temperatures illustrates how metabolism in endotherms varies with temperature (Fig. 5.5). The thermoneutral zone is the range of temperature where no additional energy is required to maintain the homeothermic condition. The lower limit of this range, $T_{lc}$, is lower temperature threshold at which the animal has to increase its metabolism to maintain body temperature, $T_b$. Beyond the upper temperature threshold, $T_{uc}$, the animal must increase its metabolism in order to cool its body. Hypothermia, as an adaptive temperature

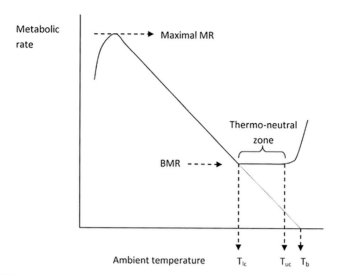

**FIGURE 5.5**   Idealized relationship between the metabolic rate of a mouse and environmental temperature. *BMR*, basal metabolic rate; *MR*, maximal rate; $T_{lc}$, lower critical temperature; $T_{uc}$, upper critical temperature; $T_b$, body temperature. *Source: Speakman, 2013. Courtesy, Frontiers.*

strategy when seasonal food resources are unavailable or temperatures become extreme, is used by hedgehogs, rodents, some bats, bears; the body temperature falls to within a few degrees of ambient, but above freezing, and the heart, respiration, and metabolic rates decrease. Conversely, when animals experience hyperthermic conditions, they may utilize physiological mechanisms to cool their bodies, such as sweating or panting. They also may display behavioral patterns to avoid direct sunlight or temperature hot spots in their environment, and they may shift from diurnal to nocturnal behavior patterns.

## 5.8 Food energy budget for an individual

The food energy available to an animal for metabolism and its disposition within its body depends upon the flow of energy along the below main pathways;

- The nonassimilated portion of food passing through the animal,
- Energy digested and incorporated into the synthesis of body tissue,
- Energy spent in locomotor activity searching for food,
- Energy devoted to reproduction,
- Respiratory heat losses, and
- Mortality and predatory losses to other trophic levels.
  The categories of energy flow through an organism are (Fig. 5.6):

  **Available energy**— the total energy available in herbage grazed or prey caught.

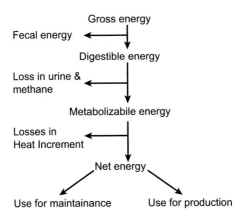

**FIGURE 5.6** Energy flow in an organism showing the categories of energy allocation and loss.

**Consumed energy**— that portion of the total available energy actually ingested by the organism; it accounts for prey killed but not eaten and all wastage during consumption whether accidental or intentional (undesirable parts such as hair).

**Digestible energy**— the gross energy of food consumed minus fecal energy; it is that portion of the energy intake subsequently available to the organism's metabolic processes.

**Metabolizable energy**— the gross energy of food consumed minus fecal energy, minus urinary energy, and minus the calorigenic effect; in other words, that portion of the digestible energy which, after processing through the biochemical machinery of the cell, is available for work.

**Net energy**— that energy available for work by the organism which may be partitioned into three fundamental requirements of the animal: Respiration for maintenance of body temperature and as respiratory heat losses during chemical transformations of substrate; Growth or incorporation of organic compound synthesized in the metabolic processes into new structure, active tissues, or storage; Reproduction as an energy cost for both synthesis of high-energy nutrient compounds and respiration by embryos while still a part of the maternal physiological system.

## 5.9 Why pork is cheaper than beef and chicken costs least of all

The answer to this question obviously must have something to do with animal energetics, otherwise why would the question be raised here? The answer lies in how efficiently these animals utilize their dietary energy intakes (Table 5.8),

**TABLE 5.8** Comparison of dietary energy utilization in the domestic pig and cow (values are % food energy ingested).

| Energetic parameter | Pig | | Cow | |
|---|---|---|---|---|
| | Usual | Range | Usual | Range |
| Digestible energy | 80 | 60–95 | 70 | 40–90 |
| Metabolizable energy | 78 | | 58 | |
| Gases | 0 | 0–0.3 | 8 | 5–12 |
| Feces | 20 | 5–40 | 30 | 10–60 |
| Urine | 2 | 1–3 | 4 | 3–7 |

and ultimately, in their food capacity and production rate limitations. All other conditions being equal, it is reasonable to suppose that those domesticated animals which produce the most biomass of comparable quality from the least food will have the less per pound monetary cost. Animal husbandry has selected from those animals, suitable for domestication, those which are energetically the most efficient. To explore our question in more detail, consider the energy budgets of the domesticated pig and cow.

Metabolizable energy is used to satisfy several energetic requirements of the animal: maintenance, growth, and reproduction. The efficiency with which the animals accomplish these tasks bears directly upon the production efficiency of the animal in terms of biomass produced per unit of food energy intake. The net efficiency of utilization of metabolizable energy for body mass gain is not a constant for each species; for the pig it is 80%, for a steer, 60%, and for a lactating cow, 70%. The pig is a more efficient converter of food energy to metabolizable energy and also a more efficient producer of biomass.

Another means by which farmers can increase the productivity of their herds is to increase the food ration, thus increasing the rate of intake of food energy and, to a limited extent, increasing the rate of biomass production. The efficiency of this operation is also not constant across species, and is essentially limited by two factors: the food capacity of the animal and the relationship between intake rate of food and food digestibility. The pig is again the winner, on both counts.

The production efficiency of farmed animals (Table 5.9) is the efficiency with which they produce protein. This efficiency can be expressed as the grams of protein produced per $10^9$ cal of dietary energy ingested.

These data represent the overall efficiency with which dietary energy is converted into food, since they include the energy cost of reproduction, rearing of

**TABLE 5.9** Rate of production and production efficiency in relation to dietary energy intake in farmed animals.

| Food product | Rate of production | Protein production (g/ $10^9$ cal ingested) | Energy efficiency (%) |
|---|---|---|---|
| Catfish[a] | 480% wt. gain in 12 weeks | – | 19.6 |
| Broiler chicken | 3.5 lbs. in 10 weeks | 13.7 | 8.9 |
| Eggs | 200 eggs per year | 10.7 | 7.5 |
| Milk | 20,000 lbs. per year | 16.3 | 6.0 |
| Pork[b] | 200 lbs. in 6 mos (2.5 lbs. feed/lb. gain) | 8.7 | 6.0 |
| Beef[b] | 1100 lbs. in 12 mos (5 lbs. feed/lb. gain) | 3.2 | 2.2 |

[a] Yousif et al. (2014)
[b] Brody (1954)

the breeding stock, and mortality, as well as that of the production itself. Therefore, these efficiencies will be lower than those reported in conventional bioenergetics texts, but they are directly comparable to production efficiencies of wild animals that include the above costs. These data also show why chicken and catfish are among the most economical sources of protein available in the supermarkets.

## 5.10 Recommended Reading

Brody, S., 1945. Bioenergetics and Growth. Reinhold, New York, p. 1023 pp. https://doi.org/10.1002/ajpa.1330040117.

Kleiber, M., 1961. The Fire of Life. John Wiley and Sons, p. 454 pp. https://www.cabdirect.org/cabdirect/abstract/19621404881.

Wiegert, R.G., 1976. Ecological Energetics. Dowden, Hutchinson and Ross, Inc., Stroudsburg, PA, 455 p. https://www.worldcat.org/title/ecological-energetics/oclc/2077851.

Wikipedia contributors. 2019. Citric acid cycle. In: Wikipedia, the Free Encyclopedia. https://en.wikipedia.org/w/index.php?title=Citric_acid_cycle&oldid=878097926.

# Chapter 6

# Species adaptations to their energy environment

## 6.1 The limits of survival

Temperature extremes and the availability of water limit the existence of life on Earth. The environmental temperatures to which organisms are exposed limit both their distribution and, at the same time, their metabolic activity. Obviously, the range of temperatures on Earth oftentimes exceeds the limits which life can tolerate. In general, life processes occur within the range of $0-40°C$, although the range tolerable for life activities is much narrower for most species. Temperatures above $41°C$ will break the interactions in many proteins and denature them. Some species may tolerate environments with more extreme temperature regimes for short periods of time, but seldom can the life processes sustain themselves beyond these limits. Some plants and animals have resistant stages of their life cycle which permit them to withstand adverse environmental conditions. Thus, some forms of bacterial spores, plant seeds, invertebrate cysts, and vertebrate gametes may survive freezing temperatures below $0°C$. Some animals live in warm springs above $40°C$, and a

**The Global Carbon Cycle and Climate Change.** https://doi.org/10.1016/B978-0-12-820244-9.00006-8

few bacterial species and some algae can tolerate hot spring environments with temperatures as high as 70°C. Why does temperature have such a constraint upon the life activities?

**Thermal properties of water.** Water is the universal solvent and indispensable for life. From the elements in the climate-space equation, it is obvious that the water balance, through evaporative cooling, can affect the organism's temperature. Therefore, it is not surprising that the temperature relations of many organisms with their environment are closely connected to their water balance. The thermal properties of water are important in determining the thermal relations of organisms. The heat conductivity of water is 0.0014 cal cm$^{-1}$ sec$^{-1}$ C$^{-1}$, which is low compared to that of other materials such as metals, but is higher than most other common solvents, e.g., olive oil, 0.000393, or ethyl alcohol, 0.00042 cal cm$^{-1}$ sec$^{-1}$ C$^{-1}$. Consequently, water as a solvent allows the organism to remain partially coupled to its thermal environment in such a way as to permit its metabolic processes to reflect thermal regimes of the environment, but sufficiently meliorated so as to permit temporary exploitation of marginal habitats.

**Specific heat of water.** The specific heat of water is high (1.0 cal g$^{-1}$ C$^{-1}$) as compared with 0.09 for copper and 0.535 for ethyl alcohol. Low thermal diffusivity results in slow warming or cooling and poor conduction of heat within the organism. Animals with great bulk are slow to cool or warm, temperature of the body being related by heat transfer to circulating body fluids. For those species subjected to extreme cold, superficial fat is used as an effective insulator, since it possesses low heat conductivity, as do most fats and oils.

**Chemical reaction rates.** The reaction rates for most chemicals are temperature-controlled, i.e., rises in temperature accelerate the chemical reaction. A fish swimming in cold water cannot go faster than energy is made available in muscles. Insects on a cool morning cannot fly until the temperature rises sufficiently to permit accelerated metabolism required for sustained flight. A means for quantitatively comparing the magnitude of the effect of temperature changes on a chemical reaction is the $Q_{10}$ approximation, also known as Arrhenius equation. The $Q_{10}$ is the factor by which the reaction velocity (rate) is increased or decreased by a respective rise or fall in temperature of 10 C:

$$Q_{10} = (R_1/R_2)^{10/t_1 - t_2} \qquad (6.1)$$

where:

R$_1$ and R$_2$ are the reaction rate constants, at temperatures t$_1$ and t$_2$ (Fig. 6.1).

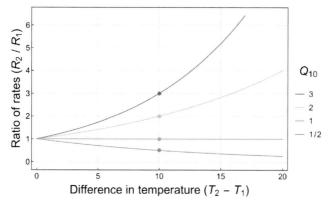

**FIGURE 6.1**  Chemical reaction rate plotted against temperature, °C, change. *Source: Wikipedia, 2018.*

It is important to recognize that the reaction rates of chemical and enzymatic-mediated reactions are not a constant, and that the $Q_{10}$ will normally vary over a temperature range. It is also important to know the temperature at which a $Q_{10}$ has been determined. Since the $Q_{10}$ is nonlinear, another useful means of displaying the change in the $Q_{10}$ coefficient is to plot the $\log_e$ of the reaction rate against temperature. Then the slope of the regression curve is the $Q_{10}$. Physical properties of solutions are less sensitive to temperature and the resulting $Q_{10}$s are between 1 and 2. Chemical reaction rates are usually doubled for each 10°C increase and thus range about 2. Enzymatic reactions are temperature accelerated more than can be accounted for on the basis of simple molecular agitation, and the $Q_{10}$ will often exceed 2. This simple difference in reaction rates for metabolic respiration (chemical) and photosynthesis (physiochemical) dictates the optimal temperature ranges over which plants can maintain a positive energy balance (Fig. 6.2).

The relationship between photosynthesis as a function of temperature is shown by the solid curve in Fig. 6.2A. At the ecosystem level of ecological complexity, temperature sensitivity of the respiration rate is constant at a $Q_{10} \approx 1.4$, globally. Thus, respiration increases exponentially with increasing temperature. However, over shorter time scales of minutes to hours, $Q_{10}$ has been shown to decrease linearly with temperature. In such situations, the temperature—respiration relationship would have the shape of the dashed curve in Fig. 6.2A. At some point high temperatures may affect the physical system that supports the basic chemical reactions which drive respiration and thereby reduce the efficiency with which the reaction can be carried out. This suggests that the temperature—respiration would bend over at high temperatures as in the dashed curve in Fig. 6.2A.

Carbon dioxide affects NEP in two ways: indirectly through temperature (both photosynthesis and respiration), and directly through a fertilization

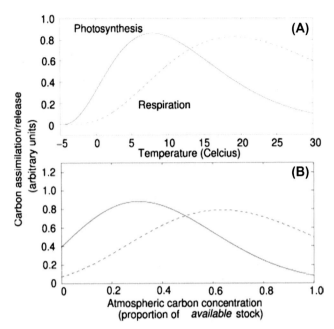

**FIGURE 6.2** Comparison of respiration and photosynthesis with temperature. *Image credit: Andereis et al., 2013. Courtesy, Environmental Research Letters and Creative Commons.*

effect. Given the uncertainty associated with changes in radiative forcing associated with $CO_2$ concentrations and changes in global and annual mean temperature, Andereis et al. (2013) made the assumption in Fig. 6.2B that annual mean temperature in the absence of feedbacks between different carbon stocks increases linearly with $CO_2$. However, the same differential response of photosynthesis and respiration to increasing temperature remains, i.e., with increasing temperatures, carbon loss through respiration proceeds more rapidly than carbon fixation through photosynthesis.

**Aquatic.** With exception of thermal springs, most natural water bodies rarely have temperatures which exceed the 35–40°C upper survival threshold for aquatic organisms. This is due to the high specific heat of water and its low heat conduction, and the great potential for evaporative cooling. At the lower extreme of the temperature spectrum, aquatic organisms can usually tolerate temperatures at or slightly below that of the medium—due in large part to the freezing point depression of their body fluids as a result of dissolved ionic and complex organic solutes. Since ice has a lower specific gravity than water, aquatic organisms will not freeze as long as they can remain beneath the floating ice. Water is densest at 4°C, which is the usual deep-water temperature during most of the year in temperate regions of the Earth.

**Terrestrial.** Terrestrial plants and animals are typically subjected to much greater fluctuations in temperature—in average values, the temperature extremes encountered, and in their frequency of change. Body temperatures cannot be allowed to "track" environmental temperatures uncontrolled, and the organisms living in terrestrial environments have evolved enumerable structural and functional mechanisms to cope with their temperature environment. These mechanisms affect the adaptability of organisms to the range of temperatures tolerable to life; temperature extremes, as for aquatic organisms, are more often than not dictated by the universal solvent and medium for life processes—water.

**Ectothermic and endothermic metabolisms.** An advantage that animals have over plants is their ability to some degree to control their body temperatures. The body temperatures of many organisms correspond directly with that of their environment. These animals are called "cold blooded" or poikilothermic animals; such animals are ectothermic, withdrawing heat from the environment, typically by sunning. Examples of ectothermic animals include amphibians, invertebrates, reptiles, and many forms of fish. A lesser number of organisms have the capability of regulating their body temperatures through physiological processes, such as respiration rate, controlled circulation, regulated evaporation, insulated covering, etc. They are call "warm blooded" or homeothermic animals. Such endothermic animals include birds and mammals and are capable of maintaining a relatively constant body temperature independent of the surrounding environment. It should not be unrecognized that homeotherms are still subject to thermal stress and that poikilotherms are not without thermal control through behavioral responses (Fig. 6.3).

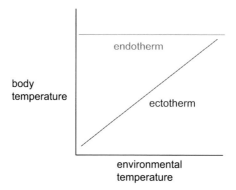

**FIGURE 6.3**   Response of ectotherms and endotherms to increasing temperature. *Image credit: Wild at Heart. Courtesy, Automatic:World Press.*

**Acclimation.** Poikilotherms exhibit alterations in chemical rate functions, behavior, and activity patterns in accordance with the temperature regimes to which they have been preconditioned. Consequently, when temperature is either raised or lowered, their accompanying $Q_{10}$ temperature adjustment may

- undershoot through hypercompensation,
- overshoot through overcompensation, or
- gradually readjust to the original rate.

These processes occur as the organism attempts to regulate its metabolic and other activity rates to correspond to those original rates to which it has become accustomed.

Following the initial rate response, a stabilized rate may last for several hours. This rate is typically a function of the present temperature and the previous temperature to which the organism had been acclimated. It is during this interval that most $Q_{10}$s are determined. If the animal remains at the altered temperature for several days, its rate functions will often show some compensation; that is, begin to approach original values—it becomes acclimated. Now, if the organism suddenly returns to its original temperature, the rate function will not return to its original level but rather to a higher or lower value depending upon the original direction of acclimation.

Change in the position (translation) of a rate/temperature curve implies a change in activity (in the thermodynamic sense) of some enzyme system. Change in the slope (rotation) indicates a change in $Q_{10}$ and, therefore, in activation energy. Factors most likely to affect each are:

| Translation | Rotation |
|---|---|
| Change in enzyme concentration | Alteration of enzymatic protein |
| Change in enzyme rations | Change in some factor |
| Change in controlling conditions | Shift of control to an alternate |
|    Ionic strength |    enzyme pathway |
|    pH | |
|    Water activity | |

In general, compensation for cold by acclimatizing reactions is more marked in aquatic than in terrestrial poikilothermic organisms, although it does occur in snails and insects. An excellent example of the practical nature of acclimatizing adaptations is cleavage of sea urchin eggs. Through acclimatization, the sea urchin can adapt to lower water temperatures and egg development proceeds at a rate proportionately higher than would be possible through a simple $Q_{10}$ response (Table 6.1).

**TABLE 6.1** The development time of sea urchin eggs as a function of temperature demonstrates how energy (heat) affects biological processes, and how acclimation to warmer summer (ˆ) temperatures, or cooler winter[(†)] temperatures, affects development. Natural populations of *Paracentrotus lividus* range between 13 °C–28°C[@].

| | Water temperature | | | | | |
|---|---|---|---|---|---|---|
| | 13°C | 16°C | 18°C | 19-22.5°C | 26°C | 34°C |
| Development range | hatching[‡] | | | no cleavage[‡] | | |
| Time for cleavage (hours) | 91†(71ˆ) | | | 25†(31ˆ) | | |
| Fertilization success (%) | 95[#] | | | >97[@,#] | | |

[‡] *(Tripneustes gratilia) Rahman, et al., 2009;*
[@] *(Paracentrotus lividus) Cohen-Rengifo et al., 2013;*
[#] *(Paracentrotus lividus) Privitera et al., 2011;*
*preconditioned at 13°C,*
[†] *preconditioned at 26°C, Vernberg & Vernberg, 1970.*

## 6.2 Adaptation to the energy environment

**Adaptive mechanisms.** Species have evolved a number of adaptive mechanisms—some physical, some chemical, some structural, some physiological, and others behavioral—to affect their coupling and uncoupling with their environment. Oftentimes parallelisms exist between plant and animal species, but generally animals, because of their mobility and self-regulating rates of metabolism, exhibit the greatest diversity of adaptive mechanisms. Some examples of the more common adaptive mechanisms are shown in Table 6.2 to illustrate how important it is for living organisms to remain in balance with their thermal environments.

**Adaptive strategies.** The examples of energy adaptations in Table 6.2 illustrate many of the adaptive strategies which plant and animal species deploy in response to their energy environments. The basic objective is that the energy gained by an organism exceeds the energy expended by a sufficient margin as to allow for energy expenditures by the organism necessary to sustain life. Analogously, these examples also imply the energetic constraints which the environment places upon species. The adaptive choices are finite, as is the extent to which any one choice can be exploited. Yet, there are a myriad of combinations of the variables affecting the total energy balance of an organism.

To illustrate, we can use the example of hummingbirds, where the energy intake is solely from nectar and insects. Energy expenditures can be itemized under the categories of maintenance, foraging, reproduction, and development;

**TABLE 6.2** Some examples of adaptive strategies of plants and animals to their energy environment.

| Adaptive Mechanism | Plants | Animals |
|---|---|---|
| **Physical** | | |
| Insulation | • Thick waxy skin of cacti<br>• Bark of trees | • Fur and hair<br>• Fatty body deposits |
| Geometry | • Small rounded leaves of desert plants<br>• Large leaves of tropical forest understory | • Body size (Bergman Principle for homeotherms)<br>• Converse Bergman for poikilotherms<br>• Size of appendages, muzzle |
| Texture | • Leaf hairs<br>• Reflectivity of surface | • Rough skin for air dwellers<br>• Smooth skin for aquatic organisms |
| **Chemical** | | |
| Pigmentation | • High absorptivity of conifers & deciduous<br>• Pigmentation | • Control of skin pigmentation<br>• Darker pigmentation of homeotherms in tropics |
| Solvent | • Tannins, oils, other organics to alter Freezing Point | • Tannins, oil, other organics; osmo regulation |
| Physiological | • Deciduousness in hot and cold climates<br>• Evapotranspiration<br>• Leaf abscission<br>• Flowering times | • Torpor, aestivation, hibernation<br>• Panting, breathing<br>• Evaporative cooling<br>• Sweating, altered fluid circulation in body<br>• Food utilization/metabolism |

| Behavioral | • Leaf angle, orientation | • Sunning to warm, |
| | | • Vibrating of wings to warm body |
| | | • Ventilation of bee hives |
| | | • Diurnal/nocturnal activity pattern |
| | | • Sheltering/denning |
| Ecological | • Altered life cycle(phenology) | • Change in rates of development |
| | • Change in GPP/NPP | • Modified reproductive strategy; Breeding seasons |
| | • Life zone preference | • Change in population dynamics |
| | | • Altered trophic structure |

then, these expenditures are compared to the parameters which influence that expenditure and the associated energy intake (Table 6.3).

The parameters that affect energy expenditures for maintenance and foraging in hummingbirds relate to their energy intake from nectar. We can hypothesize that a consumer organism should evolve in such a way that the foraging efficiency is optimized in order to provide sufficient time and energy for growth, maintenance, and reproduction. The degree to which optimization of intake versus utilization for hummingbirds occurs becomes apparent from the energy budget. The importance of different activities for such an optimization is indicated by four key strategies of hummingbird feeding:

- the optimum food sources
- the optimal time spent foraging
- the optimal foraging space, and/or
- the optimal foraging group size (i.e., competition)

**TABLE 6.3** Some aspects of an energy budget for hummingbirds. After Wolf *and* Hainsworth, *1971.*

| Energy intake | Energy expenditure |
|---|---|
| From nectar: | For maintenance: |
| • Sugar (cal) concentration (affected by temperature, humidity, photoperiod) | • Temperature and local environmental conditions |
| • Volume of nectar per flower (affected by temperature, humidity, photoperiod) | • Body size effects on metabolism |
| • Availability of flowers (affected by temperature, humidity, photoperiod) | • Torpor during stress |
| From insects: | For foraging: |
| • Caloric value | • Body size effects on metabolism |
| • Availability (numbers and size) | • Time for foraging |
| | • Energy availability (stored food reserves) |
| | • Feeding and searching strategy |
| | For reproduction |
| | • Social behavior |
| | • Energy demands of young |
| | For development |
| | • Nest environment |
| | • Energy partitioning for growth & development |

## 6.3 Phenological relationships

Temperature affects all stages of the life cycle of organisms. One of the most striking aspects of life in temperate latitudes is the changing aspect of the seasons, i.e., the climatic effects on biological activity. Although this phenomenon occurs in tropical as well as boreal environments, the alternating extremes of annual warm and cold periods places considerable stress on organisms inhabiting temperate environments. It is not surprising that climate should exert such a controlling force on biotic activity; what is striking is the responsiveness of organisms to their environmental temperature regimes. Rather sophisticated mechanisms have even developed whereby the organism can respond to anticipated temperature regimes in the environment beyond current ambient temperatures.

**Phenology.** Describing the phasing of life cycle events of plants and animals in their temporal occurrence throughout the year is termed phenology. As opposed to the astronomic or civil calendar, the phenological calendar dates the seasons of the year as distinct groups of biological events. In plants these events are called phenophases and begin with seed germination and terminate with senescence and leaf drop. Phenological relationships are important for the management of commercial crops such as soybeans (Fig. 6.4). From the model structure it can be seen that environmental temperature (heat) and solar radiation (heat) affect the rate of plant development, photosynthesis and respiration, transpiration, and seed growth (Yang, 2019).

While the sequence of phonological events in the phenological calendar is determined by the genetics of the species, the onset of each phase is often controlled by one or more environmental factors—typically temperature. Certain developmental events are responses to moisture, light, or temperature thresholds and, consequently, are often manifested with some regularity corresponding to seasonal changes. The capability to accurately predict phonological events has been utilized by climatologists and agricultural meteorologists to:

- select optimum seeding dates
- plan other agricultural practices
- predict harvest dates
- time pest control treatments

In defining the beginning, duration, and end of the growing season, from year to year, phonological observations are often better indices to the bioclimatic character of local areas than recordings made by mechanical instruments. The potential for phenology to be used as a land management practice tool is immense. This should come as no surprise to those who understand the functional coupling of plants to their environment.

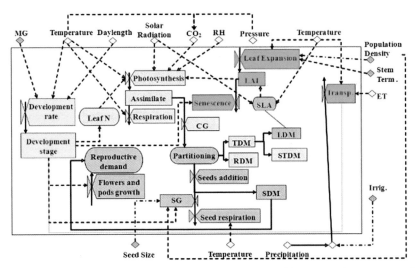

**FIGURE 6.4** The phenology, leaf expansion and senescence, and biomass growth components of a soybean simulation model interact dynamically and demonstrate how each are influenced by weather variables. *TDM*, Total above ground dry matter, *RDM*, Below ground dry matter, *LDM*, Leaf dry matter, *STDM*, Stem dry matter, *SDM*, Seed dry matter, *CG*, Crop growth, *SG*, Seed Growth, *MG*, Relative maturity group, *Stem Term*, Stem termination type (Indeterminate vs Semideterminate), *RH*, Relative humidity, *ET*, Reference evapotranspiration, *Irrig.*, Irrigation. *Image credit: Haishun Yang, University of Nebraska-Lincoln. Source: Yang, 2019.*

To illustrate how sophisticated phonological predictions can be, we take the example of floristic development (flowering) of the plants comprising a forest ecosystem in response to progressive changes in temperature as the year progresses, using the date of first flowering as the key phenophase for each of 133 species in a temperate deciduous forest. A phenogram of the oak-hickory forest association (Fig. 6.5) depicts flowering seasons with moist and dry conditions, distinguished by spring and early summer floras. The data in Fig. 6.6 are the mean dates of first flowering between 1963 and 1970. Richness of the understory is illustrated by the number of herbaceous species observed. The numbers in parentheses indicate separate species observed.

In the deciduous forest ecosystem, there are two periods of peak flowering (Fig. 6.5) during the first 6 months of the calendar year: spring in mid-April (week 16) and summer in early June (week 23). The decline in flowering following the spring peak is probably related to leafing out and canopy closure by the dominant overstory trees. Transmission of incident radiation to ground level is thereafter reduced 40% following spring flowering. The lower levels of radiation penetrating to the forest floor correlate with a canopy closure of 75%. By the time of complete leaf development (week 24) light transmission to the ground is reduced to only 11%. The relationship between the thermal environment (temperature summation) and date of first flowering (year-day)

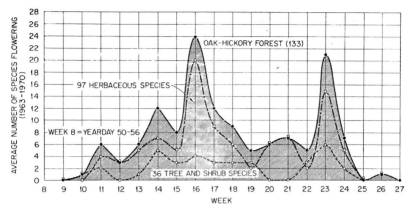

**FIGURE 6.5** Flowering phenophases in a temperate deciduous forest. *Source: Taylor, 1969. Courtesy, Oak Ridge National Laboratory.*

provides significant predictability across species (Fig. 6.6). The term *degree day* refers to the successive, cumulative daily temperature summation above a predetermined initial threshold temperature value. Silver maple (*Acer saccharinum*) and red maple (*Acer rubrum*) are the first species to begin reproductive phases at a time when the mean daily air temperature is approximately 10°C. The significant correlation ($r^2 = 0.99$) suggests that observations of mean date of first flowering spanning several years is sufficient to estimate advent of the reproductive phase for diverse species. Temperature summations, $T_s$, utilizing daily maximum and minimum temperatures can provide even more precise estimates of flowering dates from temperature records, and be used to predict the seasonality of production (Taylor, 1974).

## 6.4 Extreme environments

Before studies of biological organisms in hot springs began in the 1960s, scientists long thought that thermophilic bacteria could not survive in temperatures above 55°C (131°F). However, many bacteria were discovered that not only survived, but also thrived at higher temperatures. Brock and Freeze in 1969 reported a new species of thermophilic bacteria, which they named *Thermus aquaticus*, living in Mushroom Spring in the Lower Geyser Basin of Yellowstone National Park. The temperature optima of bacteria occurring at various temperatures along the thermal gradient (35−70°C) in a Yellowstone hot spring was studied directly in nature by measuring the rate of incorporation in the dark of $^{14}$C-glucose or $^{14}$CO$_2$ (Brock and Brock, 1968).

Bacteria in the *Deinococcus−Thermus* group have adapted by evolving a heat-resistant enzyme, Taq DNA polymerase, in their metabolic process, which is also by circumstance one of the most important enzymes in molecular biology because of its use in the polymerase chain reaction (PCR) DNA amplification technique. The gelatinous mats of blue-green algae (cyanobacteria) and bacteria in the alkaline thermal springs of Yellowstone National Park provided Richard Wiegert and his students the opportunity to study the ecology of living organisms exposed to the thermal limits of life (Weigert and Mitchell, 1973).

Another temperature extreme for life is the surface ice of Arctic glaciers. *Chlamydomonas nivalis* is a green alga that owes its red color to a bright red carotenoid pigment, which protects the chloroplast from intense visible and ultraviolet radiation. The carotenoid pigment also absorbs heat, which increases the heat content of the algae and provides it with liquid water from the surrounding melting snow (Wikipedia, 2018). Algal blooms may extend to a depth of 25 cm (10 in.), with a teaspoon of melted snow containing a million or more cells. The algae accumulate in shallow depressions in the snow. Many species feed on the alga, including protozoans such as ciliates, rotifers, nematodes, ice worms and springtails—a virtual ecosystem unto itself. Presence of the algae can reduce the albedo, or reflectivity, of the snow by as much as 13% (Griggs, 2016), thus raising surface temperature and encouraging the melt conditions. (Remember the Gaia Hypothesis from chapter 2.4?)

We have in the first six chapters discussed how the laws of thermodynamics apply to biological system; how the energy environment affects an organism's existence; how plants convert radiation energy and accumulate carbon into organic molecules; how animals process food energy and convert carbon into biomass; and how plants and animals adapt to their energy environments. Now

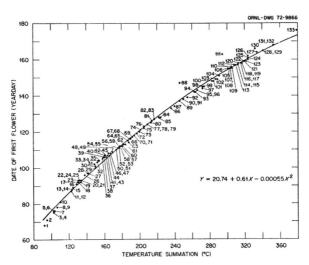

**FIGURE 6.6** Phenological degree-day summation predicting flowering for 133 species of vascular plants in an oak-hickory forest at Oak Ridge, Tennessee. *Source: Taylor, 1969. Courtesy, Oak Ridge National Laboratory.*

we will explore how energy is transformed and carbon transferred along food chains and through the trophic structure of ecosystems. We are moving from bioenergetics to greater system complexity by examining ecological energetics. Eventually we will increase spatial scales to address carbon and energy flows in the biosphere. But now let us begin with our discourse into ecological systems.

## 6.5 Recommended Reading

Lieth,, H., 1974. Phenology and Seasonality Modeling. Springer Verlag, Berlin-Heidelberg-New York, p. 444 pp. https://www.springer.com/gp/book/9783642518652.

Precht, H., Christophersen, J., Hensel, H., Larcher, W., 1973. Temperature and Life. Springer-Verlag, Berlin, Heidelberg, p. 782 pp. https://www.springer.com/gp/book/9783642657108.

Vernberg, J., Vernberg, W., 1970. The Animal and the Environment. Holt, Reinheart and Winston, 398 pp. https://www.amazon.com/Animal-Its-Environment-John-Vernberg/dp/0030796458.

# Chapter 7

# Food chains and trophic level transfers

Our overall interest has been in following radiant energy after it reaches the Earth. The transformation of energy and the production of energy-containing carbon compounds occur in the biosphere. To understand what happens to energy after it enters the biosphere requires an understanding of ecological energetics. The effective running of ecological systems depends upon the transfers of energy among organisms, and energy transfers of this type occur along food chains and move between trophic levels. Beginning with this chapter we encounter the sphere of inquiry traditionally known as ecological energetics, which embraces not only the exchanges of energy between an organism and its physical environment but also the exchanges of energy through energy-rich carbon molecules with other animals.

Although the living organism is inextricably connected energetically to its physical environment, it is to that environment that the organism's primary path of energy loss—and gain—occurs. It is the food chain which provides the main source of energy input through the chemical energy content of food molecules necessary to sustain the life processes. The food chain is the connection of animals in their photosynthesis. The successive steps in the feeding process from green plants to consumer organisms are referred to as links in the food chain. All organisms occupying the same stage in the feeding process from the green plant food base are said to belong to the same trophic (from the Greek, *trophe*, food) level. Ecologists use the concepts of food chains and trophic levels to describe and analyze the energetic relationships of plants and animals in nature.

## 7.1 Food chains

We have seen that transformation of the chemical energy of biomass into a useable form for metabolic processes occurs in three steps: (1) digestive enzymes hydrolyze macromolecules into their monomeric constituents, (2) these monomers are oxidized to one of three substances: acetic acid in the form of acetyl coenzyme A, $\alpha$-ketoglutarate, and oxaloacetate. These three compounds participate in the tricarboxylic acid cycle, where the substances are further oxidized to two $CO_2$ molecules and four pairs of H atoms, and are then further oxidized by $O_2$ with a release of energy. This final step of oxidative phosphorylation produces ATP. This is the trophic dynamic basis of all food chains. A herbivore, a predator, or a decomposer must be able to extract energy from its food. When analyzed thermodynamically, the transformation of biomass between trophic levels is always less than unity, generally closer to around 10%, because in order to enter the metabolic process of the next higher trophic level, the ingested material must first be broken down into its monomeric constituents, losing energy in both un-assimilable material and as heat of catabolism. Although the living organism is inextricably connected energetically to its physical environment, it is to that environment that the organism's primary path of energy loss — and gain — occurs.

**Links in the food chain**. In the trophic exchange of energy among organisms, regardless of the food chain, the original source of food energy derives from the transformation of radiant energy by green plants. Plants are called autotrophs (capable of making their own food). All other organisms (animals) which consume plants, or plant parts, to satisfy their food energy requirements are called heterotrophs. Each link in the food chain is described by specific terminology that not only provides which link that organism occupies but also the source of its food. Food chains are generally not very complex, typically having no more than three to five links (although parasites may add several more links) including the plant food base; but a staggering number and diversity of different kinds of food chains exist. Thus, some basic terminology aids in scientific communication of concepts. These terms carry the suffixes *phage* (from the Greek, to eat) or *vore* (from the Latin, to devour); the core common terms being:

- autotroph:                                          green plant, primary producer
- heterotroph feeding upon green plants:   herbivore, primary consumer
- heterotroph feeding upon animals:         carnivore, secondary consumer

Herbivores may feed upon foliage (grazers), seeds (granivores), wood (xylophages), plant secretions (nectivores), or dead remains (detritivores aka saprophages). Herbivores may, in turn, constitute the prey of carnivorous animals (predators) and parasites. Nonspecific feeders that consume both plant and animal tissues are called omnivores. Decomposers are the waste disposal system for dead organic material and have their own complex food web of organisms which feed upon detritus (detritivores), fungus (fungivores), and dead animals (carrion feeders). Thus, the biomass of animals in nature comprise a chain of potential energy pools connected by the flux of energy transferred during feeding, one upon the other. Such a sequence of energy exchanges is termed a food chain.

To illustrate the flow of energy along a food chain, we can conceptually compartmentalize the pools and flows of energy between successive feeders (Fig. 7.1). For consistency of notation, the energy pool represented by each organism (biomass) is shown as a box (compartment) and may change in size due to growth, reproduction, and death. There is only one pathway for intake of metabolizable food energy and that is through consumption. Four pathways of energy outflow exist (since we are following food energy, the conductive and convective heat losses by the animal are not shown). The four energy losses are: respiration (R), death ($\Delta B$), feces (F), and predation (A of the next trophic level). An understanding of the magnitude of these parameters is important, because the magnitude of these energy expenditures determines the amount of energy available to an animal and ultimately the trophic level.

The further along a food chain that an organism is removed from the primary food base (plant) of the chain, the less energy is available, and the less stable bioenergetically is its link in the food chain. This is because the transfer (conversion) of chemical energy among animals is not 100% efficient—at each transfer in the chain energy is lost through food either not consumed and food which is undigestible; between transfers energy is lost through mortality, respiration, and metabolic byproducts. The relative magnitude of these values and the efficiencies of transfer are characteristic of the individual organisms, and not unlike the energetic parameters discussed in Chapter 5. Stability is provided to the food chain through (1) regulatory feedbacks in the food chain which take the form of predator control of prey population sizes and food resource limitations, and (2) external connectivity with other food chains in a food web so as to eliminate dependence upon solely one food resource.

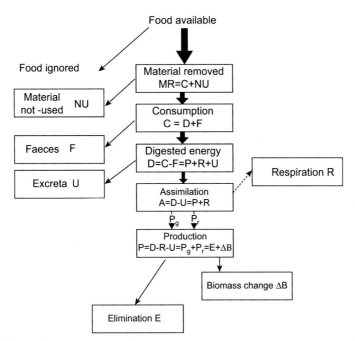

**FIGURE 7.1**   Scheme of matter and/or energy flow for a food chain or trophic level. MR, total material removed by the organism or population; NU, material removed, but not consumed; C, consumption; FU, rejecta; F, egesta; U, excreta; A, assimilation; D, digested energy/material; P, production; $P_g$, production due to body growth; $P_r$, production due to reproduction; R, respiration; $\Delta B$, changes in mass of the individual or population; E, elimination. *Nomenclature after Petrusewicz and Macfadyen, 1970.*

**Tracing food chains**. In order to analyze food chains, it is first necessary to identify the food chain—the links in nature. While simple in concept, tracing a food chain can be a challenge. Several techniques are used:

Direct observation is the simplest and quite accurate. It may be relatively easy to observe grasshoppers feeding on only one plant species, and occasionally see a spider eating one; but while adequate to establish this link, the full range of predators feeding on the grasshoppers may be far more difficult to obtain. Gut analysis often will provide quantitative data about the range of food items ingested by an animal, even though differences in digestibility may hinder quantitative budgeting of food input. Radioactive tracers are a unique tool that can be used in free-ranging animals to identify food chains but also the trophic position in the food chain. The time delay between peaks of radioactivity between consumed and consumer populations reflects the temporal delay in the flux of materials (and energy) along the food chain (Fig. 7.2). The phase differences

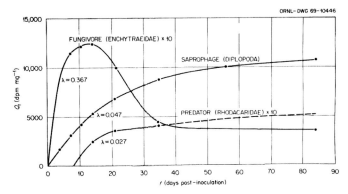

**FIGURE 7.2**  The time delays between peaks of radioactivity concentrations in trophic levels reflect the temporal delay in the flux of energy along food chains. *Image credit: Reichle and Van Hook, 1970.*

between successive peaks of radioactivity to appear in animals in a food chain can identify their trophic position and can identify successive food chain links from the food base (Crossley and Reichle, 1969). From a variety of studies, there appears to be a progressive reduction in the time between peaks as material and energy proceed along the food chain to higher trophic levels (Reichle and Van Hook, 1970). Presumably, this is related to the increasing efficiency of energy utilization at higher trophic levels.

The food chain relationship using radioactive tracers can be quantified using the intake-loss balance equation:

$$dC/dt = IA_o e^{-k_1 t} - k_2 C \qquad (7.1)$$

where:

C is the concentration of radioisotope in the consumer,
I is the consumer feeding rate in the gram plant/gram animal/time,
$A_o$ is the initial concentration of radioactivity in the plant,
$k_1$ is the loss rate of radioactivity from the tagged plant, and
$k_2$ is the loss rate of radioactivity from the consumer (loss rates measured in units of time$^{-1}$). Loss rate coefficients, k, are equal to:

$$k = 0.693/T_b \qquad (7.2)$$

where:

$T_b$ is the biological half-life of the radioisotope in that animal.

Then the radioisotope concentration in the consumer animal, as a function of time, is the solution to the above differential equation:

$$C(t) = IA_o/(k_2 - k_1)(e^{-k_1t} - e^{-k_2t}) + C_oe^{-k_2t} \tag{7.3}$$

where:

$C_o$ is the initial concentration of radioisotope in the consumer. The time required for maximum concentration to occur in the consumer is obtained by setting the derivative of the preceding equation equal to zero and algebraically solving for $t_{max}$:

$$t_{max} = (1/k_2 - k_1)\ln k_2/k_1, \quad C_o = 0 \tag{7.4}$$

The unique value of the radiotracer method in studying is that it can (and should) be used to study populations in their natural environment.

**Measuring food intake rates and assimilation.** Quantification of food intake rates, ideally for animals in their natural environment, is a crucial variable for determining energy flow along food chains. Numerous techniques have been developed, all of which are modification of gravimetric analyses, and used primarily upon captive animals which introduces uncertainties. Again, radioactive tracers offer a unique advantage. By tagging a food source and measuring the concentration of radioisotope in the food, and then measuring the radioactivity after a period of time in the consumer, the preceding equation can be solved for I, intake or the feeding rate (Reichle, 1967).

While determining the biological half-life of the radioisotope in the test animals, the food assimilation by the animal is also revealed. Animals fed upon radioisotope-tagged food acquire body burdens of isotope, Q, which when the animals are removed from the tag and fed on uncontaminated food decreases as:

$$dQ/dt = kQ, \tag{7.5}$$

For most animals the whole-body turnover of most radioisotopes consists of a two-component curve with two interacting rate coefficients, $k_1$ and $k_2$, reflecting the loss rates of the unassimilated radioisotope pool in the gut, $q_1$ and tissue pool, $q_2$; the y axis intercept of the body loss curve is the food assimilation value, $p_2$ (see Fig. 5.4). The parameters $p_1$ and $p_2$ are the proportions of each quantity of radioisotope in each compartment of the system (gut + body). Following instantaneous input of radioisotope (single feeding), digestive assimilation, $p_2$, and $q_2$ are equivalent. The identity between $p_2$ and $q_2$ ceases beyond the instantaneous input, since until the organism equilibrates with the radioisotope in its food, the tissue compartment $q_2$ continues to accrue radioisotope. Thus, beyond a single feeding, unassimilated radioisotope in the gut, $q_1$, becomes an increasingly smaller fraction of the total body radioactivity, Q, includes tissue excretion, and $q_2$ cannot be considered equivalent to digestive assimilation.

The instantaneous change in whole body radioactivity, Q, of an animal is described by the differential equation:

$$dQ = \sum_{i=1}^{n} I_{pi} - k_i Q_i \tag{7.6}$$

where:

I is a constant radionuclide ingestion rate (units of radioactivity per unit time)
k is the turnover coefficient (time$^{-1}$),
p is the proportionality constant of intake previously defined, and
n is the number of compartments in the system.

The solution to the differential equation is given by:

$$Q_t = \sum_{i=1}^{n} \frac{I_{pi}}{k_i} \left(1 - e^{-kit}\right) + Q_{oi}\, e^{-kit} \tag{7.7}$$

If the organism has not previously accumulated a body burden of radio-activity, the second term of the equation is zero and,

$$Q_t = \sum_{i=1}^{n} \frac{I_{pi}}{k_i} \left(1 - e^{-kit}\right) \tag{7.8}$$

When $t \to \infty$, $Q_t \to Q_e$ and $e^{-kit} \to 0$, Eq. (7.8) can be rearranged to solve for I (intake):

$$I = \frac{Q_e}{\sum_{i=1}^{n} \frac{p_i}{k_i}} \tag{7.9}$$

The total radioactive body burden, Q, may be defined as consisting of distinct components:

$$Q = \sum_{i=1}^{n} a_i Q_i \tag{7.10}$$

where:

$A_i$ is the fraction of total body burden in compartment i.

Therefore:

$$a_i Q = q_i \tag{7.11}$$

By evaluating Eq. (7.10) for $Q_t$ it can be shown that while the $p_i$s are constants, the $a_i$s are a function of time, t, until equilibrium, $Q_e$, is attained.

From the intake Eq. (7.3) a weighted k* may be derived for a two-compartment system at equilibrium as:

$$I = \frac{Q_e}{\frac{p_1}{k_1} + \frac{p_2}{k_2}} \tag{7.12}$$

and

$$k^* = \frac{k_1 k_2}{p_1 k_2 + p_2 k_1} \tag{7.13}$$

When $Q = Q_e$, $dQ/dt = 0$, and

$$I = Q_e \sum_{i=1}^{n} a_i k_i \tag{7.14}$$

To derive k* from Eq. (7.13) for a two-compartment system:

$$I = Q_e(a_1 k_1 + a_2 k_2) \tag{7.15}$$

and

$$k* = (a_1 k_1 + a_2 k_2) \tag{7.16}$$

The variable k* may be substituted in an abbreviated version of Eq. (7.14):

$$I = k * Q_e \tag{7.17}$$

But this intake is valid only for an equilibrium situation (an animal raised on labeled food or living in a contaminated environment). For more typical conditions of transient whole-body concentrations of radionuclide during experimental food uptake from a labeled food source, Eq. (7.8) must be used to calculate isotope (and food) intake.

Determination of the biological half-life, $T_b$, is the one experimental parameter necessary to calculate ingestion rates from the preceding Eqs. (7.8) or (7.17), isotope intake-loss balance equations. Animals accumulate measurable amounts of radioisotope after feeding for 24 h upon tagged food. When they are removed to nontagged food, animals lose radioactivity in a pattern termed a "two-component" retention curve (Fig. 5.1). Radioactivity in the animals through time, $A_t$, is:

$$A_t = A_o \left( p_1 e^{-k_1 t} + p_2^{e-k_2 t} \right) \tag{7.18}$$

where:

$A_o$ is the initial radioactivity of the animals,

$p_1$ and $p_2$ are proportions ($p_1 + p_2 = 1.0$) of radioactivity lost at rates $k_1$ and $k_2$, respectively, and

e is the base of the natural logarithm.

The rate $k_1$ represents loss of nonassimilated isotope from the gut, while $k_2$ measures excretion of assimilated radioactivity. Thus, $p_2$ is equivalent to digestive assimilation efficiency. The half-time, $T_b$, is calculated from the rate constant k.

While food chains at this point may appear to be quantitative complexities and differential equations, it is important to realize that the food chain is more than an abstract scientific concept, but instead one of the most important life-supporting ecological processes occurring in nature.

## 7.2 Population dynamics and food chains

Similar to the flow of energy through an individual, we can evaluate the result of food chain transfer of energy at the population level by examining the biomass production by that population. The biomass of organisms can, and usually does, fluctuate—oftentimes markedly—during the year. Consequently, the potential energy in biomass of a species population is variable, both within a year and between years. Consider a hypothetical population with a mean life span of 3 years, one litter of offspring per year, maturation time of 1 year, completion of growth of young in 4 months, and with a stable year-to-year reproductive rate and population size (Fig. 7.3):

**Biomass and numbers**. Biomass and production are terms frequently misunderstood and inappropriately used. Biomass is the quantity of living organic matter, often referred to as standing crop, and reported as weight per unit area. Production is the quantity of biomass produced; production per unit time is the *productivity* rate. Space and resources for most populations are finite and, therefore, and the relationship between biomass and production rate operates in a density-dependent manner. At low population densities there is a low biomass but high rate of production due to lack of competition. At high densities competition inhibits the rate of production, but biomass is nearly maximized. A complete accounting of biomass in a population requires tracking births, deaths, and growth rates throughout the growing season (Fig. 7.3).

**Productivity**. Is it better to be big or small? Relations between rates of production (*productivity)* and cumulative biomass also differ among species due to genetic difference and variations in environmental influences on phenotypic expression. A comparison of three different animal species, each with different combinations of productivity, reproduction, and metabolic rates, illustrates the ecological significance of the interplay of these parameters (Table 7.1).

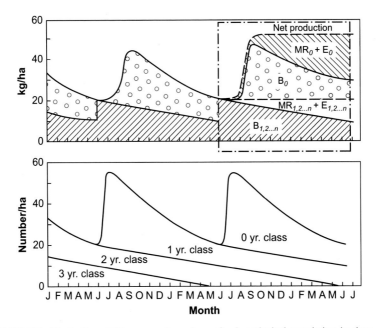

**FIGURE 7.3** Fluctuation of biomass and numbers of a hypothetical population in time. Assumptions are: a life span of 3 years, one litter per year, maturation in 1 year, completion of growth of young in 4 months, and a stable population and reproductive rate from year to year. The insert shows partitioning of biomass for net production per year. The net production exceeds the biomass peak because of the production of animals dying prior to the time of biomass peak. $B_O$, biomass of current generation; $B_{1,2...n}$, cumulative biomass from earlier generations; $E_O$, elimination and MR, material removed by predation; *Source: McCoullough, 1970, Courtesy, Springer Nature.*

**TABLE 7.1** Comparison of productivity between mouse, deer, and elephant.

| $kcal\ m^{-2}\ yr^{-1}$ | Meadow mouse | White-tailed deer | Elephant |
|---|---|---|---|
| Standing crop | 0.2 | 1.3 | 7.1 |
| Growth per unit standing crop | 2.5 | 0.5 | 0.048 |
| Net production | 0.5 | 0.65 | 0.34 |

McCullough, 1970; modified from Petrides and Swank, 1966.

The elephant, because of its longevity, large body size, and low maintenance metabolism in relation to body weight has a tremendous capacity to accumulate biomass. The elephant is not suited for production, since its rate of biomass production (growth rate) and rate of reproduction are both low. Since sustained yields are limited by production, the elephant is not a good producer. The meadow mouse represents the other extreme. Its production rate (ability to produced new biomass) is very high, but its ability to accumulate mass is very low because of the high costs of metabolism. The high reproductive capacity is offset by the inability of the mouse to maintain densities high enough to establish significant standing crop biomass. The white-tailed deer represents an intermediate example, where the species' capacity to produce new biomass (reproduction) and to accumulate biomass (growth) achieve a balance. The potential yield from animals with this type of balance has been a key factor in human's selection of wild animals for domestication—an unconscious understanding of the principles of population bioenergetics!

## 7.3 Food webs

Rarely, with the exception of domesticated crops and their specialized insect pests, does the simplistic concept of the food chain exist. Some species are so specialized, however, that they will accept only one food resource (e.g., pandas). More typically, species are polyphagous and exhibit considerable flexibility in switching food preference according to the changing seasonal availability of food (opossums). This intermeshing of food chains constitutes the food web (Fig. 7.4).

**Quantification of Food Webs**. The quantification of food webs is a laborious and complicated task, and can be a considerable challenge for ecologists (Benke and Huryn, 2017) There are two types of approaches to quantify: a flow web and an I/P web. A flow web quantifies the ingestion flow of energy to each prey and to each predator (e.g., mg m$^{-2}$ yr$^{-1}$ or kcal m$^{-2}$ yr$^{-1}$). An I/P

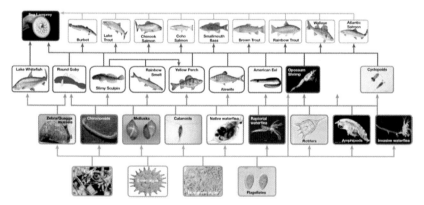

**FIGURE 7.4**  Food web showing the interactions between organisms across trophic levels in the Lake Ontario ecosystem. Primary producers are outlined in green, primary consumers in orange, secondary consumers in blue, and tertiary (apex) consumers in purple. Arrows point from an organism that is consumed to the organism that consumes it. *Source: NOAA, 2014.*

web is based upon measurement of ingestion/production (the fraction of each prey's production ingested by each predator and can sum all predatory mortality for each prey species). To construct each food web, one needs to know the production of each consumer. The I/P web also requires knowledge of the production of each food type and the fraction of each food type eaten by a consumer. Using Benke's notation:

$$RA_{ij} = G_{ij} \times AE_{ji} \qquad (7.19)$$

where:

$RA_{ij}$ is the relative amount (RA) of each food type $i$ assimilated by each taxon $j$

$G_{ij}$ is the proportion of a food type $i$ in consumer $j$'s diet, and

$AE_{ij}$ is the relative amount of food type $i$ assimilated by consumer $j$.

$AE_{ij}$ must be converted into the fractional amount assimilated as:

$$FA_{ij} = RA_{ij} \sum i = 1nRA_{ij} \qquad (7.20)$$

The actual amounts of each food type assimilated ($AA_{ij}$) and contributing to consumer $j$'s production (trophic basis of production) is determined by multiplying the fractional amount assimilated ($FA_{ij}$) by consumer $j$'s production ($P_j$) or:

$$AA_{ij} = FA_{ij} \times Pj \qquad (7.21)$$

Ingestion flows can be easily determined by dividing the production of consumer $j$ attributed to each food type ($AA_{ij}$) by the GPE for that food type. GPE is simply the product of NPE and AE. NPE is that fraction of assimilation that is converted into biomass (generally assumed to be 0.50 for all consumers). The ingestion flow from food type $i$ to consumer $j$ is therefore:

$$I_{ij} = AA_{ij}GPE_i \qquad (7.22)$$

where:

$GPE_i$ is the GPE for food type $i$.

The I/P web also uses the ingestion flow from food type $i$ to consumer $j$. However, I, the ingestion flow is divided by production of food type $i$ giving the fraction of prey production ingested by the consumer. Thus, I can be calculated as:

$$I_{ij}/P_i \qquad (7.23)$$

where:

$P_i$ is the production of the prey species (not that of the consumer or predator).

Total predation pressure of the prey is, therefore, the sum of ingestion flows from prey $i$ to all predators divided by prey production:

$$\sum j = 1nI_{ij}/P_i \qquad (7.24)$$

Thus, the I/P web is constructed using individual I/P values from consumer $j$ to food type $i$.

## 7.4 Trophic levels

The more detailed the examination of food webs, the more complex the relationships become. Diagrams of species connections become confusing tangles, and it becomes necessary to conceptualize the system. The basic abstraction of the food chain or food web is the trophic level. After each energy exchange between organisms, the energy is said to have passed to a higher trophic level.

**Trophic level pyramids**. A trophic level is each of the sequential, hierarchical levels in a food chain which is comprised of organisms that share the same function in the food chain and the same nutritional relationship to the primary sources of energy:

- Primary producer (green plants) trophic level
- Primary consumer (herbivores) trophic level
- Secondary consumer (predators) trophic level
- Tertiary consumer (apex predator) trophic level

If all organisms in each of these levels have their biomass quantified on a unit area basis for either mass or chemical energy content (i.e., grams $m^{-2}$ or kcal $m^{-2}$), we can construct a trophic level pyramid (Fig. 7.5). Ecologists speak of energy flow through trophic levels. But this is a euphemism; energy is actually moving in discrete packages of the potential chemical energy in food consumed. The concept of a pyramid of numbers, or "Eltonian Pyramid," was conceived by Charles Elton in 1927, and it has subsequently been applied to represent biomass and chemical food energy content.

The first scientific presentation of the idea that the function of an ecosystem could be represented by flows of energy through a trophic level pyramid, or food web, was by Lindeman (1942). He demonstrated that ecosystem function could be described by knowledge of two attributes of each trophic level: (1) the level of energy storage, and (2) the efficiency of energy transfer. Derived from his work is Lindeman's Law that 10% of food energy is transferred from one trophic level to another, the remaining is lost through incomplete digestion, respiration, and mortality.

**Trophic dynamic relationships**. An Eltonian pyramid can misrepresent the transfers of energy between trophic levels, i.e., trophic level efficiencies,

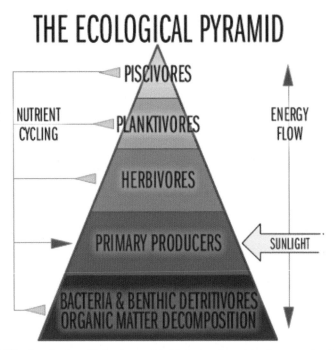

**FIGURE 7.5** A stylized trophic level pyramid with the area in each level representing biomass or chemical energy content. *Image credit: U. S. Environmental Protection Agency, 2019.*

because it represents the standing crop and does not account for the rates of productivity in each trophic level. A small herbivore biomass may occur with a large producer biomass. Or, a seemingly small biomass of primary producers may support large herbivore biomass, but if the rate of primary production is high, this is possible. Examples of each situation would be an old field in Georgia, USA (Odum, 1960) with a producer:herbivore:carnivore biomass ratio of 4700:6:1 and Silver Springs, Florida with a biomass ratio of 162:7:2:1 with the waters of the English Channel with a 1:2.5 (Fig. 7.6). The only way to make sense of these relationships is with a trophic dynamic analysis of the energetics of the different ecosystems. Since a pyramid of energy accounts for the turnover rate of the organisms, it can never be inverted.

The standing crop of biomass represents only the mass present at a point in time. To account for the trophic dynamics of the food web it is necessary to evaluate the energy balance of the system. Then the energetics, following the laws of thermodynamics, always yields an upright pyramid. This is because energy is not created and the transfer processes between trophic levels, though inefficiencies, lose energy as heat. Energetics provides a unifying concept in ecology and a means by which the mechanics of the ecosystem may be explored.

The amounts and rates of energy transfer between trophic levels in an ecosystem were represented by Lindeman (1942) as a set of mathematical

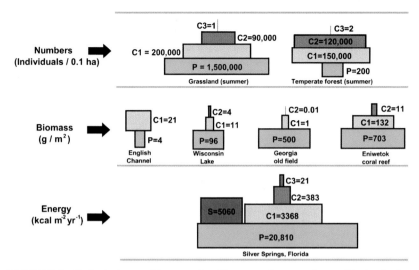

**FIGURE 7.6** Ecological pyramids comparing biomass and energy for trophic levels from different aquatic ecosystems. Notation: C1, primary consumer; C2, secondary consumer; C3, tertiary consumer; P, Producer; S, saprotroph. *Image credit: Thompsa-Own Work. Courtesy, File: EcologicalPyramids.jpg, 2016.*

equations, setting $\wedge$ as the energy content in the biomass of a trophic level (kcal $m^{-2}$), and $\lambda$ as the rate of transfer of energy between trophic levels (kcal $m^{-2}$ $yr^{-1}$). For any trophic level $\wedge_n$ we know that energy is constantly entering it from lower trophic levels $\wedge_{n-1}$ and leaving it to higher trophic levels $\wedge_{n+1}$. The flux of energy from trophic level $\wedge_{n-1}$ to trophic level $\wedge_n$ is designated by $\lambda_n$. The rate of respiratory heat loss from the trophic level is indicated by $R_n$. Therefore:

$$\Delta\wedge_n/\Delta t = \lambda_{n-1} - \lambda_{n+1} - R_n \qquad (7.25)$$

Formalizing the trophic dynamics of the system in mathematical terms focuses attention on a number of important hypotheses about energy flow through ecosystems: about the similarity/dissimilarity between different ecosystems in terms of efficiency of energy utilization, the rates of transfer between trophic levels, and the quantity of energy transferred. For instance:

- $\wedge_n/\wedge_{n-1}$ = How does the ratio of standing crops for different trophic levels and among different ecosystems? Is there a theoretical maximum of biomass (potential chemical energy) that can be supported by a given quantity of $\wedge_{n-1}$?
- $R_n/\wedge_n$ What are the metabolic costs of production? How efficient are trophic levels in maintaining their biomass? Do respiratory costs differ between trophic levels or for different ecosystems?
- $\wedge_n/\wedge_{n+1}$ How effectively do trophic levels exploit the energy available to them in lower trophic levels? Are some trophic levelsmore energetically efficient than others?

## 7.5 Trophic level efficiencies

If a commodity is both essential and scarce to an organism, it is used frugally and with economy. The greater the efficiency of use of a commodity, the more likely that the system has been honed through natural selection to use it with minimal waste. Hence, the use of various efficiency terms in trophic level and ecosystem energetics to identify emergent properties and patterns that might shed light upon the use of energy (Table 7.2).

**Energy transfer efficiencies.** Of the various ratios of energy transfers, and rates of transfer and loss, the following provide insight into the ecological processes governing energy fluxes and utilization between trophic levels. The literature is confusing on these terms and the reader is referred to Kozlovsky (1976) for clarification. Contemporary terminology uses:

$I_n$ is intake ($\lambda_{n-1}$ of Lindeman),
$A_n$ is assimilation (gross primary production or assimilation by higher trophic levels),
$R_n$ is respiration,
$NP_n$ is net production of a trophic level (standing crop and $\wedge_n$ of Lindeman) $NP = A - R$, and
P is production, the energy passed to the next higher trophic level ($\lambda_{n+1}$ of Lindeman).

**TABLE 7.2 Ecological energetic efficiencies.**

| | | |
|---|---|---|
| Assimilation Efficiency | $A_n/I_n$ | Ratio of assimilated to ingested energy in the transfer of energy between organisms in a food chain |
| Tissue growth efficiency | $NP_n/A_n$ | Ratio of standing crop of an organism or trophic level to the energy assimilated |
| Ecological growth efficiency | $NP_n/I_n$ | Ratio of standing crop energy available to be passed on to the next higher trophic level to the energy input to that trophic level |
| Respiratory coefficient | $R_n/NP_n$ | Ratio of energy expended in respiration to produce and maintain the standing crop |
| Ecological efficiency | $P_n/I_n$ | Ratio of energy passed to the next highest trophic level to that received from the next lowest trophic level |
| Trophic level production efficiency | $A_n/NP_{n-1}$ | Proportion of energy in a trophic level to that available to it at the lower trophic level |
| Trophic level intake efficiency | $A_n/A_{n-1}$ | The ratio of energy assimilated at n to the energy assimilated at n-1 |

Let us examine these efficiencies for each of the several trophic levels (Table 7.2). The assimilation efficiency is the food digestion efficiency for individuals and the intake efficiency for trophic levels, and represents how effective energy is extracted from that which is available. The tissue growth efficiency is relatively constant across herbivores and predators, indicating that a fundamental mechanism may govern what proportion of energy intake is available for production regardless of trophic position. The ecological growth efficiency increases with higher trophic position because of great assimilation efficiency (flesh is more digestible than vegetable matter). The respiratory coefficient shows no trend between trophic levels, except that predator levels have higher average values, indicating perhaps that predators may expend a greater portion of their total energy budget in locating and capturing prey. The ecological efficiency is the ratio of energy passed to the next higher trophic level to that received from the next lowest level; in other words, export as a function of input. The trophic level production efficiency ranges between 2% and 17%, with reduction at progressively higher trophic levels and reflects the Eltonian Pyramid. The trophic level intake efficiency is the efficiency of productivity (Table 7.3).

**Heterotrophic productivity**. Little has been said, heretofore, about the energetics of the heterotrophs. An excellent summary for a grassland ecosystem is provided by Heal and MacLean (1975) (Table 7.4).

Using a reference base of $100 \, \text{kcal m}^{-2} \, \text{yr}^{-1}$ produced in a grassland ecosystem, they document the energy flow through the heterotroph community. In this example, 86.5% of the energy passed to the dead organic matter pool and the saprovore system is portioned as: 71.0% as unconsumed primary production, 14.9% as feces, and 0.6% as non-predatory mortality of heterotrophic organisms. The herbivore system respires only 13.25% of the energy entering the consumer system. Production and respiration by microorganisms exceed that of animals. The high microbial production allows a production of microbivores which is greater than that of invertebrate saprovores.

For this system, 56.7% of the energy entering as dead organic matter is lost in respiration at each pass: 43.3% is recycled as organic matter. Organic matter which is not respired by organisms within the saprovore system returns to the pool of dead organic matter where it is again available for assimilation. This recycling, in conjunction with high growth efficiency in microorganisms, results in very high levels of production within the saprovore trophic level. In a forest ecosystem, 95% of total heterotrophic respiration is contributed by decomposers (Reichle et al., 1975). Placing this in the context of global C fluxes, the annual global $CO_2$ flux from soils has been estimated to average $68 \pm 4 \, \text{PgC yr}^{-1}$, with rates positively correlated with mean annual air temperature and mean annual precipitation (Raich and Schlesinger, 1992).

**TABLE 7.3** Values reported for ecological energetic efficiencies for different trophic levels (after Reichle, 1971).

| Trophic level | Assimilation efficiency, An/In | Ecological growth efficiency, NPn/In | Tissue growth efficiency, NPn/An | Respiratory coefficient, Rn/NPn | Ecological efficiency, Pn/In | Trophic level efficiency, an/NPn-1 |
|---|---|---|---|---|---|---|
| Saprovore | 0.10–0.40 | 0.05–0.08 | 0.17–0.40 | 3.7–4.6 | 0.09–0.18 | 0.11–0.17 |
| Herbivore | 0.36–0.78 | 0.08–0.27 | 0.20–0.40 | 2.16–3.06 | ~0.10 | 0.02–0.07 |
| Carnivore | 0.47–0.92 | ~0.34 | 0.10–0.37 | 1.70–4.18 | ~0.07 | ~0.02 |

**TABLE 7.4** Calculated ingestion, production, respiration, and egestion by heterotrophs in a grassland ecosystem in kcal m$^{-2}$ yr$^{-1}$ per 100 kcal m$^{-1}$ yr$^{-1}$ net annual primary production (after Heal and MacLean, 1975) Courtesy, Springer Nature

|  | Ingestion | Production | Respiration | Egestion |
|---|---|---|---|---|
| **Herbivore system** | | | | |
| Herbivores, vertebrates (Hv) | 25.000 | 0.250 | 12.250 | 12.500 |
| Invertebrates (hi) | 4.000 | 0.640 | 0.960 | 2.400 |
| Carnivores, vertebrates Cv | 0.160 | 0.003 | 0.123 | 0.031 |
| Invertebrates, Ci | 0.170 | 0.040 | 0.095 | 0.034 |
| **Saprovore system** | | | | |
| Saprovores, invertebrates (Si) | 15.153 | 1.212 | 1.818 | 12.122 |
| Microbial (Sm) | 136.377 | 54.551 | 81.826 | — |
| Microbivores, invertebrates (Mi) | 10.910 | 1.309 | 1.964 | 7.637 |
| Carnivores, vertebrates | 0.041 | 0.001 | 0.032 | 0.008 |
| Invertebrate | 0.648 | 0.155 | 0.363 | 0.130 |
| **Total % passing through** | | | | |
| Herbivore system | 15.2 | 1.6 | 13.5 | 42.9 |
| Saprovore system | 84.8 | 98.4 | 86.5 | 57.1 |

Heterotrophic regulation of ecosystem processes is essential to the stability of ecosystems processes. While heterotrophic biomass in a variety of ecosystems seems to be a constant proportion of autotrophic standing crop (Fig. 7.7), suggesting "bottoms-up resource control," herbivore populations may exert more "feedback" control on energy flow than would be indicated by the small amount of organic matter actually consumed (O'Neill and Reichle, 1976). While consumption in forest canopies may be only a few percent of net primary production, the impact on photosynthetic potential can be more substantial (Reichle et al., 1973). The role of heterotrophs as regulators is apparent from the proportion of total ecosystem respiration represented by heterotrophs. Heterotrophic respiration may account for between 35% and 55% of total ecosystem respiration (autotrophic plus heterotrophic respiration) (Reichle et al., 1975).

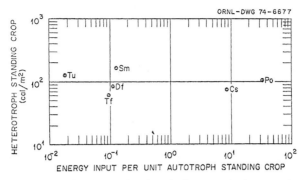

**FIGURE 7.7** Heterotroph biomass as a function of primary production per unit plant biomass. The six points represent ecosystem types: Cs, cone spring; Df, deciduous forest; Po0, pond; Sm, salt marsh; Tu, tundra; Tf, tropical forest. *Source: O'Neill and Reichle (1980). Courtesy, Oak Ridge National Laboratory and Oregon State University Press.*

Thus, a substantial portion of maintenance respiration energy of ecosystems is accounted for in heterotroph activity. While consumer organisms may play a role in rate regulation of producers, Table 7.4 indicates that heterotrophic effects are greatest in decomposer processes. In a forest ecosystem, 95% of total heterotrophic respiration is contributed by decomposers (Reichle et al., 1975).

## 7.6 Trophic structure of different ecosystems

Trophic dynamics play out in different patterns for different ecosystems, depending upon the variety of primary energy inputs and trophic structures. The flow of matter and energy through the trophic structure of an ecosystem was categorized in Fig. 7.1, where available food is consumed by herbivores and the energy content is partitioned and dissipated before resulting in production. The patterns are different for different ecosystems. Consider the successful cattleman who can anticipate a maximum yield of 30 cal m$^{-2}$ from a forage production of 5120 cal m$^{-2}$. The cattle select only about 14% of the available forage (730/5120 cal m$^{-2}$) and respire nearly 90% of that which they assimilate [244/(730−456 cal m$^{-2}$)]. Sustained yield is about 30 cal m$^{-2}$ with no loss to decomposers if the cattle remain healthy and are harvested regularly (Macfadyen, 1964). The salt marsh grasshopper studied by Smalley (1960) is another typical example of herbivores in that it is selective of the total herbage available to it, and its assimilation is fairly high (27% vs. 37% for cattle). Again, a high percentage of the assimilated energy is respired (63% vs. 89.5% for cattle). This is a characteristic feature of terrestrial food chains, and reflects the relatively high metabolic "costs" of life on land.

A third example is from the English Channel (Cushing, 1959) and reflects the fact that plankton have a high productivity rate and zooplankton are very efficient grazers upon plankton, so that very little plankton is present at any point in time (note the inverted Eltonian Pyramid in Fig. 7.6). Assimilation is lower (16%), perhaps because of the diatomaceous nature of the food base which has low digestibility. Respiration is extremely low and represents less than 1% of the energy assimilated.

**Trophic assimilation efficiencies**. The conversion of food biomass into consumer biomass or the concomitant transfer of chemical food energy is a limiting factor for food chain length in natural ecosystems. In host–parasitoid systems, higher trophic assimilation efficiencies may be expected at higher trophic levels because of the close match of host tissues and the consumer's food requirements. This may also allow longer food chains. Sanders et al. (2016) found that the efficiency of biomass transfer along an aphid-(primary) and secondary/tertiary (parasite), and found high efficiency in biomass transfer along the food chain. From the third to the fourth trophic level, the proportion of host biomass transferred was 45%, 65%, and 73%, respectively. In the following Table 7.5 are summarized the ecological energetic values for three different ecosystems. Data were reported for a fourth trophic level for Lake Mendota and Silver Springs. $D_f$ is a new term for feces or nonassimilation ($I = A + D_f$). $D_e$ is another new parameter for decomposition. L is loss from the system, a term used in aquatic systems to account for flowing water export. The reader can calculate an entire new set of ecological energy efficiencies from these data.

What should become immediately apparent are the surprising similarities of the values for the various energetic parameters among the three ecosystems, especially the higher trophic levels; the one disparity being gross primary production (I), respiration (R), and net production (NP) for Lake Mendota's trophic level 1. Not surprising when one realizes that the Lake Mendota ecosystem operates at a significantly lower ambient temperature than the Georgia salt marsh and the Florida spring. Yet, even here we see the effects of temperature on ecosystem energetics. What we see are consistent patterns in energy transfer in different ecosystems. These patterns become significant as we scale up to larger spatial scales and begin to understand how larger units of the living landscape function.

Our overall objective has been to follow radiant energy, after it reaches the Earth, and then its capture by biological systems. The transformation of energy and the production of energy-containing carbon compounds occur in the biosphere. As we have seen, the flux of energy depends upon the transfers among organisms, and transfers along food chains and between trophic levels. Ecological energetics requires us to consider carbon and energy fluxes at a larger spatial scale and at the ecosystem level of complexity. This requires an understanding of ecosystem metabolism and productivity.

**TABLE 7.5** Ecological energetic efficiencies for three different ecosystems (cal cm$^{-2}$ yr$^{-1}$): Lake Mendota (Lindeman, 1942); Silver Springs (Odum, 1957); Salt Marsh (Teal, 1962).

| Parameter | Trophic level 1 | | | Trophic level 2 | | | Trophic level 3 | | |
|---|---|---|---|---|---|---|---|---|---|
| | Lake | Spring | Marsh | Lake | Spring | Marsh | Lake | Spring | Marsh |
| I | 118,872 | 1,700,000 | 600,000 | – | – | – | – | – | – |
| A | 480 | 20,810 | 6,380 | 41.6 | 33,68 | 767 | 2.3 | 383 | 59 |
| $D_f$ | 118,392 | 1,679,190 | 563,620 | 10.4 | – | – | 0.3 | – | – |
| R | 107 | 11,977 | 28,175 | 15 | 1,890 | 596 | 1.1 | 316 | 48 |
| NP | 373 | 8,833 | 8,205 | 26.6 | 1,478 | 171 | 1.3 | 67 | 11 |
| P | 52 | – | – | 2.6 | – | – | 0.3 | – | – |
| $D_e$ | 321 | – | – | 24 | – | – | 1.0 | – | – |
| L | – | – | 3548 | – | – | 112 | – | – | 11 |

## 7.7 Recommended reading

Lindeman, R., 1942. The trophic dynamic aspect of ecology. Ecology 23 (4), 399−418. DOI: 10.2307/1930126. https://www.jstor.org/stable/1930126.

Macfadyen, A., 1964. Energy flow in ecosystems and its exploitation by grazing, pp. 3−20. In: Crisp, D.J. (Ed.), Grazing in Terrestrial and Marine Environments, British Ecological Soc. Symp., Blackwell, Oxford, 322 pp.

Odum, H.T., 1957. Trophic structure and productivity of Silver springs, Florida. Ecological Monographs 25, 55−112. https://doi.org/10.2307/1948571.

Phillipson, J., 1966. Ecological Energetics., St. Martin's Press, New York, 57 pp. https://www.worldcat.org/title/ecological-energetics/oclc/220312888.

Wiegert, R.G. (Ed.), 1976. Ecological Energetics. Dowden, Hutchinson and Ross, Inc., Strouds-burg, PA, 455 pp. https://www.worldcat.org/title/ecological-energetics/oclc/2077851/.

# Chapter 8

# Energy flow in ecosystems

The fundamental, self-sustaining, energy-processing system on Earth today is the ecosystem. This generalization is an ecological example of cyles in steady-state systems and is consistant with Morowitz's Generalization VII (Morowitz, 1968) which states that sustained life, under present-day conditions, is a property of an ecological system rather than a single organism or species. A one-species ecological system is never found in nature. The carbon cycle requires at least one primary producer and a means of returning carbon to the $CO_2$ pool. A system with only herbivores would perish without a food source. A system with only primary producers cannot persist unless autolysis produces

The Global Carbon Cycle and Climate Change. https://doi.org/10.1016/B978-0-12-820244-9.00008-1
**119**

$CO_2$ at a sufficient rate; this does not occur in nature. Decomposers thus plays a uniquely important role in the ecosystem.

The First Law of Thermodynamics states that whatever amount of energy enters a system, the same amount ultimately leaves. In other words, over sufficiently long periods of time, thermal physics and chemistry dictate the basic principle of energy flux through the organism, trophic levels, and the entire ecosystems of which these components are a part. Just as metabolism integrates the functional processes of plants and animals, so does metabolism integrate the processes in ecosystems. Understanding that ecosystems are functional units of the landscape allows the ecosystem's metabolism to be translated to larger geographic scales. Thus, by quantifying ecosystem energetics, we begin to develop the tools to understand the processes underlying the biosphere's energy and carbon dynamics.

## 8.1 Ecosystem energetics

The ecologist defines an ecosystem, functionally, as an energy-processing system (Reichle et al., 1975). Ecosystem energetics is primarily concerned with the quantity of incident radiation per unit area of the ecosystem and the efficiency with which this energy is converted and the net ecosystem productivity. An energy balance for the ecosystem can be obtained by quantifying the efficiency with which all the autotrophs in the system convert solar energy into the chemical energy of plant protoplasm, the efficiency by which this energy is utilized by heterotrophs, the growth and mortality of system components, and losses through decomposition. A typical approach would be to quantify the energy fluxes with gas exchange chambers for particular species, and then summing according to the percentage composition that species population contributes to the ecosystem. In that way, one obtains a total value for the whole ecosystem as well as an understanding of the contribution of different species populations and structural components. Alternatively, recent instrumentation advances in field gas spectrometry, eddy covariance, and Lidar technologies permit fluxes to be measured at different points within and for the whole ecosystem.

**Thermodynamic properties of ecosystems**. Although the quantity of energy which ultimately leaves an ecosystem is equivalent to that which entered it, many energetic transformations occur in between. Accepting the premise that over sufficiently long time scales of many years, ecosystems can be considered as equilibrium systems, and that over even longer time scales net production is zero, then the Second Law of Thermodynamics also applies. Energy transformations proceed from an ordered state to a random, low-energy form (heat). Thus, an ecosystem can be envisaged as an autonomous system behaving in a thermodynamic sense mush as does an individual organism. The flux of energy through an ecosystem can be defined from the Gibbs free energy equation (Eq. 5.1) as:

$$\Delta H = \Delta G + T\Delta S \qquad (8.1)$$

where:

H is equivalent to the metabolizable energy entering the system or, the heat content (enthalpy) assimilated by the ecosystem

G is the net free energy available for work within the ecosystem, e.g., maintenance and growth, including that dissipated as respiratory heat, and

$T\Delta S$ is the specific dynamic action (calorigenic effect) of the system which represents energy not available for maintenance or growth of the ecosystem.

The metabolism of an ecosystem is the transformation of radiant energy into chemical energy and its successive transfers and transformations throughout the complex structure of the system. The transformation of solar radiant energy to the chemical energy of biomass by plants conforms to the laws of thermodynamics:

| Solar energy assimilated by autotrophs | = | Chemical energy of plant biomass | = | Heat energy of autotrophic respiration |

In general, excluding episodic outbreaks, a dynamic equilibrium exists between the net primary production and the amount of food consumed by heterotrophs. In Chapter 4 we learned that individual heterotrophs do not assimilate all of the food which they consume. For herbivores as much as 90% of the total food intake may not be assimilated and passes the body as feces. In carnivores, where digestive assimilation efficiencies are higher, as much as 75% of the prey eaten may be assimilated (Phillipson, 1966), although 30% to 50% in more normal. Thus, according to the Laws of Thermodynamics we can state for heterotrophs:

| Chemical energy intake by by heterotroph(s) | = | Chemical energy assimilated by heterotroph(s) | + | Chemical energy of feces produced by heterotroph(s) |

and

| Chemical energy assimilated by heterotroph(s) | = | Chemical energy of heterotroph biomass (including reproduction) | + | Heat energy of respiration |

The storage of chemical energy in heterotrophic tissues (biomass) is termed secondary production. The body tissues and feces of heterotrophs eventually serve as food for other heterotrophs and decomposers, respectively. At each trophic level transfer of energy heat is evolved. The entire process conforms with the laws of thermodynamics and may be expressed as:

Solar energy entering the ecosystem = Heat energy leaving the ecosystem

## 8.2 Ecosystem production equations

**Productivity terminology**. The quantity or mass of a group or organisms per unit area is referred to as the standing crop, i.e., the standing crop of trees in a forest. The entire mass (weight) of all organisms in that unit area is termed biomass, i.e., the biomass of the forest. Most of the energy captured in gross production of ecosystems is depleted by green plant respiration, and nearly all the resulting net productivity of ecosystems is usually consumed by the respiration of consumer organisms and decomposers. Production is the amount of biomass produced (g m$^{-2}$) or of radiant energy transformed (cal m$^{-2}$). Productivity is the rate (quantity per unit time) of production (g m$^{-2}$ yr$^{-1}$ or cal m$^{-2}$ yr$^{-1}$). These relationships are described by the following terms and equations:

> **(Total) Standing crop (TSC)** is the quantity of biomass present at any given point in time (e.g., g m$^{-2}$). The synthesis of plant biomass by autotrophs is designated as primary production.
> **Gross primary production (GPP)** is the amount of chemical energy as biomass that primary producers create in a given period of time.
> **Autotrophic respiration ($R_A$)** is the cellular respiration and maintenance energy expenditures by the tissues of the green plant producers.
> **Net primary production (NPP = GPP-$R_A$)**. Gross primary production minus autotrophic respiration is net primary production and represents the potential food energy available to heterotrophs. This is the measure of energy captured minus the costs of energy transformation and green plant maintenance metabolism.
> **Heterotrophic respiration ($R_H$)** is that of consumer organisms and represents the collective metabolism by all consumer trophic levels. In the ecosystem this comprises all herbivores, carnivores, and saprovores. The largest component by far being that by saprovores reprocessing dead organic matter.
> **Ecosystem respiration ($R_E = R_A + R_H$)** is the sum of all respiration within the ecosystem by all trophic levels.
> **Net ecosystem production (NEP = NPP-$R_H$ = GPP-$R_E$)** is gross primary production minus all respiratory losses by the ecosystem. This is the net increment of mass, and its carbon or energy content, added to the ecosystem.

## 8.3 Measurement of pools and fluxes

Since ecosystem productivity is based upon the photosynthetic conversion of radiant energy and $CO_2$ into carbon-based substrates, measurement of production by green plants is the first step. Productivity of higher trophic levels is merely the reapportionment of the photosynthate base. The methods of measurement are various and extensive and not necessary to address in detail. A summary will illustrate the effort and techniques required to measure biomass:

**Standing crop of vegetation.** The biomass, or standing crop, of an ecosystem is the mass present at any given time, and while it is an approximation of the system's production, it should not be confused with the productivity of the system (Chapin, 2006). The variety of methods to measure primary production in terrestrial ecosystems has been reviewed by (Whittaker and Marks, 1975; Bonham, 1989). The basic approaches are, for terrestrial ecosystems:

For herbaceous and grassland terrestrial ecosystems, and in littoral communities of wetland ecosystems, standing crop can be simply measured by harvesting and weighing the vegetation per unit area or quadrats (e.g., White, 1978). These harvest method data may provide an estimate of net production. In ecosystems where there is an accumulation of biomass from year to year, such as forests, this approach is not as practical.

For woody vegetation the harvest method can again be used in forest ecosystems, but a variant called dimension analysis is simpler. This method involves harvesting a sufficient number of trees sufficient to cover the entire range if individuals composing the stand and developing regression equations of the relation between a tree dimension (e.g., basal area, dbh, etc.) and the mass of that tree. Then the mass value for each tree, w, can be multiplied times the number of trees with similar dimensions, n, and individual of all trees summed to obtain a value, W, for the entire stand being $W = \sum(wn)$.

In aquatic ecosystems algal biomass in both moving and stillwater ecosystems can be estimated in three ways:

**(1)** by quantifying chlorophyll *a,*
**(2)** by measuring carbon biomass as ash-free dry mass, or
**(3)** by measuring the particulate organic carbon in a sample.

The chlorophyll *a* procedure measures the photosynthetic pigment common to all types of algae, while (2) and (3) measure the carbon in a filtered water sample (Hambrook-Berkman and Canova, 2007). Measurements of aquatic plant biomass can be measured using variants of the quadrat harvest techniques used for terrestrial ecosystems, with variations to accommodate the more difficult-to-sample aquatic habitat (see, Downing and Anderson, 2011).

**Animal biomass**. The techniques for measuring the biomass of the variety of animals from invertebrates to vertebrates, living in habitats both aquatic and terrestrial, including soils and sediments, are extensive. Nearly as many techniques exist comparable to the diversity of animal types. A review here is neither necessary nor appropriate. The reader is offered references for further inquiry: aquatic invertebrates (DiFranco, 2014); terrestrial invertebrates (Johnson et al., 2012); aquatic vertebrates (Randall and Minns, 2000); novel new means of estimating aquatic vertebrate biomass using DNA (Takahara et al., 2012); and terrestrial vertebrates (Petrusewicz and Macfadyen, 1970).

**Soil and litter microbial biomass**. Measuring the biomass of other components of the ecosystem can be challenging, soil bacteria and fungi being examples. Since they account for much of the heterotrophic respiration, great care must be given to obtaining these values. There is a variety of both direct and indirect methods (Islam and Wright, 2015). In direct methods, microorganisms are determined by colony forming units counted on agar media. Biomass of bacteria and fungi is then calculated from cell dry weight and volume relationships.

Chloroform ($CHCl_3$) Fumigation Incubation (Jenkinson and Powlson, 1976) has been the standard indirect method because of its simplicity and accuracy. In the $CHCl_3$ fumigation incubation procedure, the fumigant lyses the soil microbes, and the resulting increase in $CO_2$ evolution from the fumigated soil is compared to that from unfumigated soil, The $CO_2$ evolved is measured by NaOH absorption followed by acid-base titration to yield the carbon content of the soil microbial biomass.

Other indirect methods, with the carbon measurement technique in parentheses are: Microwave (MW) Irradiation Incubation ($CO_2$ is absorbed by NaOH); Rehydration Method ($K_2Cr_2O_7$ oxidation); Freeze-dried Soil Extraction (extracted C is analyzed colorimetrically); Adenosine Triphosphate Extraction (luciferin-luciferase light reactive system); Phospholipid Fatty Acids Extraction (capillary gas chromatography with flame ionization detection); Substrate-Induced Respiration ($CO_2$ respiration). The extraction methods, particularly MW soil extraction, are widely used because they provide simple, rapid, and precise measurement of soil microbial biomass.

**Respiration.** Respiration is the chemical breakdown of complex organic substances, such as carbohydrates and fats that takes place in the cells and tissues of animals and plants, during which energy is released and carbon dioxide produced (internal respiration). The process in living organisms of taking in oxygen from the surroundings and giving out carbon dioxide from catabolic metabolism is termed external respiration.

Since there is a definite relationship, through the respiratory quotient (see Eqs. 5.10−5.12), between the amount of oxygen consumed and the amount of carbohydrate produced, oxygen consumption can be used to determine the rate of production. In aquatic ecosystems, application of this technique uses the light bottle/dark bottle technique (Fig. 8.1). Two bottles are filled with water samples containing a typical collection of plants and animals. The $O_2$ content of each is measured. One bottle is fitted with an opaque black cover, the other left exposed to light; both are then returned to the natural habitat. In the light bottle, oxygen is produced by photosynthesis and some is used in respiration (NPP). In the black bottle, no photosynthesis takes place and no $O_2$ is produced, but some is consumed in respiration as in the light bottle. Therefore, the sum of $O_2$ produced in the light bottle, plus that used in the dark bottle, is equal to the total $O_2$ production in photosynthesis, GPP. This is a measure of the gross production of the autotrophs in the water.

Similar in principle to the $O_2$ uptake method, $CO_2$ analysis also provides a measure of the metabolic activity of the system. This involves enclosing parts

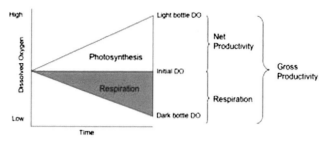

**FIGURE 8.1** Oxygen production during the light bottle:dark bottle experiment. *Image credit: Scott Deeken at John Burroughs School.*

or all of the ecosystem, and measuring $CO_2$ concentrations in known volumes of air in air supplied or removed. Enclosed leaves or branches of a forest give autotrophic respiration values, inverted chambers on the forest floor measure heterotrophic respiration by decomposers. $CO_2$ concentrations can be measured chemically by KOH precipitation or by using infrared gas analyzers. Measurement of $CO_2$, as well as $O_2$ and $CH_4$, gas concentrations is made using gas chromatography.

There are other techniques to measure the biological activity of the ecosystem, or the rate of metabolism, based upon the premise that the technique "monitors" some metabolic parameter of the ecosystem which is directly or indirectly related to productivity capacity.

- Radioisotope uptake. $^{14}C$ and $^{32}P$, among other isotopes, are used to measure net productivity when a given amount is introduced into the system and its rate of incorporation can be measured. Knowing the biochemistry of the isotope, can provide information on metabolic activity.
- Disappearance of resources. A variant of the isotopic method, this involves measuring the rate of utilization of nutrients added to the system in the form of measured quantities of fertilizer for plant growth, or substrate for decomposer activity.
- Biological indicators. Structural or compositional measures of the ecosystem can be used to estimate production, such as leaf area index or chlorophyll content as surrogates for photosynthetic capacity.
- Gas exchange. Instead of measuring the amount of mass at a point in time, an alternative estimate of production can be obtained by measuring its metabolic activity. This can be accomplished by measuring the $O_2$ and $CO_2$ fluxes of the system. Empirical relations can be developed to related fluxes to standing crops.

**Eddy covariance**. Over the past several decades, this has become the most significant advancement in the measurement of the fluxes of respiratory gases for entire ecosystems (Baldochi et al., 1988). Eddy covariance is a micrometeorological method that is currently popular to directly observe the exchanges of gas, energy, and momentum between ecosystems and the atmosphere. Eddy covariance can be used to measure:

- Carbon dioxide fluxes,
- Methane fluxes,
- Water loss, evapotranspiration,
- Instantaneous water use efficiency, and
- Instantaneous radiation use efficiency.

The Eddy covariance method is used in terrestrial ecosystems within and above the vegetation zone for total system fluxes, and in the benthic zone for measuring oxygen fluxes between the seafloor or lake sediments with overlying water. The Eddy covariance calculates the covariance of fluctuations in the vertical wind velocity and in the physical quantity to be measured (Liang et al., 2012). This method is also able to directly measure the carbon, water, and heat flows between plant communities and the atmosphere. Since photosynthesis stops at night, the measured nighttime flux represents the total respiration of the ecosystem, consisting of both autotrophic plant respiration and heterotrophic soil respiration. $Q_{10}$ correction (See Chapter 5, Eq. 6.2) for temperature differentials in reaction rates (Eq. 8.2) of nighttime and daytime fluxes, allow the 24-hour ecosystem respiration (ER) to be calculated:

$$R_1 = R_2 Q_{10} (T_1 - T_2/10) \qquad (8.2)$$

where:

$R_1$ is the energy flux at temperature $T_1$, to which the processes are being normalized, typically mean 24-hr temperature,
$R_1$ is the energy flux at the observed temperature $T_2$,
T is temperature in °C, and
$Q_{10}$ is the factor by which the rate changes for every 10°C.

Using the $O_2$ fluxes to measure respiration, R, similar to the light bottle:dark bottle method discussed above for aquatic systems, the sum of the net $O_2$ flux produced in the daytime (NEP), plus that used during the night $(R_A + R_H)$, is equal to the total $O_2$ produced in photosynthesis (or GPP). The daytime energy flux is the net energy flux for the ecosystem. The sum of the daytime and nighttime energy fluxes is the gross ecosystem energy flux.

Through the AmeriFlux Network—a collection of long-term, eddy flux measurement stations located across the Western Hemisphere—carbon, water, and energy flux measurements are gathered, processed, and shared with the scientific community (AmeriFlux, 2018). AmeriFlux is a network of sites, established by the U.S. Department of Energy, which is measuring ecosystem $CO_2$, water, and energy fluxes in the northern, central, and southern U.S. Ecosystem types included in the AmeriFlux network represent the major climatic regions and biomes, including tundra, grasslands, savanna, crops, and conifer, deciduous, and tropical forests. Research currently being conducted is intended to:

- Quantify carbon sources and sinks for diverse terrestrial ecosystems and evaluate how these sources and sinks are influenced by disturbance, land use, climate, nutrients, and pollutants;
- Advance understanding of the processes associated with photosynthesis, respiration, and carbon storage in ecosystems;
- Collect observations that promote understanding and modeling of the current global carbon budget; and
- Enable improved predictions of future atmospheric carbon concentrations.

**Remote sensing**. While all of these techniques may be useful for ecosystems with less massive structure (grasslands and tundra), or structural sub-components (forest floor herbaceous layers, or soils), larger systems present a more difficult challenge. New technologies now offer new approaches (DeFries et al., 1995). Respiration values for entire ecosystems, which here-tofore had to be obtained by measurements like those summarized previously on subcomponents of the ecosystem, and then added together—each step compounding associated errors of measurement—now can be measured directly. Values for gas exchange and production can be obtained remotely for ecosystems over increasingly larger segments of the landscape.

For measuring productivity, satellite-based remote sensing is a relatively new technology that provides less labor-intensive estimates of standing crop biomass, while not as detailed for individual species, integrative over the entire ecosystem. The LANDSAT Thematic Mapper measures reflectance in each waveband for a pixel ($30 \times 30$ m) of land. Measurement of NPP is then made by correlating LANDSAT's measured differential absorption of red light by chlorophyll in the 800—1200 nm waveband, assuming that "greenness" cor-relates with the leaf area index (LAI, $m^2m^{-2}$) of vegetation which has been shown to be a reliable measure of photosynthetic activity. Thus, a remotely sensed measure of NPP is obtained. Similarly, NOAA's Coastal Zone Color Scanner (CZCS) on the Nimbus-7 Satellite measures reflection from chloro-phyll in the ocean surface waters (Balch et al., 1992).

Traditionally, freshwater ecosystems have been challenging to measure with satellite remote sensing because they are small and spatially complex, require high fidelity spectroradiometry, and are best described with biophys-ical variables derived from high spectral resolution data. The U.S. National Research Council's guidance to NASA regarding missions for the coming decade includes a polar orbiting, global mapping, hyperspectral satellite mission, the Hyperspectral Infrared Imager (HyspIRI). Using retrieval of multiple biophysical variables, such as phycocyanin and chlorophyll-a which drive primary productivity, it will be possible to make quantitative measure-ments of ecosystem change, both terrestrial and freshwater (NASA, 2018).

All modern total solar incidence (TSI) satellite instruments use active cavity electrical substitution radiometry. This technique applies measured electrical heating to maintain an absorptive blackened cavity in thermal equilibrium, while incident sunlight passes through a precision aperture of a calibrated area, and the aperture is modulated with a shutter. Measurement uncertainties of <0.01% are required to detect long-term solar irradiance variations, because expected changes are in the range $0.05-0.15$ W m$^{-2}$ per century. The Solar Radiation and Climate Experiment/Total Irradiance Measurement (SORCE/TIM) TSI values are lower than prior measurements by the Earth Radiometer Budget Experiment (ERBE) on the Earth Radiation Budget Satellite (ERBS). Similar measurements have been made by VIRGO on the Solar Heliospheric Observatory (SOHO), and the ACRIM instruments on the Solar Maximum Mission (SMM) Upper Atmosphere Research Satellite (UARS) and ACRIMSat (Stephens et al., 2012).

The Atmospheric Research Measurement (ARM) program is the U.S. Department of Energy's scientific user facilities which provide the climate research community with strategically located, ground-based remote sensing observatories. ARM is designed to improve the understanding, and representation, in climate and Earth system models, of clouds and aerosols as well as their interactions and coupling with the Earth's surface (ARM, 2018). The Southern Great Plains site in Oklahoma, established in 1992, provides a wide variability of climate, cloud type, and surface flux properties, as well as seasonal variation in temperature and specific humidity. The North Slope of Alaska sites, established in 1997, provide data about cloud and radiative processes at high latitudes, which have been identified as one of the most sensitive regions to climate change.

Recognizing the importance of long-term records of ecosystem properties and exchanges with the atmosphere, the U.S. National Science Foundation, over a 30-year timeframe, is funding NEON (the National Ecological Observatory Network). With automated instrumentation, NEON collects data to characterize the causes and impacts of environmental change for 81 terrestrial and freshwater ecosystems within 20 ecoclimatic domains, making the data available to the scientific community (NEON, 2018).

**Measurement of carbon in pools and fluxes**. Carbon occurs in many different materials in many different forms. In energetics, one is interested in the carbon content of organic molecules. Carbon content may be categorized as:

- *Total Carbon (TC)*—all the carbon in the sample, including both inorganic and organic carbon,
- *Total Inorganic Carbon (TIC)*—often referred to as inorganic carbon (IC), carbonate, bicarbonate, and dissolved carbon dioxide ($CO_2$),

- *Total Organic Carbon (TOC)*—material derived from decaying vegetation, bacterial growth, and metabolic activities of living organisms or chemicals, and
- *Elemental Carbon (EC)*— charcoal, coal, and soot. Resistant to analytical digestion and extraction, EC can be a fraction of either TIC or TOC depending on the analytical approach.

Carbon mass may be calculated from the proportional composition of carbon in the substrate if its molecule composition is known, i.e.,

$$
\begin{array}{c}
\text{proportional} \\
\text{mass of} \\
\text{element}
\end{array}
=
\begin{array}{c}
\text{mass of element} \\
\text{in compound} \\
\hline
\text{total mass of} \\
\text{compound}
\end{array}
\tag{8.3}
$$

with the proportion of carbon times the weight of the compound yielding the mass of carbon present.

Measurement of $CO_2$, as well as $O_2$ and $CH_4$, gas concentrations are made using gas chromatography and thermal conductivity (e.g., Haskin, 2013; Emerson). Carbon equivalencies are calculated from $CO_2$ using the molecular ratio of $C/CO_2 = 0.27$.

The energy content of carbonaceous compounds is determined with a bomb calorimeter. Differential scanning calorimeters, isothermal microcalorimeters, titration calorimeters, and accelerated rate calorimeters are among the most common types (Wikipedia, 2018). A bomb calorimeter is a type of constant-volume calorimeter used in measuring the heat of combustion of a substance, determined by burning a sample and measuring a temperature change in the surrounding water. For chemical reactions, to find the enthalpy change per mole of a substance A in a reaction between two substances A and B, the substances are separately added to a calorimeter and the initial and final temperatures are recorded. Multiplying the temperature change by the mass and specific heat capacities of the substances gives a value for the energy given off or absorbed during the reaction.

## 8.4 The carbon cycle in ecosystems

All ecosystems fueled by radiant energy have basically the same trophic structure and patterns of internal energy transformation, exceptions being those dependent upon chemosynthesis for their energy source. Within the ecosystem energy flows through it, obeying the laws of thermodynamics, while carbon is recycled; thus, the terms "energy flow" and "carbon cycle." The concept of an ecosystem carbon cycle will be displayed in the following examples of a freshwater spring, the sea, and a forest— each using a different style of representation of the ecosystem: the spring uses H.T. Odum's analog flow diagram template, and the forest is represented by the box model of Oak Ridge National Laboratory's systems ecologists (Shugart et al., 1976) melded with the European forest mineral portraits (Duvigneaud, 1971). The uniqueness of Silver Springs and the Oak Ridge forest is that they are among the very few ecosystems for which all of the major carbon pools and fluxes have been measured empirically and concurrently.

**Freshwater ecosystems**. In most freshwater lakes and streams, inorganic forms of carbon are more abundant than organic forms. These inorganic carbon compounds exist as the equilibrium products of the chemical reaction between $CO_2$ and $H_2O$:

$$CO_2 + H_2O \leftrightarrow H_2CO_3$$

$$H_2CO_3 \leftrightarrow H^+ HCO_3^=$$

Inland waters with low content of dissolved solids range from solutions approaching distilled water (arctic springs and oligotrophic lakes) to saline brines (estuaries and tidal pools) with total carbonate exceeding several moles per liter. The normal freshwater range for total $CO_2$ ($CO_2 + HCO_3^- + CO_3^=$) is 50 µmol to 10 mmol.

Dissociation equilibrium and kinetics of the chemistry of hydration and dehydration of $CO_2$ are affected by pH, temperature, and salinity. Atmospheric $CO_2$ at 0.033% by volume dissolves in water to yield an unhydrated concentration of $CO_2$ similar to that in air. The hydration of $CO_2$ to $H_2CO_3$ is slow and at, equilibrium, yields a concentration of $H_2CO_3$ 0.0025 times that of $CO_2$. Dissociation of $H_2CO_2$ to $HCO_3^-$ and $CO_3^-$ is instantaneous. If $CO_2$ is at equilibrium in a solution buffered to constant pH, the $CO_2$ concentration and $H_2CO_3$ are independent of pH, whereas $HCO_3^-$ and $CO_3^-$ increase with pH until saturation kinetics are achieved. Such equilibria are strongly influenced by temperature and salinity.

Both oceanic and freshwater systems are in close equilibrium with atmospheric $CO_2$. In marine habitats, the inorganic-carbon pool is essentially $HCO_3^-$ and amounts to 2 mmol C or about 50 times that of the atmosphere. In freshwaters, the total C is much more dependent upon pH in relation to [$HCO_3^-$] and [$CO_3^-$]. In marine systems, the inorganic-carbon fluxes are basically determined by inorganic ionization equilibria. In freshwaters, respiration and other fluxes of $CO_2$ can contribute significantly to overall carbon cycling. Complex reaction products of $CO_2$ may also occur. Large quantities of C may exist as carbonate and $CaCO_3$. The solubility product of calcium carbonate is low ($0.48 \times 10^{-8}$) and $CaCO_3$ precipitates may occur in stable colloidal form. In either situation, this inorganic form of carbon is both an important sink and a source term for carbon in the aquatic cycle.

Another important source of carbon in aquatic ecosystems is dissolved organic carbon (DOC) and particulate organic carbon (POC), usually in the ratio of DOC to POC of 10:1. Dissolved and particulate organic carbon forms typically exceed many times the amount of carbon in the entire fauna, plankton, and bacteria. It is a characteristic of most aquatic ecosystems, and in marked contrast to terrestrial ecosystems, that most of the organic carbon budget in the system exists in the form of nonliving detritus. In the carbon budget of aquatic ecosystems, organic carbon inputs occur through two pathways:

- Allochthonous carbon from outside the system, typically from terrestrial ecosystems in the form of DOC in ground water or POC from particulate detritus, and
- Autochthonous carbon products produced in the littoral and pelagic zones.

  Major pools of carbon are:

- The dominant pool of dissolved organic carbon, DOC, and
- A benthic deposition of particulate organic carbon, POC, where most of the decomposition in the aquatic ecosystem occurs.

**Lentic ecosystem—ponds and lakes**. An example of the ecosystem for the aquatic carbon cycle (Fig. 8.2) is the classic study of Silver Springs (Odum, 1957). Aquatic ecosystems, with the absence of woody vegetation characteristic of many terrestrial systems, typically have low standing crops but high rates of productivity. This is a characteristic of many aquatic ecosystems. This is illustrated at Silver Springs where the standing crop biomass of producers, largely the grass, *Sagittaria,* and other macrophytes and algae, was

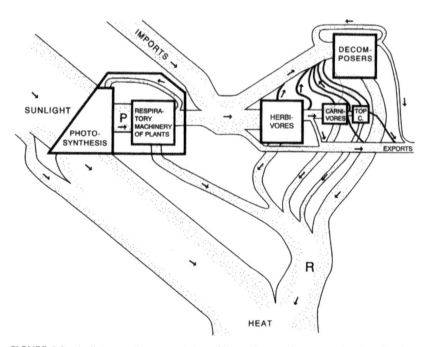

**FIGURE 8.2**   A diagrammatic representation of the pathways of energy and carbon flux in a freshwater ecosystem: Silver Springs, Florida. Carbon values given in Table 8.1. *Image credit: Odum, 1957. Courtesy, John Wiley & Sons.*

**TABLE 8.1** Comparison of the carbon budgets of five aquatic ecosystems: Spartina Salt Marsh, GA (Teal, 1962); Silver Springs, FL (Odum, 1957); oligotrophic Lake Eckarfjärden, Sweden (Andersson and Kumblad, 2006); Lake Washington, WA (Eggers et al., 1978); eutrophic Lake Lawrence, MI (Wetzel and Rich, 1973). Units are: fluxes in kg C m$^{-2}$ yr$^{-1}$, standing crop in kg C m$^{-1}$).

| Carbon budget component | Salt marsh[a] | Silver springs | Lake Eckarfjärden | Lake Washington | Lake Lawrence Total | Lake Lawrence Littoral |
|---|---|---|---|---|---|---|
| Standing crop | – | 0.404 | 0.056 | – | – | – |
| Gross primary production (GPP) | 8.93 | 3.200[b] | – | 0.216[b] | – | 0.1251 |
| Autotrophic respiration ($R_A$) | 6.88 | 1.920 | 0.06? | 0.059 | – | 0.1175[e] |
| Net primary production (NPP = GPP – $R_A$) | 2.05 | 1.280 | 0.421 | 0.157 | 0.191[d] | 0.0076 |
| Heterotrophic respiration ($R_H$) | 1.12 | 1.090 | 0.13[c] | 0.048 | – | – |
| Net ecosystem production (NEP = NPP - $R_H$) | 0.93 | 0.190 | 0.292[b] | 0.109[b] | – | – |
| Ecosystem respiration ($R_E = R_A + R_H$) | 8.00 | 3.00 | 0.20[b] | 0.107[b] | 0.160[d] | – |

[a]calories converted to grams using 4072 cal g$^{-1}$ for Spartina (Golley, 1961).
[b]calculated from author's data.
[c]benthic bacteria (0.126 kg C m$^{-2}$ yr$^{-1}$) treated as heterotrophs.
[d]Wetzel and Rich (1973).
[e]benthic respiration.
?= questionable value.

404 g C m$^{-2}$. Gross primary production (GPP), summarized in Table 8.1 was 3.2 kg C m$^{-2}$. The gross production efficiency is:

$$GPP/TSC \sim 8.0$$

Net primary production (NPP) was 1.28 kg C m$^{-2}$. The aquatic ecosystem, with a relatively small standing crop biomass of producers, exhibited a relatively high productivity. On an annual basis, NPP amounts to approximately three times the mean standing crop biomass of producers and the net production efficiency is:

$$NPP/TSC \sim 3.2$$

Heterotrophic biomass, including all consumers and decomposers, was 18.5 g C m$^{-2}$ — or about 4.6% of producer biomass. Although most of the heterotrophic biomass and respiration occurs within the decomposer community, Fig. 8.2 visually illustrates how the carnivore trophic level acts as a "speed brake" upon herbivorous consumption and promotes a "steady state" in primary producer biomass.

Autotrophic respiration ($R_A$) accounted for 1.92 kg C m$^{-2}$. The efficiency of photosynthate conversion, NPP/GPP is about 40%, with most of the assimilated carbon in photosynthesis being lost through respiration, with an effective production of:

$$NPP/GPP \sim 0.4$$

Heterotrophic respiration, $R_H$, of 1.09 kg C m$^{-1}$, when summed with $R_A$ yields a total ecosystem respiration, $R_E$, of 3.0 kg C m$^{-2}$ yr$^{-1}$. Heterotrophic respiration was nearly 40% of $R_E$ and with over two-thirds of heterotrophic respiration was due to decomposer metabolism. This clearly demonstrates the important role that decomposers play in the carbon cycle, with a respiratory allocation of:

$$RH/RE \sim 0.40$$

Also note the importance of allochthonous carbon inputs to the metabolism of decomposers which comes from outside the system. The magnitude of allochthonous carbon input distinguishes aquatic ecosystems from those of terrestrial ecosystems.

**Lotic ecosystem—streams and rivers**. In order to appreciate the dynamics of allochthonous inputs to standing water bodies, and their export beyond the system, one must examine the movement of organic carbon in flowing water ecosystems (streams and rivers). This is not easily done, but a stream can serve as an example. Streams are open ecosystems strongly connected with their watershed. The unidirectional flow of water connects streams and rivers through the transport of dissolved and particulate material (Cummins, 1975). Much of the carbon resource available in a stream at a specific location is delivered from upstream sources. Nutrients in streams are subject to downstream transport, thus a carbon cycle turns into a carbon spiral (Fig. 8.3).

Spiraling length ($S$) is the mean distance required for a nutrient atom to complete one cycle from its dissolved inorganic form in the water column, through a particulate phase, and finally through a consumer phase to be returned to the water column in a dissolved inorganic form. The cycle begins with the availability of the nutrient atom in the water column, including the distance that it is transported downstream in the water ($S_w$) until its uptake (U) and assimilation by the biota, and whatever additional distance the atom travels downstream within the biota ($S_B$) until that atom is eventually remineralized and released back to the water column. F is the downstream flux of organic carbon in water ($F_W$) and in the biota ($F_B$) in grams per sec. per meter width of the stream.

One measure of how efficiently the stream utilizes its carbon input is the stream organic carbon turnover length, $S_{OC}$, defined as the distance traveled by a carbon atom between its entry into the stream and its ultimate oxidation in respiration (Fisher, 1977). Thus, $S_{OC}$ values are a function of the rate at which the organic carbon pool moves downstream ($V_{OC}$) and the rate of its oxidation in respiration ($K_{OC}$):

$$S_{OC} = V_{OC}/K_{OC} \tag{8.4}$$

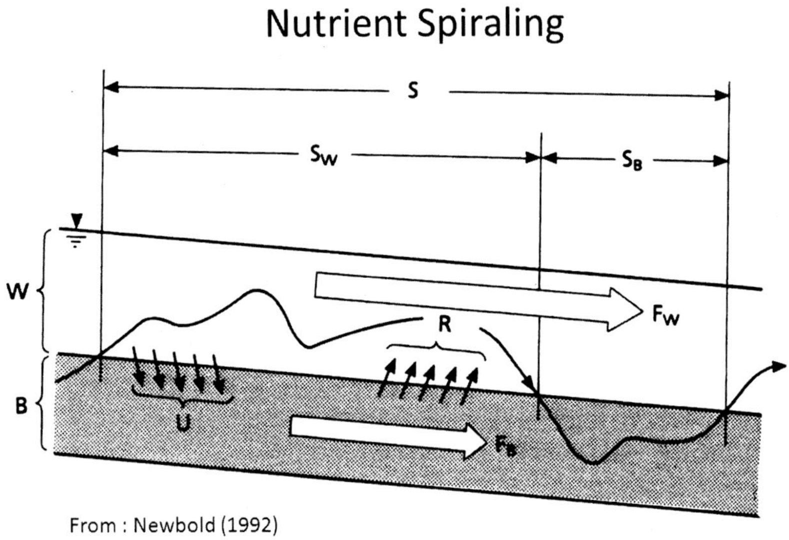

## Nutrient Spiraling

From : Newbold (1992)

**FIGURE 8.3** Conceptual representations of stream spiraling and uptake length affecting carbon metabolism in flowing waters. *Image credit: Newbold et al., 1982. Courtesy,* Oak Ridge National Laboratory and Springer Nature.

where: $S_{OC}$ is the turnover (spiraling) length of organic carbon ($m^{-1}$), $K_{OC}$ is the rate of oxidation of the organic carbon pool ($g\ C\ sec^{-1}$), and $V_{OC}$ is the rate of downstream movement of the organic carbon pool ($g\ C\ m^{-1}\ sec^{-1}$).

Before carbon turnover length there were other terms referring to organic carbon processing in lotic ecosystems (see Lisboa et al., 2016): "ecosystem efficiency" (EE) and "stream metabolism index" (SMI). EE was defined by Fisher and Likens (1973) as the ratio between the energy utilized, $R_E$, and the total energy input, I, to the ecosystem ($R_E/I$ ratio). The ratio derived from the production/respiration, P/R, diagram created by Odum (1956) to evaluate autotrophy or heterotrophy in ecosystems. However, the transition between autotrophy and heterotrophy in streams is not properly expressed by the P/R ratio because it does not show the extent of production that comes from instream primary producers, and the extent that comes from external input. The EE index allowed the assessment whether a stream was accumulating or losing organic carbon. However, it showed some limitations that hampered comparing different open ecosystems (Fisher, 1977) because it did not measure the degree to which respiration rates were in balance with inputs (Newbold et al., 1982).

This concept was improved later when Fisher (1977) developed the SMI index, as a loading factor that indicates the rate of increase or decrease of total organic matter concentration per unit ecosystem. SMI was defined as simply the ratio of observed respiration to respiration required for zero loading in the system, which can be used to compare total processing efficiencies in streams of any size (Fisher, 1977). Newbold et al. (1982) showed that organic carbon turnover length and SMI are complementary, and proposed an equation relating the two indices:

$$SMI(x) = 1 - exp(-x/S_{OC}) \tag{8.5}$$

where:

SMI is the stream metabolism index,
X is the distance from headwaters to the studied reach (m), and
$S_{OC}$ is the turnover length of carbon ($m^{-1}$).

Both SMI and $S_{OC}$ indices use similar properties of a stream, but the difference relies on the fact that the ecosystem efficiency, EE, and SMI are each length dependent, while the $S_{OC}$ can be calculated for any given point of the stream (Newbold et al., 1982). While SMI measures the processing of organic carbon from lateral input, $S_{OC}$ calculates the relative processing of all the carbon that is within the ecosystem originating from upstream input.

$$R_E = \frac{\sum (18:00/06:00)(R_{OC}) + (D)(k)}{\Delta t} \tag{8.6}$$

where:

$R_E$ is the respiration (mg $O_2$ liter$^{-1}$ day$^{-1}$),
$\sum$ sum of hourly rates (mg $O_2$ liter$^{-1}$ day$^{-1}$),
$R_{OC}$ is the rate of change of $O_2$ per hour ($\Delta O_2$ hr$^{-1}$),
D is the $O_2$ deficit (mg L$^{-1}$),
k is the reaeration coefficient, base (day-1), and
$\Delta t$ is the time interval (to return a rate per time) (hr$^{-1}$).

For $S_{OC}$ calculations, respiration rates must be transformed firstly in g $O_2$ m$^{-2}$ day$^{-1}$ (dividing by depth), then converted to units of carbon using a respiratory coefficient (RQ) of 0.85:

$$gC = gO_2(RQ)(12/32) \tag{8.7}$$

where:

12 is the atomic weight of carbon, and
32 is the atomic weight of oxygen.

Since $S_{OC}$ in Eq. 8.4 requires only heterotrophic respiration, the $R_H$ equivalence (proportionality) is calculated from $R_E$ using a coefficient of $p = 0.38$, with the formula:

$$R_H = R_E - (p)(GPP) \tag{8.8}$$

Thus, there are several ways of quantifying the carbon cycle in flowing water ecosystems. The terms carbon spiraling and carbon turnover appear synonymous, apart from the condition that spiraling only makes sense in terms of length (m), but conceptualizes a carbon cycle moving downstream, while turnover is used both as a rate (day$^{-1}$), representing $K_{OC}$ (rate of oxidation), and as length (m), representing $S_{OC}$ (turnover length) of carbon oxidation. The shorter the turnover length the richer the biota and the more intense is the stream's metabolism.

The carbon metabolism of 14 rivers in the western and midwestern U.S., with discharges ranging from 14 to 84 m$^3$ s$^{-1}$, was measured by Hall et al. (2016). GPP ranged from 0.6 to 22 g $O_2$ m$^{-2}$ d$^{-1}$ and $R_E$ tracked GPP, suggesting that autotrophic production was the source of much of the carbon for riverine $R_E$ during the summer. Gas exchange from these 14 rivers correlated with river slope and velocity; carbon turnover lengths, $S_{OC}$, ranged from 38 to 1190 km, with the longest turnover lengths in high-sediment, arid-land rivers, where GPP and $R_E$ were lowest. The mean ratio of carbon turnover length to river length was 1.6, in river segments between major tributaries or lakes, demonstrating that rivers can mineralize much of their organic carbon load along their length at baseflow. Carbon mineralization velocities ranged from

$0.05$ to $0.81$ m d$^{-1}$, and were not different from small streams. With high GPP relative to $R_E$, combined with generally short organic carbon spiraling lengths, $S_{OC}$, rivers can metabolize organic carbon very efficiently.

**Marine ecosystems**. The ocean plays an important role in the global carbon cycle (Fig. 8.4). The ocean is called a carbon "sink" because it takes up more carbon from the atmosphere than it releases. Carbon moves in and out of the ocean daily, but it is also stored there for thousands of years. The ocean behaves like a biological pump for carbon. Surface turbulence helps the mixing of $CO_2$ with surface waters through a bicarbonate chemistry, and phytoplankton take up carbon through photosynthesis. Carbon gets incorporated into marine organisms as organic matter or structural calcium carbonate. Remains of primary production, dead cells, shells, and other parts sink into deep water and eventually into the deeper ocean layers as sediments. This action of organisms moving carbon in one direction is often described as a "biological pump." Over millions of years, chemical and physical processes may turn these sediments into rocks. This part of the carbon cycle can lock up carbon for millions of years.

Carbon dioxide is the basis of photosynthesis by ocean phytoplankton (Fig. 8.4). Carbon dioxide dissolves in the ocean to form carbonic acid

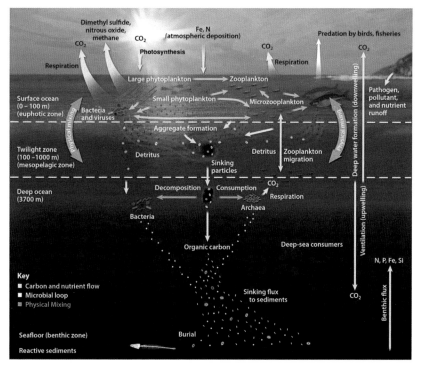

**FIGURE 8.4** The biogeochemical cycle of carbon in the ocean ecosystem. *Image credit: Oak Ridge National Laboratory and U.S. Department of Energy.*

($H_2CO_3$), bicarbonate ($HCO_3^-$), and carbonate ($CO_3^{2-}$). There is about 50 times as much carbon dissolved in the oceans as exists in the atmosphere. Of the many chemical reactions involving carbon in ocean waters, several are fundamental to the carbon cycle:

$$\text{Respiration} \leftrightarrow \text{Photosynthesis:}$$

$$6CO_2 + 6H_2O \leftrightarrow 6O_2 + C_6H_{12}$$

Carbon dioxide and ocean water:

$$CO_2 + H_2O \leftrightarrow H_2CO_3$$

Carbonic acid chemistry:

$$H_2CO_3 \leftrightarrow H^+ + HCO_3^-$$

$$HCO_3^- \leftrightarrow H^+ + CO_3^{2-}$$

Weathering of rock into the ocean:

$$2Ca^{2+} + CO_2^2 \leftrightarrow 2CaCO_3 \leftrightarrow \text{Calcareous sedimentation}$$

Gaseous $CO_2$ in the atmosphere dissolves easily in water, accentuated by turbulent mixing. Carbon dioxide reacts with the water molecules according to the equations below (see Feely et al., 2001). When carbon dioxide mixes with water it is partially converted into carbonic acid, which ionizes to form hydrogen ions ($H^+$), bicarbonate ions ($HCO_3^-$), and carbonate ions ($CO_3^{2-}$). Seawater can assimilate much more $CO_2$ than fresh water, because bicarbonate and carbonate ions have been discharged over aeons by rivers into the sea from the weathering of carbonaceous rocks on land. The carbonate reacts with $CO_2$ and $H_2O$ to form bicarbonate, which leads to a further uptake of $CO_2$ and a decline of the $CO_3^{2-}$ concentration in the ocean. All of the $CO_2$-derived chemical species in the water, collectively, carbon dioxide, carbonic acid, bicarbonate and carbonate ions, are referred to as dissolved inorganic carbon (DIC). The carbonic acid-carbonate equilibrium determines the number of free protons ($H^+$ ions) in the seawater and thus its pH value. This is why rapid infusion of large quantities of $CO_2$ into surface waters, until buffered, can cause ocean acidification.

$$CO_2(atm) \leftrightarrow CO_2(aq) + H_2O \leftrightarrow H_2CO_3 \leftrightarrow H^+ + HCO_3^- \leftrightarrow H^+ + CO_3^{2-}$$

carbonic acid      bicarbonate      carbonate

$$Ca^{2+} + CO_3^{2-} \leftrightarrow CaCO_3$$
Calcium carbonate

Oceanic waters are in close equilibrium with atmospheric $CO_2$. Dissociation equilibrium and kinetics of the chemistry of hydration and dehydration of $CO_2$ are affected by pH, temperature, and salinity. Atmospheric $CO_2$ at 0.033% by volume dissolves in water to yield an unhydrated $CO_2$ concentration similar to that in air. The hydration of $CO_2$ to $H_2CO_3$ is slow and, at equilibrium, yields a concentration of $H_2CO_3$ 0.0025 times that of $CO_2$. Dissociation of $H_2CO_2$ to $HCO^-_3$ and $CO^-_3$ is instantaneous. If $CO_2$ is at equilibrium in a solution buffered to constant pH, the $CO_2$ concentration and saturation kinetics are achieved. Such equilibria are strongly influenced by temperature and salinity.

Over time scales of hundreds of years or more, and ignoring the infusion of $CO_2$ from fossil fuel combustion, carbon chemistry in the ocean is independent of the atmosphere. It is the ocean that dictates the atmosphere's content of $CO_2$ (Broecker, 1973). (We will consider the anthropogenic emissions of $CO_2$ from fossil fuels in Chapter 11). The ocean can be viewed as a two-layered system: a warm surface ocean and a deep cold ocean, separated by a thermocline going from 20°C to 2°C. There is about 20% more carbon dissolved in deep water than in surface water. About 10% to 29% of the organic carbon produced in surface water is not reprocessed there and falls into the deep sea. Of organic tissue produced in the surface, 99% is catabolized and about 1% becomes kerogen and is deposited in sediments. Of the calcium carbonate formed in ocean waters, 85% undergoes dissolution in the deep sea and about 15% is buried in sediments. Over geologic time, ocean sediments have become a huge sink for carbon. Carbon has a long residence time of about 100,000 years in the ocean (Broecker, 1973).

The photosynthetic fixation of $CO_2$ in surface waters by phytoplankton drives the oceanic carbon cycle. The range of productivity in marine ecosystems is about the same as that of terrestrial ecosystems (Dunbar, 1975). The extremes of productivity are represented by coral reef and the Arctic Ocean. Open ocean ecosystems are most nearly analogous to grassland ecosystems (Fig. 8.5). Both systems are intermediate in productivity; rates of production are high, but the biomass of the producer plant community is low due to constant cropping by consumers. Estimates of net primary production in different latitudes of the world are mg C $m^{-2}$ $day^{-1}$: Gulf of Guinea, 365 g C $m^{-2}$ $yr^{-1}$; Long Island Sound, 190 g C $m^{-2}$ $yr^{-1}$; and the Sargasso Sea, 70−145 g C $m^{-2}$ $yr^{-1}$. The normal range of primary production in the oceans is between 50 and 150 g C $m^{-2}$ $yr^{-1}$ (Riley, 1970), although productivity in inshore regions and areas of upwelling can be up to ten times greater than oceanic values.

There is a general consensus that the total amount of carbon in primary production of the world's oceans is:

$$NPP = \sim 50 \text{ GtCyr}^{-1} (5 \times 10^{16} \text{grams carbon}), \text{and}$$

And that the average worldwide rate of production is:

$$NPP = \sim 140 \text{ gCm}^{-2}\text{yr}^{-1}.$$

Field et al. (1998) estimated total ocean NPP of 48.5 Gt C yr$^{-1}$ to be about 46.2% of the world's net primary production of 104.9 Gt C yr$^{-1}$ ($1.05 \times 10^{17}$ g carbon). Productivity in the upper layers of the oceans is dependent upon the supply of nutrients which, in turn, is dependent upon instability of the water column. An exception would be where outflows of nutrients from land occur where rivers discharge. Vertical instability in the water column is caused by winds, storm turbulence, tides, and vertical exchange in wintertime in temperate and subarctic regions. The influence of temperature and solar insolation (latitude) is far less than that of vertical instability (Dunbar, 1975), although the effects of future climate change on ocean circulation could have unforeseen consequences.

Even though the total amounts of carbon incorporated in primary production are about equal between the oceans and land, the amounts per area are greater on land than in the ocean. Average NPP on non-ice-covered land is 426 g C m$^{-2}$ yr$^{-1}$, while in the ocean it is 140 g C m$^{-2}$ yr$^{-1}$. The lower NPP per unit area of the ocean largely results from competition for light between phytoplankton and their strongly absorbing medium, seawater. Only about 7% of the incident radiation, i.e., photosynthetically active radiation (PAR), is absorbed by the phytoplankton; the remainder is absorbed by water and dissolved organic matter. In contrast, terrestrial plants absorb about 31% of the PAR incident on land. Even though primary producers in the ocean are responsible for nearly half the total world NPP, they represent only 0.2% of the global producer biomass. Thus, the turnover time of plant organic carbon in the ocean (average 2−6 days) is about a 1000 times faster than on land.

**Terrestrial ecosystems.** In contrast to aquatic ecosystems, inorganic carbon sources are typically unimportant in the terrestrial carbon cycle. Carbon is inorganic form generally occurs as calcium and magnesium carbonates, and such forms of carbon contribute less than 5% of the total carbon content of living biomass. Fluxes of release and uptake of carbon occur directly with the atmosphere as gaseous $CO_2$ and, unlike in aquatic ecosystems, dissolved carbon forms are relatively unimportant. Beyond these dissimilarities, other than the differences in total standing crops of producers, much of the trophic structure and carbon allocations are similar between terrestrial and aquatic ecosystems.

Forest ecosystems are ubiquitous around the world and will be discussed in more detail in Chapter 9, but here will serve as an example of the carbon cycle

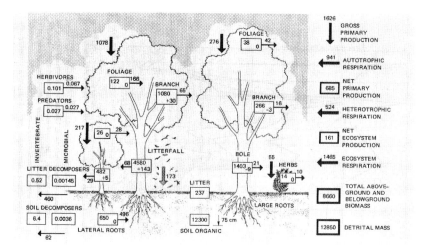

**FIGURE 8.5** The carbon cycle in a mesic deciduous forest in Tennessee. Trees, left to right, represent understory, dominant *Liriodendron tulipifera*, and all other overstory trees. Decomposers are separated by surface litter and soil zones. Heterotrophs are invertebrates only for both herbivores and carnivores; values do not include vertebrates. All values are in g C m$^{-2}$ for biomass (boxes, upper left standing crop; lower right, annual increment) and in g C m$^{-2}$ yr$^{-1}$ for fluxes (*arrows*). *Image credit: Reichle et al., 1973. Courtesy, Oak Ridge National Laboratory and Cambridge University Press.*

$$GPP/TSC = 0.19 \tag{8.9}$$

in a terrestrial ecosystem. Forest ecosystems utilize energy in creating a structure, and in maintaining it, which allows them to vertically stratify and most effectively intercept incident radiation. Major types of forests are boreal, temperate, and tropical forests, both evergreen and deciduous; swamp, scrubland, alpine, and savannah transitions (chaparral). Although there are many differences, most of the true forests are distinguished by a substantial woody biomass, complex structure, and high species diversity. An example of the carbon cycle in a temperate deciduous forest will illustrate the carbon cycle in forests (Fig. 8.5), while Tables 8.3 and 8.4 will illustrate not only the range in carbon cycle parameters across different forests, but also their similarities in net ecosystem production (NEP).

Our model terrestrial carbon cycle is that of a mesic, temperate deciduous forest (Fig. 8.5). This box and arrow representation of the carbon cycle in the forest is a compartment model (Shugart et al., 1976). Boxes represent pool sizes of carbon, i.e., biomass production values. Values (g C m$^{-2}$) in the upper left of boxes are standing crops; those in the lower right are annual increments. The arrows are the fluxes; in this example they are rates expressed as g C m$^{-2}$ yr$^{-1}$. Downward arrows represent system inputs in photosynthesis; lateral

**TABLE 8.2** Mean values and ranges for GPP, $R_E$, and NEP for aquatic ecosystems Values are g $O^2$ $m^{-2}$ $day^{-1}$).

| Ecosystem | Parameter | Mean | Range |
|---|---|---|---|
| Stream (n = 215) | GPP | 2.4 | 0.0–16.3 |
| | $R_E$ | 6.7 | 0.0–30.0 |
| | NEP | −4.2 | −27.0–7.3 |
| Wetland/Pond (n = 13) | GPP | 3.3 | 0.7–5.7 |
| | $R_E$ | 5.8 | 1.9–8.9 |
| | NEP | 12.5 | −6.0–1.3 |
| Lake (n = 72) | GPP | 5.8 | 0.1–35.2 |
| | $R_E$ | 6.1 | 0.1–32.1 |
| | NEP | −0.3 | −7.0–9.9 |
| Estuary (n = 4) | GPP | 10.8 | 1.2–28.1 |
| | $R_E$ | 13.6 | 1.2–33.4 |
| | NEP | −2.7 | −9.0–5.2 |
| All (n = 350) | GPP | 4.2 | 0–35.2 |
| | $R_E$ | 7.4 | 0–33.4 |
| | NEP | −3.1 | −26.6–9.8 |

**Source:** Hoellein et al., 2013. *Courtesy, John Wiley & Sons.*

arrows are respiratory fluxes. The ecosystem totals are obtained by adding the subcomponent values. The much larger standing crop biomass in woody structural components of forests results in gross production efficiency of: several orders of magnitude lower than that of our Silver Springs aquatic example of GPP/TSC = 8.0. The forest also has a similarly lower net production efficiency:

$$NPP/TSC = 0.08 \tag{8.10}$$

also, several orders of magnitude lower than Silver Springs with an NPP/TSC $\sim$ 3.2. This means that the forest's effective production efficiency:

$$NPP/GPP = 0.42 \tag{8.11}$$

is nearly identical to that of Silver Springs at NPP/GPP $\sim$ .36, with different structural composition, but with very similar effective production efficiencies. This is because the autotrophic respiratory "costs" $R_A$ in the forest were much less at only 942 g C $m^{-2}$ $yr^{-1}$. Heterotrophy in the forest amounts to

**TABLE 8.3** Comparison of the carbon budgets of eight terrestrial ecosystems: Spruce Forest, Sweden (Karlberg et al., 2007); Mesic Tulip Poplar forest, TN (Reichle et al., 1973); Oak-Pine forest, NY (Woodwell and Botkin, 1970); Tropical Rain Forest, Thailand (Tan et al., 2010); Shortgrass Prairie, CO (Andrews et al., 1974); Tundra (after Reichle, 1975); Agricultural ecosystems values from L. Ryszkowski (Reichle, 1981). Units are: fluxes in kg C $m^{-2}$ $yr^{-1}$, standing crop in kg C $m^{-2}$).

| Carbon budget component | Spruce forest[a] | Mesic forest[b] | Oak-pine forest[c] | Tropical rain forest[d] | Shortgrass prairie[e] | Tundra[h] | Potato field[i] | Rye field[i] |
|---|---|---|---|---|---|---|---|---|
| Total standing crop (TSC) | 9.930 | 8.660 | 5.96 | 14.747 | ~0.002[g] | — | — | — |
| Gross primary production (GPP) | 1.870 | 1.626 | 1.28 | 2.601 | 0.641 | 0.240 | 1.286 | 1.006 |
| Autotrophic respiration ($R_A$) | 1.320 | 0.941 | 0.68 | 1.721 | 0.218 | 0.120 | 0.431 | 0.342 |
| Net primary Prod. (NPP = GPP-$R_A$) | 0.550 | 0.685 | 0.60 | 0.880 | 0.423 | 0.120 | 0.849 | 0.664 |
| Heterotrophic respiration ($R_H$) | 0.450 | 0.524 | 0.32 | 0.521 | 0.294 | 0.108 | 0.500 | 0.310 |
| Net Ecosys. Prod. (NEP = NPP - $R_H$) | 0.100 | 0.161 | 0.28 | 0.359 | 0.129 | 0.012 | 0.355 | 0.354 |
| Ecosystem respiration ($R_E$ = $R_A$ + $R_H$) | 1.770 | 1.465 | 1.00 | 2.242[b] | 0.512 | 0.228 | 0.931 | 0.652 |
| Soil carbon | 7.990 | 12.850 | — | 9.238 | ~3.5[f] | — | — | — |

[a]Karlberg et al., 2007.
[b]C values reported by Reichle et al. (1973), Harris et al. (1975).
[c]Woodwell and Botkin (1970).
[d]Tan et al. (2010).
[e]Andrews et al. (1974).
[f]Pepper et al. (2005).
[g]Bokhari (1978).
[h]Reichle et al. (1973), 1975, pers. communication from P. C. Miller, L. L. Tieszen, P. I. Coyne and J. J. Kelly in Bowen, Ed., 1972 Tundra Biome Symposium.
[i]Ryszkowski, personal communication.

$524 \text{ g C m}^{-2} \text{ yr}^{-1}$, or about 36% of the total heterotrophic respiration; this heterotrophic respiration is largely due to the metabolic activity of decomposers, with a respiratory allocation of:

$$R_H/R_E = 0.36 \tag{8.12}$$

Large quantities of organic carbon in the forest are below ground. With roots amounting to $\sim 9\%$ ($780 \text{ g C m}^{-2}$, revised estimate by Harris et al., 1975) of aboveground tree biomass ($8660 \text{ g C m}^{-2}$, Table 8.2), and $12,850 \text{ g C m}^{-2}$ in detritus, 56% of the organic carbon in the forest is nonliving ($8660 + 780/8660 + 12,850$). This nonliving organic carbon pool is a very important component of the forest carbon cycle; a large pool, with a slow turnover time, lends stability to both the carbon cycle and the forest's nutrient cycles.

**Turnover times of carbon.** The movement of carbon in and out of the various pools can be described by its turnover time. Resident time and turnover time are synonymous measures. Turnover time is calculated by dividing the quantity of nutrient present in a particular nutrient pool or reservoir by the flux rate for that nutrient element into or out of the pool. Turnover time thus describes the time it takes to fill or empty that particular nutrient reservoir. The turnover rate is the reciprocal of the turnover time. Turnover time of the soil organic matter pool in his forest ecosystem is:

$$\frac{\text{Pool size soil organic}}{R_H \text{ soil decomposers}} = \frac{12,300 \text{ g Cm}^{-2}}{62 \text{ g Cm}^{-2} \text{ yr}^{-1}} = 198.4 \text{ years} \tag{8.13}$$

equivalent to a turnover rate of soil organic material of 0.5% $\text{yr}^{-1}$. Just as the nonliving, belowground carbon mass is a significant portion of the total ecosystem's carbon mass, so is the respiration by heterotrophs, $R_H$, of $524 \text{ g C m}^{-2} \text{ yr}^{-1}$ and that by the living roots. Living root respiration accounts for between 35% and 53% of the total $R_A$ in the forest (Reichle et al., 1973; Edwards and Harris, 1977).

**Chemosynthetic ecosystems.** Chemosynthesis is the biological conversion of carbon-containing molecules (usually carbon dioxide or methane) and nutrients into organic matter using the oxidation of inorganic compounds (e.g., hydrogen gas, hydrogen sulfide) or methane as a source of energy, instead of radiant energy from the sun, as in photosynthesis. Chemoautotrophs are primarily bacteria and are found in rare ecosystems where sunlight is not available, such as in those associated with dark caves or hydrothermal vents at the bottom of the ocean. Many chemoautotrophs in hydrothermal vents use hydrogen sulfide ($H_2S$), which is released from the vents as a source of their chemical energy.

$$12H_2S + 6CO_2 \rightarrow C_6H_{12}O_6 + 6H_2O + 12S$$

This allows chemoautotrophs to synthesize complex organic molecules, such as glucose. While not significant as ecosystem fluxes of carbon and energy in the global context, they form the food base for rare and interesting food webs and are possible sites of where early life may have originated and evolved.

## 8.5 Comparison of carbon metabolism among ecosystems

The parameters of the carbon budget for a variety of aquatic and terrestrial ecosystems, summarized in Tables 8.1−8.3, provide the basis for comparing the carbon metabolism of ecosystems—similarities and dissimilarities—with the ultimate objective of developing predictive properties of the carbon budget for all ecosystems.

**Aquatic ecosystems.** Aquatic ecosystems are characterized by having low standing crop biomasses and relatively high rates of primary production (Table 8.1). The aquatic ecosystems have a range in order of magnitude in GPP from relatively warm Silver Springs to colder, but eutrophic, Lake Lawrence of $0.125-3.2$ kg C m$^{-2}$ yr$^{-1}$. Oligotrophic lakes are deep and cold; eutrophic lakes tend to be shallow and warm; their metabolism varies accordingly.

High productivity by an ecosystem does not necessarily denote a high standing crop, e.g., NPP of $1.2$ kg C m$^{-2}$ yr$^{-1}$ in Silver Springs with a standing crop biomass of only $0.4$ kg C m$^{-2}$ (Odum, 1957) or grass in a Spartina marsh (Teal, 1962). Nor does a relatively high productivity of a young pine plantation at $1.260$ kg m$^{-2}$ yr$^{-1}$ with a standing crop of only $5.6$ kg C m$^{-2}$ compare to that of a climax Douglas fir forest having a productivity of only $999$ g m$^{-2}$ yr$^{-1}$ but with a biomass two times greater at $20$ kg m$^{-2}$ (Woodwell and Whitaker, 1968).

Autotrophic respiration, $R_A$, ranged between $0.118$ and $1.92$ kg C m$^{-2}$ yr$^{-2}$, paralleling their respective autotrophic biomasses. Net primary production, NPP, of the temperate climate aquatic ecosystems ranged between $0.19$ and $1.28$ kg C m$^{-2}$ yr$^{-1}$. Heterotrophic respiration, $R_H$, in these aquatic ecosystems ranged between $0.48$ and $1.09$ kg C m$^{-2}$ yr$^{-1}$. Net ecosystem respiration, $R_E$, was between $0.5$ and $0.2$ kg C m$^{-2}$ yr$^{-1}$. The net ecosystem production, NEP, of these aquatic ecosystems was between $0.11$ and $3.0$ kg C m$^{-2}$ yr$^{-1}$. Differences between the four aquatic ecosystems (excluding the salt marsh) are more comparable.

Autotrophic respiration as a function of gross primary production, or the respiratory cost of production, RA/GPP, was between $0.27$ and $0.60$; for the salt marsh it was $0.77$. The apportionment of total ecosystem respiration between autotrophs and heterotrophs, $R_H/R_A$, the respiration ratio, was $0.57-0.81$; the salt marsh was $0.16$. Heterotrophic respiration (largely decomposers in all these ecosystems) was a significant percentage of total ecosystem respiration; $R_H/R_E$

x 100 was 36%−65%; for the salt marsh it was 14%. The relative respiration, RE/NEP, was roughly 1−15; the salt marsh was 8.6.

Net ecosystem production, NEP, is a consequence of the balance between GPP and $R_E$. These are three fundamental metrics of ecosystems required in models of global carbon cycles (sources and sinks). The ecosystem carbon balance, GPP:RE, thus, is an important relationship. Correlation analyses of GPP (g $O_2$m$^{-2}$ day$^{-1}$) against $R_E$ (g $O_2$ m$^{-2}$ day$^{-1}$) for 215 streams ($r^2 = 0.23$), 15 wet lands/ponds ($r^2 = 0.02$), 72 lakes ($r^2 = 0.84$), and 47 ($r^2 = 0.87$) estuaries show promise (Hoellein et al., 2013). Respiration in terms of $O_2$ consumed can be converted to carbon (from $CO_2$) released by using appropriate RQs (Eqs. 5.10−5.14) and Eq. (8.13). While there are few detailed, total ecosystems carbon balances with all parameters measured, for Odum's Silver Springs and Lake Washington (Table 8.3) the range was GPP:$R_E$ = 2.0 ≤ 3. Hoellein et al.'s GGP: $R_E$ average for lakes was GPP:$R_E$ = 0.95; Odum's 1956 synthesis of lake and ponds had GPP:$R_E$ = 1 ≤ 2. The values for the Odum's Silver Springs and Lake Washington (Table 8.3) were in the range GPP:$R_E$ = 2.0 ≤ 3. While the values for GPP:$R_E$ are close, greater precision will require better understanding of the internal processes (i.e., temperature, nutrient loading, solar insolation, and heterotrophy).

**Terrestrial ecosystems**. Terrestrial ecosystems are characterized by having a larger standing crop biomass, TSC, than aquatic ecosystems, but with more moderate rates of primary production (Tables 8.3 and 8.4). Natural terrestrial ecosystems also ranged an order of magnitude in gross primary production, GPP, from the tropical rain forest at 2.6 kg C m$^{-2}$ yr$^{-1}$ to the cold tundra with 0.240 kg C m$^{-2}$ yr$^{-1}$. Monocultured pine stands in plantations can attain 4.12 kg C m$^{-2}$ yr$^{-1}$. Autotrophic respiration, $R_A$, ranged between 0.12 and 2.00 kg C m$^{-2}$ yr$^{-1}$, paralleling their respective autotrophic biomasses. Net primary production, NPP, of the terrestrial ecosystems ranged between 0.12 and 0.88 kg C m$^{-2}$ yr$^{-1}$. Heterotrophic respiration, $R_H$, in these terrestrial ecosystems ranged from 0.108 to 0.694 kg C m$^{-2}$ yr$^{-1}$. Ecosystem respiration, $R_E$, of the terrestrial ecosystems was between 0.512 and 2.76 kg C m$^{-2}$ yr$^{-1}$. Agricultural fields show the results of human "engineering" in monoculture ecosystems, with the highest net primary production 0.849 kg C m$^{-2}$ yr$^{-1}$ in potato fields and 0.664 kg C m$^{-2}$ yr-$^1$ in rye fields, with respective net ecosystem production of 0.355 kg C m$^{-2}$ yr$^{-1}$ and 0.354 kg C m$^{-2}$ yr$^{-1}$ (Table 8.3).

The NPP/GPP, effective production, ratios suggest that those ecosystems dominated by annual autotroph communities (prairie and tundra), without substantial energy or carbon in perennial woody biomass, have the highest yields relative to total carbon fixed. Strikingly, the shortgrass prairie ecosystem and the two agricultural ecosystems all have the same NPP/GPP ratios of 0.66. Differences between the three forest ecosystems can be more easily compared. Autotrophic respiration as a function of gross primary production, or the production efficiency, $R_A$/GPP, in the forest ecosystems were between 0.53 and 0.71; similarly, the prairie and tundra were 0.34 and 0.40, respectively,

**TABLE 8.4** Comparative metabolic parameters for six different forest ecosystems. All values above the dotted line are in kg C m$^{-2}$ and kg C m$^{-2}$ yr$^{-1}$; values below the dotted line are dimensionless indices.

| Carbon budget parameter | Pine plant[a] USA | Sub. Alpine[b] Conif. | Beech[c] Japan, Denmark | Rain fforest[c] Ivory Coast | Oak forest[d] Britain | Oak-Hornbeam[e] Poland |
|---|---|---|---|---|---|---|
| Total standing crop (TSC) | 7.062 | 15.905 | — | — | 7.466 | (1.377) |
| Gross primary production (GPP) | 4.124 | 1.910 | 1.175 | 2.675 | 2.330 | (0.133) |
| Autotrophic respiration (R$_A$) | 2.068 | 1.375 | 0.500 | 2.000 | 1.412 | (0.051) |
| Net primary Prod | 2.056 | 0.535 | 0.675 | 0.675 | 1.918 | (0.082) |
| Heterotrophic respiration (R$_H$) | 0.694 | 0.331 | 0.675 | 0.675 | 0.564 | — |
| Ecosystem Resp. (R$_E$ = R$_A$ + R$_H$) | 2.762 | 1.706 | 1.175 | 2.675 | 1.976 | — |
| Net Ecosys. Prod. (NEP = NPP-R$_H$) | 1.362 | 0.204 | — | — | 0.354 | — |
| R$_A$/GPP | 0.50 | 0.72 | 0.43 | 0.75 | 0.61 | 0.38 |
| NPP/GPP | 0.50 | 0.28 | 0.57 | 0.25 | 0.39 | 0.62 |
| R$_A$/NPP | 1.00 | 2.57 | 0.74 | 2.96 | 1.54 | 0.62 |
| R$_H$/R$_A$ | 0.34 | 0.24 | 1.35 | 0.34 | 0.40 | — |
| NEP/GPP | 0.33 | 0.11 | — | — | 0.15 | — |

[a]from Kinerson et al. (1977).
[b]from Kitazawa (1977).
[c]from Macfadyen (1970) (calculation assumes NEP = 0).
[d]from Satchell (1973).
[e]from Medwecka-Kornas et al. (1974) (values in parentheses are in 10$^{-6}$ cal, i.e., multiply by 10$^6$ for cal).
**Source:** Reichle (1981).

with the agricultural fields at 0.34. When the production efficiency is compared across ecosystems, a striking similarity among all ecosystems appears. $R_A$/NPP, maintenance efficiency, illustrates how these systems allocate their carbon reserves between maintenance respiration and production. Intuitively this allocation should be related to the quantity and persistence of nonphotosynthetic biomass characteristic of each ecosystem. Maintenance metabolism costs are lowest in the prairie and agricultural systems ($\sim 0.51$) and progressively increases with the tundra and xeric forest, reaching maximum values in the mesic deciduous forest. The apportionment of total ecosystem respiration between autotrophs and heterotrophs, $R_H$/$R_A$, the respiration allocation, was 0.30−0.56 in the forests, 0.90 in the tundra, 1.35 in the prairie, and 1.16 and 0.91 in the potato and rye fields, respectively. Heterotrophic respiration (largely decomposers in all these ecosystems) was a significant percentage of total ecosystem respiration; $R_H$/$R_E \times 100$ was 23% −32% for the forests and 47%−57% for the prairie and tundra, and 48%−54% for the agricultural systems. The relative respiration, $R_E$/NEP, was roughly 3.5 to 9. The ecosystem carbon balance, GPP:$R_E$ = 1.15 (n = 6) for the natural terrestrial ecosystems in Table 8.3.

The difference between the two temperate forests (Table 8.3), a mature mesic deciduous tulip poplar forest and a young oak-pine forest, is exhibited in their net ecosystem production, the detritus pools, and decomposer activity. The tulip poplar forest annually loses 522 g C m$^{-2}$ yr$^{-1}$ through decay, the oak-pine forest decomposes only 320 g C m$^{-2}$ yr$^{-1}$. This would be expected, since the oak-pine is a successional forest, still developing an organic soil, and is drier. For the tulip poplar forest, this amounts to 4.1% annual turnover of the detrital pool of 12,850 g C m$^{-2}$. About 44% of the carbon in the spruce forest and 60% of the carbon in the tulip poplar forest is in the detritus pool. With NEP values about one half that of the oak-pine forest, the spruce and tulip poplar are closer to equilibrium in soil organic matter. The ecosystem productivity, NEP/GPP, of the temperate and tropical forests was (0.10−0.22), and in the prairie and agricultural systems (0.20−0.35); and much lower for the spruce forest (0.05) and tundra (0.05) ecosystems of cooler, more northern latitudes. Additional data (Table 8.4) on the carbon metabolism of geographically diverse forest ecosystems provides further information for additional forest ecosystem types in different latitudes, with associated ecosystem production and metabolic efficiencies.

The efficiency of conversion of radiant energy into photosynthate, while lower than that of aquatic ecosystems, is higher for forests (2.0%−3.5%) than for perennial herbaceous communities or annual and crop systems, 1%−2% and <1.5%, respectively (Kira, 1975). The relatively higher gross photosynthetic efficiency may be attributed in large part to maintenance of a large foliar surface area within the forest canopy. Support of this geometrically complex photosynthetic surface area has its metabolic costs, however, and maintenance of a large aboveground and belowground structural support and supply system, and its associated metabolic demands, draws heavily on the GPP of forest ecosystems.

Ecosystems, regardless of type or complexity, all appear to exhibit common properties of persistence and growth. Ecosystems tend to display consistent patterns of optimization toward maximum persistent biomass. The specific nature of the mechanisms underlying this phenomenon is complex, but ecosystems have evolved regulatory mechanisms whereby a secure energy base is established in the presence of a fluctuating environment. Autotrophic components of ecosystems display one of two basic properties (or combinations thereof): (1) small individuals with rapid turnover, or (2) large individuals with slow turnover. Combinations of these properties serve to provide homeostasis, with attributes of both rapid response and long-term stability in the photosynthetic base.

Concomitantly, ecosystems have also developed mechanisms for energy storage as a basis for homeostasis. This carbon/energy reservoir is characteristically large, with slow response time, but is crucial for nutrient storage and reutilization. Many ecosystems, across a wide range of internal structure and external environmental conditions, possess an energy reservoir in the form of organic detritus. These patterns in energy metabolism and carbon allocation for a variety of ecosystem types show similar patterns in net ecosystem production, NEP, i.e., storage in detritus. This storage also minimizes the energetic costs of maintaining essential elements available for reutilization by the ecosystem.

**Carbon turnover in soils and sediments**. The soil organic matter content of ecosystems is an important sink in the ecosystem's carbon cycle, because of its size and slow turnover. In a mature, steady-state ecosystem carbon does not usually accumulate from year to year; deep ocean sediments and peat accumulation in northern latitudes, where environmental temperature restrain $R_H$, are notable exceptions. The importance of these "sinks" in the global carbon cycle will become apparent in Chapter 10, but first an understanding of the residence times of carbon in various parts of ecosystems is important. The residence time is equal to the energy or carbon in biomass divided by that of the net productivity (Wang, 2017):

$$R_t = (C \text{ in TSC biomass}/C \text{ net productivity}), \qquad (8.14)$$

$$R_t = (\text{total soil } C/\Delta \text{ soil } C), \text{ or} \qquad (8.15)$$

$$R_t = (\text{tot soil } C/R_H \qquad (8.16)$$

where:

$R_t$ is carbon turnover time (years),
C is TSC biomass (kg C m$^{-2}$),
NPP is net primary production (veg) or net increment (soil) (kg C m$^{-2}$ yr$^{-1}$), and

$R_H$ is heterotrophic respiration.

In the carbon cycle there are carbon "sinks" in ocean sediments and terrestrial soil organic residues. Boecker (1974) states that the overall turnover time in the ocean is 100,000 years, but this reflects the long-term storage in carbonaceous sediments. The turnover time of purely organic carbon in detritus is much shorter, but it still has the longest turnover times of any of the parts (carbon pools) of ecosystems. Median residence times of soil carbon in forests range from 12 to 19 years in tropical—subtropical systems to 21 for warm-temperate forests (Wang, 2017) to mean values of 25 years for temperate forests (Reichle, 1973), to $\geq 100-200$ for boreal forests and tundra (Fig. 8.5).

## 8.6 Net ecosystem production and net ecosystem exchange

As a result of eddy covariance measurements, a new term enters our lexicon, net ecosystem exchange; NEE, is the net $CO_2$ exchange with the atmosphere, i.e., the vertical and lateral $CO_2$ flux from the ecosystem to the atmosphere (Baldochi, 2003). NEE approaches NEP (but is opposite in sign) when fluxes

**TABLE 8.5** Comparison carbon fluxes of five forest ecosystem using eddy covariance: WB=Walker Branch; TN, MMSF = Morgan Monroe State Forest, IN; HF=Harvard Forest, MA; UMBS=University of Michigan Biological Station, MI; WC=Willow Creek, WI. Units are: fluxes in kg C m$^{-2}$ yr$^{-1}$, standing crop in kg Cm$^{-2}$.

| Carbon budget Component | WB Oak-Maple | MMSF Beech-Maple | HF Oak-Maple | UMBS Maple | WC Aspen |
|---|---|---|---|---|---|
| Standing crop | 11.283 | 11.820 | 12.860 | 7.860 | 8.530 |
| Net primary production(NPP) | 0.727 | 1.049 | 0.565 | 0.639 | 0.511 |
| Heterotrophic respiration ($R_H$) | 0.475 | 0.604 | 0.400 | 0.566 | 0.405 |
| Net ecosystem production (NEP = NPP - $R_H$) | 0.252 | 0.354 | 0.165 | 0.073 | 0.106 |
| Net ecosystem exchange (NEE) | 0.577 | 0.236 | 0.200 | 0.167 | 0.220 |
| ΔC (by biometrics) | 0.264 | 0.320 | 0.175 | 0.212 | 0.186 |
| Soil carbon | 7.89 | 10.80 | 8.80 | – | 21.60 |

**Source:** Curtis et al., 2002. *Courtesy, Elsevier.*

due to extreme atmospheric stability, or extremely soil $CO_2$ respiration, are small (Chapin et al., 2006), i.e., when NEE = NPP-$R_H$, if $R_H$ approaches zero (Chapin et al., 2006).

Comparison of NEP and NEE estimates were made for five intensely studied North American temperate forests (Table 8.5) using three different methodologies (Curtis et al., 2002). The three techniques were metabolic pool and flux analysis for NEP, biometric changes in pool sizes for $\Delta C$, and meteorological/eddy covariance for NEE. The average NEP for the five forests in Table 8.5 is in very close agreement with the average for the three forests in Table 8.4, 190 g C m$^{-2}$ yr$^{-1}$ and 173 g C m$^{-2}$ yr$^{-1}$, respectively. Assuming the means of the three techniques most closely approximated the "true value," in the Oak-Maple and Beech-Maple forests NEP was closest, while NEE was closer in the remaining three forests. NEP (carbon accumulation) was higher in the younger Oak-Pine successional forest on sandy soil (280 g C m$^{-2}$yr$^{-1}$) than the older tulip poplar forest (160 g m$^{-2}$ yr$^{-1}$) on a well-developed mull soil (Table 8.4). In the aquatic ecosystems NEP was higher in the two oligotrophic water bodies, Silver Springs (190 g C m$^{-2}$ yr$^{-1}$) and Lake Eckarfjärden (292 g C m$^{-2}$ yr$^{-1}$, Table 8.1).

None of the measures of net ecosystem production (i.e., net carbon storage in the ecosystem) are perfect approximations. The GPP-$R_A$ expression underestimates NPP, because it does not include losses other than $R_A$ during production (i.e., herbivory, root exudations). The biometric, bottoms-up calculation suffers from compounding of measurement errors. While NEP = GPP-$R_H$ underestimates NEP, NEE must overestimate carbon storage, if NEE only approaches NPP when $R_H$ approaches zero. Eddy correlation-derived NEE is most accurate for short-term estimates, since errors of measurement due to environmental variability such as weak turbulence and precipitation, plus integration of daytime-nighttime measurements and calculation of annual values, all introduce compounding errors. Estimation of NEP and NEE should converge as methodologies are refined.

The conceptual framework used by ecologists and biogeochemists must allow for accurate and clearly defined comparisons of carbon fluxes made with disparate techniques across a spectrum of temporal and spatial scales. Ultimately, eddy covariance must become the methodology of choice for terrestrial ecosystems, especially forested landscapes, because it is the only practical means to address ecosystem carbon flux and net exchange over large spatial scales. NEP is the net carbon accumulation by ecosystems, calculated from the balance between gross primary production (GPP) and ecosystem respiration ($R_E$). But, use of this calculation has often ignored other carbon fluxes from ecosystems (e.g., leaching of dissolved carbon and losses associated with disturbance). To avoid conceptual ambiguities, the historical definition of NEP should be accepted as the net carbon accumulation by ecosystems, but it explicitly incorporates all the carbon fluxes from an ecosystem, including autotrophic respiration, heterotrophic respiration, losses associated with disturbance, dissolved and particulate carbon losses, volatile organic compound emissions, and lateral transfers among ecosystems. Net biome productivity, NBP, which has been proposed to account for carbon loss during episodic disturbance, is equivalent to NEP at regional or global scales (Randerson et al., 2002).

**Ecosystem sequestration efficiency**. I suggest a new parameter, the Ecosystem Sequestration Efficiency (ESE Index):

$$ESE = NEP/GPP \qquad (8.17)$$

which is approximated by:

$$ESE \sim NEP/TSC \text{ (biomass)} \qquad (8.18)$$

In the context of the carbon cycle, the ESE represents the capacity of the ecosystem to store carbon, i.e., withdraw carbon from the active short-term carbon cycle. For the four forests in Table 8.3, and the five forests in Table 8.5 (Eq. 8.18), ESE is solved at a value of 0.02 (range 0.01−0.05) using TSC, and a value of 0.11−0.19 for the three forests in Table 8.4 using GPP and Eq. 8.17), surprising consistent values. If proven reliable, this term, at least for forests, would enable prediction of NEP from easily measured total standing crop biomass.

The ecosystem carbon balance of $GPP:R_E = 0.95$ (n = 72), for lakes (Hoellein et al.) differs only slightly from the value for $GPP:R_E = 1.16$ (n = 6) for the terrestrial ecosystems in Table 8.3 and the salt marsh in Table 8.2. Could the $GPP:R_E$ ratio be a parameter of value for predicting ecosystem carbon flux? Perhaps some of the errors of methodology for calculating NEP and NEE would cancel if the ratio of $GPP:R_E$ were used. GPP:RE should be the same as NPP:RA; therefore, estimates of RE could be obtained from eddy covariance calculations of NPP.

Sometimes it is difficult to see the forest for the trees. One way in which to gain an appreciation for the dynamics of carbon in different ecosystems, and at the same time recognize the relative mobility of carbon in different ecosystems, is to look at the turnover times of carbon. The turn over time for carbon in an entire ecosystem is the same as that in Eq. (8.14) for its component parts:

$$\text{Ecosystem} = \text{total standing crop biomass (TSC)} \qquad (8.19)$$

Turnover time annual exchange with atmosphere (NEE) is:

$$\text{Turnover rate} = 1/\text{Turnover time} \qquad (8.20)$$

The annual atmospheric exchange requires use of NEE for the flux term for an ecosystem. Thus, the turnover times for the five temperate forests in Table 8.5, using NEE, range between 20 and 64 years (average 44 years). Eq. (8.14) works for an ecosystem pool but not for the whole ecosystem, because

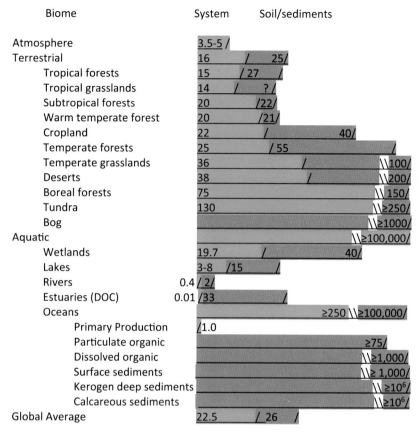

| Biome | System | Soil/sediments |
|---|---|---|
| Atmosphere | 3.5-5 / | |
| Terrestrial | 16 | / 25/ |
| Tropical forests | 15 | / 27 / |
| Tropical grasslands | 14 | / ? / |
| Subtropical forests | 20 | /22/ |
| Warm temperate forest | 20 | /21/ |
| Cropland | 22 | / 40/ |
| Temperate forests | 25 | / 55 / |
| Temperate grasslands | 36 | / \\100/ |
| Deserts | 38 | / \\200/ |
| Boreal forests | 75 | \\ 150/ |
| Tundra | 130 | \\≥250/ |
| Bog | | \\≥1000/ |
| Aquatic | | \\≥100,000/ |
| Wetlands | 19.7 | / 40/ |
| Lakes | 3-8 /15 | / |
| Rivers | 0.4 / 2/ | |
| Estuaries (DOC) | 0.01 /33 | / |
| Oceans | | ≥250 \\≥100,000/ |
| Primary Production | /1.0 | |
| Particulate organic | | ≥75/ |
| Dissolved organic | | \\≥1,000/ |
| Surface sediments | | \\≥ 1,000/ |
| Kerogen deep sediments | | \\ ≥10⁶/ |
| Calcareous sediments | | \\≥10⁶/ |
| Global Average | 22.5 | / 26 / |

**FIGURE 8.6** Approximate turnover times in years representative for carbon in major world ecosystem types: vegetation in green and soils/sediments in brown, approximate average times in years derived from the sources below. *Sources: Broecker, 1974; Carvalhais et al., 2014, 2016; Reichle et al.,1973; Wang et al., 2017; Karlberg, 2007; Trumbore and Harden, 1997; Odum, 1957; Andersson and Kumblad, 2006; Whitfield, 1972; Satchell, 1971; Gore and Olson, 1967; Ryszkowski, 1975; Raich and Schleisinger, 1992; Findlay et al., 1992; Shimel et al., 2007; Hedges, 1992; Eswaren et al., 1993; Siegenthaler and Samiento, 1993; Meyer and Edwards, 1990; Sinsabaugh and Findley, 1995; Sanderman et al., 2017; Chmiel et al., 2015; Cristian et al., 2011.*

of the ecosystem's carbon storage parameter of detritus. The residence times of energy in various ecosystems have a wide range (Fig. 8.6).

Fig. 8.6 should be self-explanatory. The green/blue portion of each bar is the turnover time for the ecosystem; the darker extension is the turnover time for the soil carbon. As example, consider the temperate forest (but now include previously undocumented 733 g c m$^{-2}$ yr$^{-1}$ annual small root mortality and 340 g C m$^{-2}$ small root biomass already included in the soil carbon pool, Edwards and Harris, 1977). The forest's soil carbon pool of 12,850 g C m$^{-2}$ ÷ 173 g C m$^{-2}$ yr$^{-1}$ litterfall plus small root mortality of

733 g C m$^{-2}$ yr$^{-1}$ = 14.2 years turnover time. In all cases, ecosystem turnover times are shorter than their longer-subcomponent of soil carbon (soil carbon acts as a "brake" on the rate of the systems' overall carbon turnover). The importance of root biomass is illustrated by the grasses in a Missouri tallgrass prairie that incorporated 67 to 41 μCi of $^{14}$Cm$^{-2}$ from a single exposure to 151 μCi of $^{14}$CO$_2$m$^{-2}$ for 6 h. The root systems had incorporated over 50% and the roots of mature vegetation as much as 85% of the assimilated $^{14}$C remaining in the plant biomass 8 weeks after tagging, illustrating the accumulation of food reserves in underground storage organs (Dahlman and Kucera, 1968).

Soil turnover times of carbon are averages for the soil profile, but there are some components of recalcitrant C in the soil with turnover times ≥100 years. In terrestrial forests and grasslands, the tropical systems have shorter turnover times than their temperate counterparts, and the turnover times are correlated with mean annual temperatures and rainfall. Grasslands have longer turnover times than their latitudinal forest counterparts, related to the activity of their respective decomposer communities. Of course, there are many other singular ecosystems, e.g., montane forests, evergreen rain forests, croplands, estuaries, coral reefs, etc., which may not fit precisely into this template.

It is worth noting that freshly fallen surface detritus in terrestrial ecosystems decomposes much more rapidly than recalcitrant soil organic matter, with turnover time values of ∼1 year in deserts, 2−5 years in grasslands, ∼3 years in woodlands, and ∼0.15 years in the tropics.

Understanding, and being able to quantify, these terms will be important to predicting how organisms will respond to environmental changes, especially temperature changes. By knowing how ecosystems alter biochemical fluxes in response to environmental change, or conversely how altered environmental conditions change their biochemical fluxes, may eventually enable predictive models of energy flux within major groups of organisms and, ultimately, between them and the ecosystems of which they are a part.

## 8.7 Emergent properties of ecosystems

Emergent properties of systems are the phenomena that emerge from the collaborative function of the system but do not belong to any one part of that system (Mayr, 1982). The bubble model applied to ecosystems (Ponge, 2005) presents the vegetative canopy of an ecosystem as a "skin" for energy exchange; the ecotone between ecosystems as a transition barrier for invasive species, and trunks as a skeleton for water exchange (Gilbert and Henry, 2015). The energetic properties and CO$_2$ flux patterns of ecosystems are summarized in Chapter 7, which result from the interactions of constituent parts, further illustrate and quantify these emergent properties. Jorgensen (2008) has presented a number of emergent properties of ecosystems, many of which involve the thermodynamic and energetic properties of ecosystems which we have discussed:

**(1)** All ecosystems are open systems, both receiving and discharging energy, $CO_2$ and matter; thermodynamically, this is a prerequisite for ecological processes. If ecosystems could be isolated, they would be at thermodynamic equilibrium without life and without gradients. The laws of thermodynamics are rooted in Prigogine's use of "thermodynamics far from thermodynamic equilibrium." The openness, explains Prigogine and Stengers (1997), is why the system can be maintained far from thermodynamic equilibrium without violating the Second Law of Thermodynamics (Jorgensen, 2008).

**(2)** Thermodynamically, carbon-based life has a viability domain between $250°K$ and $350°K$. Within this range there is a good balance between ordering and disordering processes.

**(3)** Mass, including biomass and energy, is conserved.

**(4)** Carbon-based life on Earth has a characteristic basic biochemistry which all organisms share. They have the same elementary chemical composition. This principle allows one to identify stoichiometric relationships to ecology.

**(5)** No ecological part can exist in isolation. Interconnectivity has synergistic effects on components. The sum is greater than the parts.

**(6)** All ecosystem processes are irreversible (another way of expressing the Second Law of Thermodynamics). Living organisms need energy to maintain, grow, and develop. This energy is lost as heat to the environment and cannot be recovered again as usable energy within the system.

**(7)** Biological processes use captured energy to move further from thermodynamic equilibrium and maintain a state of low entropy and high energy relative to its surroundings and to thermodynamic equilibrium; another way of stating that ecosystems can grow.

**(8)** After initial capture of energy across the system boundary, ecosystem growth and development becomes possible by (a) increase in size (biomass), (b) increase in the network (cycling), or (c) an increase in information imbedded in the system. All three growth and development forms imply that the system is moving away from thermodynamic equilibrium, and all three are associated with an increase of (a) the eco-energy stored in the ecosystem, (b) the energy flow in the system (power), and (c) the ascendancy. When cycling increases, the energy storage capacity, the energy-use efficiency, and space-time differentiation all increase. When the information increases, the feedback control becomes more effective, the system gets bigger, which implies that the specific respiration decreases, and there is a tendency to replace $v$ strategists with $k$ strategists. Note that the first growth form corresponds to the eco-energy of organic matter of $18.7$ kJ g$^{-1}$. The three forms of growth and development are in accordance with H. T. Odum's trends of ecosystem development. A typical growth and development sequence with increased biomass (form 1) has a positive feedback allowing even more additional solar energy capture until a limit of around 75% of the available solar energy is reached. Thereafter, the ecosystem continues to grow and develop by increasing network interaction (form 2) and improving energy efficiencies (form 3).

**(9)** An ecosystem receiving solar radiation will attempt to maximize eco-energy storage or maximize power such that if more than one possibility is offered, then in the long run the one which moves the system furthest from thermodynamic equilibrium will be selected. The eco-storage and energy flow increase during all three growth and development forms. When an ecosystem evolves it can apply all three forms in a continuous Darwinian selection process. The nested space-time differentiation in organisms optimizes thermodynamic efficiency as expressed here, because it allows the organism to simultaneously exploit equilibrium and nonequilibrium energy transfers with minimal dissipation.

## 8.8 Recommended Reading

Broecker, W.S., 1973. Factors controlling $CO_2$ content in the oceans and atmosphere. In: Woodwell, G.M., Pecan, E.V. (Eds.), Carbon and the Biosphere, CONF-720510. National Technical Information Office, Springfield, VA, pp. 32−50, 392 pp. https://www.biodiversitylibrary.org/bibliography/4036#/summary/.

Chapin, F.S., Woodwell, G.M., Randerson, J.T., Rastetter, E.B., Lovett, G.M., Baldocchi, D.D., Clark, D.A., Harmon, M.E., Schimel, D.S., Valentini, R., Wirth, C., Aber, J.D., Cole, J.J., Goulden, M.L., Harden, J.W., Heimann, M., Howarth, R.W., Matson, P.A., McGuire, A.D., Melillo, J.M., Mooney, H.A., Neff, J.C., Houghton, R.A., Pace, M.L., Ryan, M.G., Running, S.W., Sala, O.E., Schlesinger, W.H., Schulze, E.D., 2006. Reconciling carbon-cycle concepts, terminology, and methods. Ecosystems 9, 1041−1050. https://doi.org/10.1007/s10021-005-0105-7.

Jorgensen, S.E., 2008. Ecosystem theory: fundamental laws in ecology. In: Encyclopedia of Ecology. Elsevier, pp. 1697−1701 pp. https://www.elsevier.com/books/ecosystem-ecology/jorgensen/978-0-444-53466-8.

Odum, H.T., 1956. Primary production in flowing waters. Limnology & Oceanography 1 (2), 102−117. https://doi.org/10.4319/lo.1956.1.2.0102.

Reichle, D.E. (Ed.), 1981. Dynamic Properties of Forest Ecosystems. Springer Verlag, Berlin-Heidelberg-New York, p. 683 pp. https://www.worldcat.org/title/dynamic-properties-of-forest-ecosystems/oclc/1025081165.

Reichle, D.E., Dinger, B.E., Edwards, N.T., Harris, W.F., Sollins, P., 1973. Carbon flow and storage in a forest ecosystem. In: Woodwell, G.M., Pecan, E.V. (Eds.), Carbon and the Biosphere. National Technical Information Service, Springfield, Virginia, pp. 345−365, 392 pp. https://www.biodiversitylibrary.org/bibliography/4036#/summary.

Wetzel, R.G., Rich, P.H., 1973. Carbon in freshwater systems. In: Woodwell, G.M., Pecan, E.V. (Eds.), Carbon and the Biosphere, CONF-720510. National Technical Information Office, Springfield, VA, pp. 241−263, 392pp. https://www.biodiversitylibrary.org/bibliography/4036#/summary/.

# Chapter 9

# Ecosystem productivity

Just as carbon is fundamental to the chemistry of life, so is the carbon metabolism of ecosystem types fundamental to the biogeochemistry of the biosphere. As Morowitz (1968) so aptly stated, "The surface of the earth does not gain net energy from the sun, but remains at a constant total energy, reradiating as much energy as is taken up. The subtle difference is that it is not

*The Global Carbon Cycle and Climate Change.* https://doi.org/10.1016/B978-0-12-820244-9.00009-3
**157**

energy per se that makes life go, but the flow of energy through the system." Energy enters the trophic system as photons and is transformed into energetic covalent bonds. Thus is summarized energy flow in the biosphere and the production of biomass, idealized as:

$$CO_2 + N_2 + H_2O + H_2SO_4 + H_2PO_4. \rightarrow \text{Biomass} + O_2 \quad (9.1)$$

which describes the life process of primary production with the by-product of oxygen contributing to an aerobic environment.

Translating the carbon dynamics of different ecosystems, to the carbon cycle of the biosphere requires information about the standing crop and fluxes on a biome-wide, landscape-scale, basis. To accomplish this requires knowledge of the global distribution of carbon in biomass. This requires knowledge of the mean biomass, mean productivities, and geographic areas occupied by each ecosystem type. Biomes are defined as the largest easily recognizable geographical subsection of the biosphere distinguished by climate conditions (Fig. 9.1).

The quantity of solar energy entering the Earth's atmosphere is approximately $15.3 \times 10^8$ cal m$^{-2}$ yr$^{-1}$. Much of this is scattered by dust particles, or is used in the evaporation of water. The average amount of radiant energy per unit area per unit of time that is actually available to plants varies with geographic location, but in Great Britain the figure is of the order of $2.5 \times 10^8$ cal m$^{-2}$ yr$^{-1}$; in Michigan, USA, it is $4.7 \times 10^8$ cal m$^{-2}$ yr$^{-1}$; and in Georgia, USA, $6.0 \times 10^8$ cal m$^{-2}$ yr$^{-1}$. As much as 95%−99% of this energy is immediately lost from plants in the form of sensible heat and heat of evaporation. The remaining 1%−5% is used in photosynthesis and is transferred into the chemical energy of plant tissues. The total primary production of the Earth was calculated by Rodin et al. (1975) as being $2.33 \times 10^{14}$ kg yr$^{-1}$ of dry organic material ($1.05 \times 10^{17}$ g C yr$^{-1}$); at ~ 9.9 cal g$^{-1}$ C (Table 9.1) that is equivalent to $1.04 \times 10^{18}$ cal, or $4.35 \times 10^{18}$ J represented by the Earth's biomass.

**Units of measurement**. Throughout this chapter values reported in different mass and energy units in various scientific studies will be converted to standardized values using the conversions in Table 9.1. Mass measures will continue following international (System International, SI, protocol). New mass notations merely reflect new units to encompass larger values at global scales. Energy values will begin shifting from calories, used in previous chapters traditionally preferred in ecological energetic studies, to Joules as the now accepted SI notation for use in global energy applications.

Movement of carbon in the global cycle is a function of the pool size (biomass) of carbon in a biome, the metabolic activity of ecosystems in that biome, and the area which that biome occupies (Table 9.2). Big pools with low rates may have the same net effect as small pools with high rates; large pools with high rates, such as forests and marine biomes, play the biggest roles in the global cycle. The energy values in columns five and six in Table 9.3 give mean energy equivalents for organic carbon in plants of different ecosystem types and energy fixed in net primary production (Whittaker and Likens, 1973). Column six gives the energy equivalents in Joules.

**TABLE 9.1** Conversion factors of units of measure for mass and energy values.

| | |
|---|---|
| 1 g carbon combustion value | = 9.9 k calories |
| 1 g dry wt. biomass | = 4.46 k calories |
| 1 g dry-weight plant material | = 0.45 g carbon |
| Kilo | = $10^3$ |
| Metric ton | = $10^6$ g |
| Giga | = $10^9$ |
| Peta[a] | = $10^{15}$ |
| 1 calorie | = 4.184 J |
| 1 J | = 0.230 calories |
| 1 day | = 0.0027 years |
| 1 ha | = 10,000 $m^2$ |
| Metric tons km-$^2$ | = grams m$^{-2}$ |

[a]*How many ways can you say $10^{15}$ g? Gigaton, Petagrams, billion metric tons, $10^{12}$ kg, $10^9$ metric tons.*

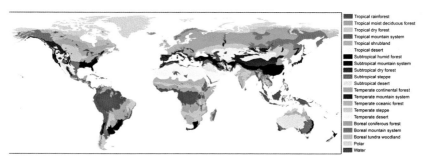

**FIGURE 9.1** The global distribution of biomes, or "ecofloristic zones" mapped by the United Nations Food and Agricultural Organization. Source: Ruesch and Gibbs, 2008. *Courtesy: Carbon Dioxide Information and Analysis Center (CDIAC).*

World net photosynthetic efficiency (energy of NPP/energy of sunlight at the Earth's surface) has been estimated to be 0.13% (Leith, 1975). Efficiency of gross primary production for light in the visible spectrum is about 0.5%. The magnitude of world energy fixation in gross primary production is about $1.4 \times 10^{18}$ kcal yr$^{-1}$ (Cook, 1971).

## 9.1 Terrestrial ecosystems

Climate is the main factor determining the distribution of vegetation (ecosystem) types around the world, primarily temperature and precipitation (Fig. 9.2), as developed from Whittaker, 1975. Latitudinal and seasonal differences in solar radiation are both a direct (photoperiod) and indirect (climate influence) determinant on plant distribution. Mean annual temperatures correlate with the distribution of plant species, and temperature affects both the rates of photosynthesis and respiration. Temperature extremes determine the length of the growing season and annual primary production. Precipitation, both quantity and availability during the growing season, directly affects production, and ecosystems' adaptations to water abundance, or lack thereof, is a determinant of the general character of ecosystem types, i.e., forests versus grassland versus desert. Nutrients, nitrogen, and phosphorous, being keenly important, also influence productivity. Excellent global maps of plant life

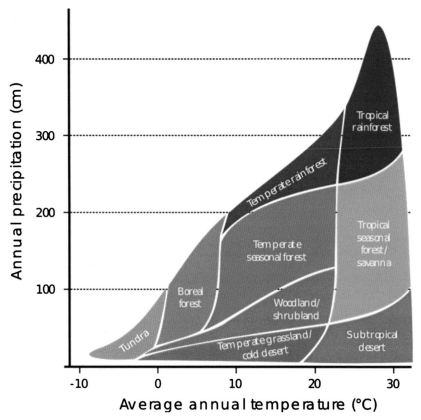

**FIGURE 9.2** Ecofloristic zones (biomes) as determined by mean annual temperature and annual precipitation. *Image credit: Navarras. Courtesy: Wikimedia Commons, 2017.*

zones are now available from satellite-derived imagery (Fig. 9.1), which will remove much discrepancy in earlier estimates. Such maps play a key role in computer models of the carbon cycle by providing the areal parameter over which particular ecosystem metabolic parameters apply.

**Productivity of forests**. The largest amounts of ecosystem biomass are found in forests. Forests are distinguished by their structural woody biomass which is a multiyear accumulation of NPP and part of the system's NEP. Forest types include tropical rainforests characterized by high rainfall, with normal annual rainfall between 175 cm to over 200 cm $yr^{-1}$ and mean monthly temperatures (8°C −30°C range) exceeding 18°C during all months of the year. Tropical rainforests have a range of productivity 720−1800 g C $m^{-2}$ $yr^{-1}$ (Olson, 1975), averaging 900−925 g C $m^{-2}$ $yr^{-1}$ (Tables 9.2 and 9.3), with a total global value of 2.44 × $10^{17}$ g C in its standing crop biomass (Table 9.2).

Tropical dry forests have much lower productivity, ranging between 320 and 760 g C $m^{-2}$ $yr^{-1}$. Estimation of aboveground net primary productivity in secondary tropical dry forests using the Carnegie−Ames−Stanford approach (CASA) model was 3.23−7.59 Mg C $ha^{-1}$ $yr^{-1}$ (Cao et al., 2016).

The areas in which deciduous forests are located receive about 75−150 cm of precipitation spread fairly evenly throughout the year. Temperate forests (both deciduous and coniferous) range in productivity between 540 and 585 g C $m^{-2}$ $yr^{-1}$, with a standing crop value of 9.2−9.5 × $10^{16}$ g C globally (Tables 9.2 and 9.3). Boreal forests average 355−360 g C $m^{-2}$ $yr^{-1}$ with a global carbon mass of 1.1−2.2 × $10^{16}$ g C (Tables 9.2 and 9.3).

**Productivity of grasslands and savannas**. From 25 to 90 cm (mean 75 cm) of precipitation falls annually across grasslands. A steppe is a dry grassland. The average annual temperature in grassland is 1.4°C, with an annual temperature range in the summer of up to 40°C and during the winter down to −40°C. A tropical savanna is a grassland located in semiarid to semihumid climate regions of subtropical and tropical latitudes, with average temperatures remaining at or above 18°C and rainfall between 750 and 1270 mm (50 in) a year. They are widespread in Africa and parts of India, northern S. America, Malaysia, and Australia. The productivity of tropical and subtropical grasslands and savannas ranges from 270 to 2070 g C $m^{-2}$; temperate steppe 270−585 g C $m^{-2}$; temperate meadows 315−1440 g C $m^{-2}$; subarctic meadows, 135−225 g C $m^{-2}$ (Coupland, 1975). The mean global value of grasslands is estimated to be 650 C $m^{-2}$ (Table 9.2). Global values of C contained in the primary production of grassland ecosystems range between 33.3 and 75 × $10^{15}$ g C, depending upon ecosystem classification categories (Tables 9.2 and 9.3).

**Productivity of deserts**. A landscape is considered to be a desert if it receives very little precipitation, less than 10 inches, or 25 cm, of rain a year. Some

**TABLE 9.2** Summary of global area, annual net primary production (NPP), plant carbon content, and soil carbon content in broadly categorized terrestrial ecosystems.

| Ecosystem | Area ($10^{12}$ m$^2$) | NPP (g C m$^{-2}$)[a] | NPP (Pg C yr$^{-1}$)[a] | Plant C (g m$^{-2}$)[a] | Plant C (Pg)[a] | Soil C (g m$^{-2}$)[a] | Soil C (Pg)[a] |
|---|---|---|---|---|---|---|---|
| Forest, tropical | 14.8 | 925 | 13.7 | 16,500 | 244 | 8300 | 123 |
| Forest, temper & plantation | 7.5 | 670 | 5.0 | 12,270 | 92 | 12,000 | 90 |
| Forest, boreal | 9.0 | 355 | 3.2 | 24,450[c] | 22[c] | 15,000 | 135 |
| Woodland, temperate | 2.0 | 700 | 1.4 | 8000 | 16 | 12,000 | 24 |
| Chaparral | 2.5 | 360 | 0.9 | 3200 | 8 | 12,000 | 30 |
| Savanna, tropical | 22.5 | 790 | 17.7 | 2930 | 66 | 11,700 | 264 |
| Grassland, temperate | 12.5 | 350 | 4.4 | 720 | 9 | 23,600 | 295 |
| Tundra, arctic & alpine | 9.5 | 105 | 1.0 | 630 | 6 | 12,750 | 121 |
| Desert, semidesert, scrub | 21.0 | 67 | 1.4 | 330 | 7 | 8000 | 168 |
| Desert, extreme | 9.0 | 11 | 0.1 | 35 | 0 | 2500 | 23 |
| Perpetual ice | 15.5 | – | – | – | – | – | – |
| Lake and stream | 2.0 | 200 | 0.4 | 10 | 0 | – | – |
| Wetland | 2.8 | 1180 | 3.3 | 4300 | 12 | 72,000 | 202 |
| Peatland, northern | 3.4 | – | – | – | – | 133,800 | 455 |
| Cultivated & perm crop | 14.8 | 425 | 6.3 | 200 | 3 | 7900 | 117 |

| | | | | | | |
|---|---|---|---|---|---|---|
| Human area | 2.0 | 100 | 0.2 | 500 | 1 | 5000 | 10 |
| TOTAL | 150.8 | 391 | 59.0 | 3220 | 486 | 13,640 | 2057 |

[a] Assuming that phytomass is 45% C.

[b] Values for all biomes except northern peatlands are for the top 1 m of soil only, whereas Nepstad et al. (1994) report that stores of C below 1 m depth exceed those in the top 1 m in an Amazonian forest. Values for all biomes except wetlands and northern peatlands exclude surface litter. Surface litter and standing dead plants may contain from 50 to more than 200 Pg C globally, with large amounts in some forest ecosystems (see references in Ajtay et al. (1979); Amthor 1995).

[c] The value for boreal forest is based on Botkin and Simpson (1990) who claim that living phytomass in boreal forests is only 24% the value given in Ajtay et al. (1979).

Sources: Amthor, 1995 using Ajtay et al.; 1979; Post et al., 1982; Botkin and Simpson, 1990; Gorham, 1995; FAO, 1997.

**TABLE 9.3** Primary production and biomass estimates for the biosphere.

| Ecosystem type | Area $10^6$ km$^2$ = $10^{12}$ m$^2$ | Mean net primary productivity g C m$^{-2}$ yr$^{-1}$ | Total NPP $10^9$ metric tons C yr$^{-1}$ = $10^{15}$ g C yr$^{-1}$ | Combustion value kcal g$^{-1}$ C | Net energy fixed $10^{15}$ kcal yr$^{-1}$ | Net energy fixed $10^{15}$ joules yr$^{-1}$ | Mean plant biomass kg C m$^{-2}$ | Total plant mass, $10^9$ metric tons C = $10^{15}$ g C |
|---|---|---|---|---|---|---|---|---|
| Tropical rain forest | 17.0 | 900 | 15.3 | 9.1 | 139 | 582 | 20 | 340 |
| Tropical seasonal forest | 7.5 | 675 | 5.1 | 9.2 | 47 | 197 | 16 | 120 |
| Temperate evergreen forest | 5.0 | 585 | 2.9 | 10.6 | 31 | 0 | 16 | 80 |
| Temperate deciduous fores | 7.0 | 540 | 3.8 | 10.2 | 39 | 163 | 13.5 | 95 |
| Boreal forest | 12.0 | 360 | 4.3 | 10.6 | 46 | 192 | 9.0 | 108 |
| Woodland & shrubland | 8.0 | 270 | 2.2 | 10.4 | 23 | 96 | 2.7 | 22 |
| Savanna | 15.0 | 315 | 4.7 | 8.8 | 42 | 176 | 1.8 | 27 |
| Temperate grassland | 9.0 | 225 | 2.0 | 8.8 | 18 | 75 | 0.7 | 6.3 |
| Tundra and alpine meadow | 8.0 | 65 | 0.5 | 10.0 | 5 | 21 | 0.3 | 2.4 |

| | | | | | | | |
|---|---|---|---|---|---|---|---|
| Desert scrub | 18.0 | 32 | 0.6 | 10.0 | 6 | 25 | 0.3 | 5.4 |
| Rock, ice, and sand | 24.0 | 1.5 | 0.04 | 10.0 | 0.3 | 12.6 | 0.01 | 0.2 |
| Cultivated land | 14.0 | 290.0 | 4.1 | 9.0 | 37 | 155 | 0.5 | 7.0 |
| Swamp and marsh | 2.0 | 1125 | 2.2 | 9.2 | 20 | 84 | 6.8 | 13.6 |
| Lake and stream | 2.5 | 225 | 0.6 | 10.0 | 6 | 25 | 0.01 | 0.02 |
| **Total continental** | **149** | **324** | **48.6** | **9.5** | **459** | **1933.6** | **5.55** | **827** |
| Open ocean | 332.0 | 57 | 18.9 | 10.8 | 204 | 854 | 0.0014 | 0.46 |
| Upwelling zones | 0.4 | 225 | 0.1 | 10.8 | 1 | 4 | 0.01 | 0.004 |
| Continental shelf | 26.6 | 162 | 4.3 | 10.0 | 43 | 180 | 0.005 | 0.13 |
| Algal bed and reef | 0.6 | 900 | 0.5 | 10.0 | 5 | 21 | 0.9 | 0.54 |
| Estuaries | 1.4 | 810 | 1.1 | 9.7 | 11 | 46 | 0.45 | 0.63 |
| **Total marine** | **361** | **69** | **24.9** | **10.6** | **264** | **1105** | **0.0049** | **1.76** |
| **Full total** | **510** | **144** | **73.2** | **9.9** | **723** | **3038.6** | **1.63** | **829** |

Source: Whittaker and Likens, 1973. All values in columns 3 to 8 expressed as carbon on the assumption that carbon = 0.45 × dry matter. Conversions to joules by author.

locations may get less than 1 cm, or less than one-half of an inch of moisture per year. Deserts usually have a large diurnal and seasonal temperature range, with high or low, depending on location, daytime temperatures in summer up to 45°C and low nighttime temperatures in winter down to 0°C (Fig. 9.2). Deserts average between 32 and 67 g C m$^{-2}$ yr$^{-1}$, with a global total biomass of 7 × 10$^{15}$ g C (Table 9.2).

**Productivity of tundra.** Tundra is delimited by permafrost substrate. Precipitation in tundra is about 25 cm yr$^{-1}$. There is a wide range of production in tundra, from as low as 6 g C m$^{-2}$ to as high as 1150 g C m$^{-2}$ (Wielgolaski, 1975). The global NPP ranges between 65 and 120 g C m$^{-2}$, for a global value of 6 × 10$^{15}$ g C in the global biomass (Table 9.3) and 1.4−1.8 × 1018 g C in soils of northern permafrost regions (Tarnocai et al., 2009).

**Productivity of cultivated lands.** Most cultivated lands generally occur within temperate forest and grassland regions. Productivity in agricultural lands is around 290 g C m$^{-2}$ yr$^{-1}$ (Table 9.3) to as high as 425 g C m$^{-2}$ yr$^{-1}$ if permanently cultivated lands are included (Table 9.2). Highly intensive sugar cane cultivation in Java has attained productivity as high as 3450 g C m$^{-2}$ yr$^{-1}$ (Crisp, 1975). The global pool of carbon in cultivated land is $\sim$7 × 10$^{15}$ g C (Table 9.3).

## 9.2 Freshwater ecosystems

This category of ponds and lakes, streams and rivers, and springs, while not major components of the global carbon budget, have a relatively high rate of production and a significant indirect stimulation of oceanic production through their export of nutrients.

**Productivity of lakes and reservoirs.** Solar energy has the main influence on production globally, but over narrow latitudinal ranges nutrients assume greater importance. Chlorophyll $a$ is a good indication of production. GPP can vary widely depending upon nutrient status, ranging from 50 to 300 mg C m$^{-2}$ day$^{-1}$ in oligotrophic lakes to 0.6−8 g C m$^{-2}$ day$^{-1}$ in eutrophic lakes (Likens, 1975), to 75−250 g C m$^{-2}$ yr$^{-1}$ reported by Lund in 1970 (Crisp, 1975) and $\sim$800 g C m$^{-2}$ yr$^{-1}$ reported by Bray et al., in 1959 (Crisp, 1975) in cattail beds in Minnesota. Although lakes and reservoirs occupy only 1.7% of continental land area worldwide, they have a relatively high rate of productivity $\sim$225 g C m$^{-2}$ yr$^{-1}$ with a global NPP of $\sim$0.6 × 10$^{15}$ g C (Table 9.2). Across a spectrum of lake and reservoir types, from tropics to the arctic, primary productivity ranges from 5000 to 13,000 kcal m$^{-2}$ yr$^{-1}$, with outliers as low as 30 kcal m$^{-2}$ yr$^{-1}$ (Brylinski and Mann, 1973).

**Productivity of rivers and streams.** The net primary productivity of rivers and streams is about 225 g C m$^{-2}$ yr$^{-1}$, $\sim$2400 kcal m$^{-2}$ yr$^{-1}$, nearly equal to that of grasslands (Table 9.3). Total area represented globally is small, but that underestimates their contribution to the global flux of carbon and energy, through their discharges of nutrients to coastal marine areas. They play an important role in the productivity of the coastal zone.

**Productivity of swamps and marshes.** The highest mean rates of net primary production rates occur in this ecosystem type, averaging 1125 g C m$^{-2}$ yr$^{-1}$. Due to a global area of only $2.0 \times 10^{12}$ m$^{-2}$, swamps and marshes yield a net annual global net production of $2.2 \times 10^{15}$ g C m$^{-2}$ or about 3% of total net primary production (Table 9.3); the global standing crop of carbon by all swamps and marshes worldwide is of the order of $13.6 \times 10^{15}$ g C (Table 9.3). **Productivity in sewage treatment ponds.** Studies by Goluake and colleagues in California have recorded primary production values of 4500 g dry wt. m$^{-2}$ yr$^{-1}$ in sewage treatment ponds in California (Crisp, 1975).

## 9.3 Marine ecosystems

**Productivity of the open ocean.** Open ocean ecosystems are most analogous to grasslands, being intermediate in the ranges of ecosystem productivities. The rate of production (Table 9.4) is modest at $\sim$130 g C m$^{-2}$ yr$^{-1}$ Knauer, 1993), with a range of 50–150 g C m$^{-2}$ yr$^{-1}$ (Riley, 1972; Crisp, 1975; De Vooys, 1979), but the biomass is small at 1.4 g C m$^{-2}$ (total of $0.46 \times 10^{15}$ g C total in the open ocean). Notable examples of relatively high ocean productivity are: 70–145 g C m$^{-2}$ yr$^{-1}$ in the Sargasso Sea, 190–470 g C m$^{-2}$ yr$^{-1}$ in Long Island Sound, and 365 g C m$^{-2}$ yr$^{-1}$ in the Gulf of Guinea.

**Productivity of the continental shelf.** Estimates of the rates of productivity on the continental shelf average 162 g C m$^{-2}$ yr$^{-1}$ (Whittakwr and Likens, 1973; Table 9.3), less than that in upwelling zones of 225–420 g C m$^{-2}$ yr-$^{1}$,

**TABLE 9.4** Net primary productivity in the ocean (Knauer, 1993).

| Ocean region | % Ocean | Area 10$^{12}$ m$^2$ | Mean productivity g C m$^{-2}$ yr$^{-1}$ | Total global NPP 10$^{15}$ g C |
|---|---|---|---|---|
| Open ocean | 90 | 326 | 130 | 42 |
| Coastal zone | 9.9 | 36 | 250 | 9.0 |
| Upwelling area | 0.1 | 0.36 | 420 | 0.15 |
| Total | | 362 | | 51 |

Courtesy: Springer Nature.

and production (g C yr$^{-1}$), and can be up to 10 times, or more, greater than open ocean values. Extremely high values in well-lit, warm shallow waters can reach 1750 to 3836 g C m$^{-2}$ yr$^{-1}$ (Crisp, 1975). Both upwelling zones and estuaries benefit from nutrient enrichment from ocean circulation and river discharges, respectively.

**Productivity of coral reefs.** By far the highest rates, and greatest variation, of primary productivity occur in coral reefs. Coral reefs function almost independently of the open ocean with far higher rates of productivity. NPP ranges widely, from 190 g C m$^{-2}$ yr$^{-1}$ in the Marshall Islands to 6307 g C m$^{-2}$ yr$^{-1}$ in the Hawaiian Islands, compared to 20−40 g C m$^{-2}$ yr$^{-1}$ in the open ocean (de Vooys, 1997). Coral reefs occupy $1.2 \times 10^5$ km$^2$ of the $360 \times 10^6$ km$^2$ of open ocean area, with a net primary production of $0.30 \times 10^{15}$ g C yr$^{-1}$.

**Productivity of estuaries.** Estuaries are the nursery beds for the secondary production of the continental shelf. They are characterized by warm, shallow, nutrient-enriched waters favorable for primary production, with NPP ranging from 160 to 450 g C m$^{-2}$ yr$^{-1}$ for estuarine phytoplankton (de Vooys, 1997), with rates of NPP for estuarine ecosystems (Whittaker and Linkens, 1973) averaging between 675 g C m$^{-2}$ yr$^{-1}$ (Woodwell et al., 1973) and 810 g C m$^{-2}$ yr$^{-1}$ (Table 9.3). Along with coral reefs averaging 900 g C m$^{-2}$ yr$^{-1}$ (Whittaker and Likens, 1973), they have among the higher rates of primary production of the biomes, comparable to marshes, swamps, and tropical rain forests.

**Productivity of upwelling zones.** The highest rate of productivity in the oceans occurs in upwelling zones where nutrient-rich waters are brought to the sunlit surface layers. Various estimates range from 225 to 420 to g C m$^{-2}$ yr$^{-1}$ (Tables 9.3 and 9.4).

**Productivity of salt marshes.** Aboveground NPP in salt marshes in eastern United States is high, ranging from 133 to 1153 g C m$^{-2}$ yr$^{-1}$ (Hatcher and Mann, 1975), 338−1800 g C m$^{-2}$ yr$^{-1}$ (Odum, 1974), with other records as high as 2050 g C m$^{-2}$ yr$^{-1}$ (Smalley, 1960 and Teal, 1962). Net dry matter production by *Spartina alterniflora* in Georgia, USA, salt marshes is as high as 3300 g dry wt. m$^{-2}$ yr$^{-1}$ (Woodwell et al., 1973). Fluxes to the ocean from salt marshes average $\geq 100$ g C m$^{-2}$ yr$^{-1}$ (Nixon, 1980).

## 9.4 Secondary production

Secondary production is that portion of the energy flow in an ecosystem which includes the ingestion and assimilation of energy, and the expenditure of that energy in metabolism, by consumer organisms. Data from 20 terrestrial animal populations showed that ∼29% of the energy assimilated by invertebrates is converted to net secondary production, while the equivalent value for birds and mammals is only 2%. The difference depends upon the capacity for homeothermy (Golley, 1968). Globally, herbivores consume $3 \times 10^{15}$ g C $\text{yr}^{-1}$ (Whittaker and Likens, 1973). Ecological theory has assumed that the ecosystem is in a steady-state condition where primary production is balanced by heterotrophic utilization (Odum, 1963). Theory also assumes that ecological succession trends toward a balanced system from a situation where organic matter is accumulating more rapidly than it is being produced. Therefore, analysis of quantitative relationships between primary and secondary productivity should be useful in interpreting what is going on in the entire ecosystem (Evans, 1967).

Table 9.5 shows the rather consistent pattern of secondary productivity across many different ecosystems. Lindeman's trophic dynamic theory projects a 10% efficiency in energy transfer across trophic levels. Values for lakes fall between 9% and 20%. The remaining productivity estimates are for only a part of the consumer trophic level, and are in the range of 1% or less. The forest values are based upon invertebrate biomass, and not productivity, and therefore could be significant underestimates of NSP. Secondary production in managed agricultural systems can much higher, approaching Lindemann's theoretical values.

## 9.5 Global biome-scale production

Rodin et al. (1975) estimated the total "phytomass" of the Earth to be $2.4 \times 10^{18}$ 2.2 g dry wt., or its equivalent of $1.20 \times 10^{18}$ g C. Whittaker and Likens (1973) estimated it to be $0.89 \times 10^{18}$ g dry wt., or $8.29 \times 10^{17}$ g C. This gives a range for global biomass of $0.89 \times 10^{18} - 1.2 \times 10^{18}$ g C (Table 9.6).

Annual global production estimates range from $1.31 - 2.33 \times 10^{18}$ for biomass; $0.59 - 1.05 \times 10^{17}$ for carbon; and $7.2 - 10.3 \times 10^{17}$ for calories. Whittaker and Likens (1973) estimated global net primary production to be $0.73 \times 10^{17}$ g C $\text{m}^{-2}$ $\text{yr}^{-1}$; Lieth (1975) estimated $0.80 \times 10^{17}$ g C $\text{m}^{-2}$ $\text{yr}^{-1}$; Rodin et al. (1975) estimated $1.05 \times 10^{17}$ $\text{yr}^{-1}$; and Atjay et al. (1979) estimated global NPP to be $2.25 \times 10^{17}$ g C $\text{yr}^{-1}$. This gives a range of global NPP of $0.73 - 2.25 \times 10^{17}$ g C $\text{yr}^{-1}$, with an average of these four estimates at $1.21 \times 10^{17}$ g C $\text{m}^{-2}$ $\text{yr}^{-1}$ (Tables 9.3 and 9.4). Lieth (1975) used a temperature-light-moisture algorithm, to arrive at a value for terrestrial global NPP of $0.63 \times 10^{17}$ g C $\text{yr}^{-1}$, illustrating the close dependence of NPP upon these climatic variables.

Not only do the biomes differ in their biomass and carbon content, but they also differ in their rates of primary productivity, i.e., the production of mass or conversion of energy per unit area per unit time. Ranking of the net primary productivity of the biomes illustrates these differences (Table 9.6). The contributions of each of the biomes in terms of biomass (pools) and fluxes of

**TABLE 9.5** Secondary production (NSP) by consumers in different ecosystems. Values are for specific consumer groups, except where indicated by "A" = productivity for the entire animal trophic level.

| Ecosystem type | Heterotroph k cal m$^{-2}$ yr$^{-1}$ | Net primary prod. k Cal m$^{-2}$ yr$^{-1}$ | Net secondary prod. | NSP/NPP |
|---|---|---|---|---|
| Tropical rain forest[a] | A | 87 kcal m$^{-2}$ day$^{-1}$ | 1.5 kcal m$^{-2}$ day$^{-1}$ | 0.02 |
| **Mesic forest** | | | | |
| Leaves | Invert herbivore[b] | 200 g C m$^{-2}$ yr$^{-1}$ | 0.115 g C m$^{-2}$ | 0.001 |
| Leaves | Vert herbivore[c] | 200 g C m$^{-2}$ yr$^{-1}$ | <0.5 g C m$^{-2}$ | 0.01 |
| Litter | Invert saprovore[d] | 237 g C m$^{-2}$ yr$^{-1}$ | 0.280 g C m$^{-2}$ | 0.001 |
| Oak forest[e] | A | 2194 | 170 | 0.008 |
| Grassland[f] | Invert herb | 1075–1360 | 10–13 | 0.01 |
| | Vert herb | 747 | 3.1 | 0.004 |
| Old field[g] | Vert herbivore | 17.50 | 0.5 | 0.03 |
| Meadow[h] | A | 2075 | 287 | 0.014 |
| Rye field[i] | A | 4170 | 450 | 0.011 |
| Potato field[i] | A | 2660 | 487 | 0.018 |
| Saltmarsh[j] | Invert | 6585 | 81.0 | 0.01 |
| Silver Springs[k] | A | 8428 | 1383 | 0.16 |
| Oligotrophic lakes[l] | A | 250–1000 mg C m$^{-2}$ day | 50–100 mg C m$^{-2}$ day | 0.10–20 |

| Lake Mendota[m] | A | 4800 | 416 | 0.09 |
| Ocean[n] | | 36,380 | 596 | 0.02 |

e,h,i use TSC for NPP and NPP consumed for NSP, thus values for NSP/NPP may be < by up to $10^{-1}$ due to trophic level transfer inefficiency. "A" is entire NSP trophic level (Kozlovsky, 1976).

[a]Odum, 1970, after Lugo, 2004.
[b]Reichle and Crossley, 1967.
[c]Reichle et al., 1973.
[d]McBrayer and Reichle, 1971
[e]Medwecka-Kornas, 1973.
[f]Wiegert, 1964, 1965, Evans, 1951, 1964.
[g]Golley (1960)
[h]Breymeyer, 1971.
[i]Trojan, 1967.
[j]Teal, 1962, Smalley, 1959, 1960.
[k]Odum, Fig. 7.8.
[l]Andersson, Table 8.1.
[m]Lindeman (1942)
[n]Fig. 8.5. Grams dry wt., b-d, converted to g C using C content of 0.50.

**TABLE 9.6** Various estimates of total global production in carbon and energy units.

| Parameter mass (dry wt.) | Carbon (g) | Energy (cal) | Source |
|---|---|---|---|
| **Biomass** | | | |
| $2.40 \times 10^{18}$ g | $1.08 \times 10^{18}$ g C | $1.69 \times 10^{22}$ cal | Rodin et al., 1975 |
| $0.89 \times 10^{18}$ g | $8.29 \times 10^{17}$ g C | $0.82 \times 10^{22}$ cal | Whittaker and Likens, 1975 |
| | $5.60 \times 10^{17}$ g C | | Olson et al., 1983 |
| | $5.59 \times 10^{17}$ g C | | Houghton and Skole, 1990 |
| $1.08 \times 10^{18}$ g | $4.86 \times 10^{17}$ g C | | Amthor et al., 1998 |
| **NPP** | | | |
| $2.33 \times 10^{17}$ g yr$^{-1}$ | $1.05 \times 10^{17}$ g C yr$^{-1}$ | $1.03 \times 10^{18}$ cal yr$^{-1}$ | Rodin et al., 1975 |
| $1.62 \times 10^{17}$ g yr$^{-1}$ | $0.73 \times 10^{17}$ g C yr$^{-1}$ | $7.23 \times 10^{17}$ cal yr$^{-1}$ | Whittaker and Likens, 1975 |
| | $0.65 \times 10^{17}$ g C yr$^{-1}$ | | Olson et al., 1983 |
| | $0.60 \times 10^{17}$ g C yr$^{-1}$ | | Houghton and Skole, 1990 |
| $1.31 \times 10^{17}$ g yr$^{-1}$ | $0.59 \times 10^{17}$ g C yr$^{-1}$ | | Amthor et al., 1998 |

carbon and calories (or Joules) depend upon their pool sizes, fluxes (rates of production and respiration), and also their percentage composition of the biosphere (i.e., their proportion of the Earth's total organic carbon or energy content). Table 9.4 is a summary of NPP by ecosystem type, the surface area of the Earth occupied by each, and their annual NPP. Table 9.7 ranks these ecosystem types by the rates of primary production for the major biomes. These simple tables belie the detail and sophistication with which modern science can bring to bear on the subject. Detailed tabulations of the pools and fluxes of ecosystem types (or biomes), which have been traditionally built from "the bottom up" through years of on-the-ground research and measurement, can now be remotely sensed in exquisite detail (Fig. 9.3).

**TABLE 9.7** Ranking of the net primary productivity of the biomes based upon the values reported by the references cited in Chapter 8 and Tables 9.2 and 9.3

| Biome | NPP (g carbon $m^{-2}$ $yr^{-2}$) |
|---|---|
| Reefs and algal beds | 190-6300 |
| Swamps and marshes | 190-1180 |
| Salt marshes | 133-2050 |
| Tropical rain forests | 880—925 |
| Estuaries | 675—810 |
| Temperate evergreen forests | 585—680 |
| Temperate deciduous forests | 540—700 |
| Tropical dry forests | 320—670 |
| Boreal forests | 355—360 |
| Tropical grasslands | 315—790 |
| Cultivated land | 290—425 |
| Woodland/shrubland, chaparral | 270—360 |
| Prairies | 225—430 |
| Aquatic freshwater, lakes, rivers | 200—430 |
| Marine upwelling zones | 225—420 |
| Continental shelf | 160—250 |
| Open ocean | 50—150 |
| Tundra | 65—120 |
| Desert | 32—67 |
| Extreme desert | 1.5—11 |

Examination of the production of ecosystem types (Table 9.3) and the map of global biomass (Fig. 9.3), the standing crop of organic biomass, begins after the frozen arctic deserts in the tundra with $\sim 0.3$ kg C $m^{-2}$. A significant increase occurs in the great circumpolar boreal forests of taiga with 9 kg C $m^{-2}$. The temperate mixed and deciduous forest follows with 3.5—16 kg C $m^{-2}$. Further to the south, in the temperate dry steppes, biomass decreases to 0.7 kg C $m^{-2}$. It falls further still in the desert shrub woodland, 0.3 kg C $m^{-2}$. Further south, above and below the equatorial belt, biomass in the subtropical and tropical woodlands increases again to 16 kg C $m^{-2}$. Peak biomass is

## Global Above- and Below-ground Living Biomass Carbon Density

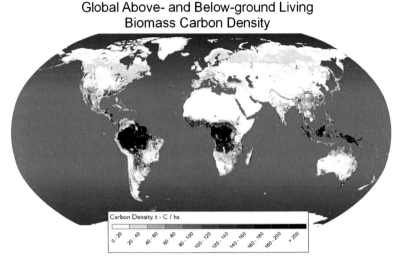

**FIGURE 9.3** IPCC Tier-1 Global Biomass Carbon Map (above and below-ground) for the Year 2000 in metric tons carbon per hectare ($100 \, g \, m^{-2}$). *Source: Ruesch and Gibbs, 2008; Courtesy, Carbon Dioxide Information and Analysis Center (CDIAC).*

reached in the great tropical rain forests of East Asia, Africa, and South America, where biomass reaches $20 \, kg \, C \, m^{-2}$. NPP follows a similar geographical pattern.

The calculations of pools and fluxes of carbon in Table 9.3 required data on global areas of the different biomes (Fig. 9.3, Table 9.2), biomass of major ecosystem types (Table 9.6), and rates of NPP (Table 9.5). It should be apparent that biomes with high vales in all three parameters, such as tropical and temperate forests, or systems with moderate NPP rates but huge areas, such as the oceans, can have a dominate influence on the biosphere's carbon cycle. It is also why there is so much concern about the clearing of tropical forests; besides they being the home of great diversity of species.

## 9.6 Factors affecting global productivity

**Climate**. The effects of climate on primary productivity can be most clearly seen in terrestrial environments. Primary production is greatest in the tropics and declines with increasing latitude to lowest values in boreal forests and shrub tundra. There is a clear tendency toward an increase in the amount of green, photosynthetic biomass as one proceeds from northern latitudes to southern ones in the deciduous forests, which is associated with the duration of the growing season. The largest quantity of photosynthetic tissue recorded ($1.26-5.0 \times 10^3$ g dry wt.) has been for tropical forests (Rodin et al., 1975).

The quantity of photosynthetic tissue in terrestrial ecosystems is correlated with the amount of light, heat, and moisture. The effects of moisture on ecosystem distribution become evident when one compares the production of photosynthetic tissue in such different types of vegetation as savannas and coniferous forests (Table 9.6). NPP decreases along a precipitation gradient from forests to grasslands to lowest values in deserts. NPP declines with elevation, reflecting the influence of declining temperature and shorter growing season (Table 9.8).

**Nutrients**. Notice in Table 9.6 that three ecosystem types approach, with one even exceeding, the tropical rain forest's productivity of $\sim 900$ g C m$^{-2}$ yr$^{-1}$. These ecosystem types are swamps and marshes, NPP$\sim 1125$ g C m$^{-2}$ yr$^{-1}$, ocean reefs, $\sim 900$ g C m$^{-2}$ yr$^{-1}$, and estuaries, $\sim 810$ g C m$^{-2}$ yr$^{-1}$. Swamps, marshes, and estuaries have a common feature: the flush of nutrients from upland watersheds stimulate their biological productivity. The continental shelf at $\sim 162$ g C m$^{-2}$ yr$^{-1}$ also benefits from nutrient input from the land and has a productivity nearly three times that of the open ocean. The important nutrients for ocean productivity encompass the entire sweet of elements necessary for metabolism, but paramount among them are nitrogen and phosphorus, which are often the elements limiting productivity.

In the open oceans near the continents, and beyond the coastal plain where Fe in sea water is sufficient to support NPP, Fe is a limiting factor on NPP. This is particularly evident in the central Pacific Ocean, where 98% of new NPP is supported by Fe in wind-blown dust from the deserts of central China. Thus, some scientists have suggested that ocean fertilization might be an effective means of carbon sequestration, increasing oceanic NPP, and thus increasing uptake of excess atmospheric $CO_2$ (Martin and Gordon, 1988; Martin et al., 1988).

**River discharge**. The U.S. National Center for Atmospheric Research estimated global water discharge from the continents to the oceans, based upon data from 991 rivers, to be $37.29 \pm 0.66 \times 10^6$ m$^3$ yr$^{-1}$ (Dai and Trenberth, 2002). The predictive modeling of river nutrient export from watersheds by the Global Nutrient Export from Watersheds (NEWS) model estimated export from 5761 watersheds distributed globally as a function of land use, nutrient inputs, hydrology, and other factors (Seitzinger, 2005). The Global NEWS system models predicted global export by rivers in the mid-1990s to be $367 \times 10^{12}$ g yr$^{-1}$ organic carbon, $66 \times 10^{12}$ g yr$^{-1}$ nitrogen, and $11 \times 10^{12}$ g yr$^{-1}$ phosphorus. With a C:N:P ratio (Redfield Ratio) for marine algae of 106:16:1 (Ekholm, 2008), and calculating from open ocean C productivity values in Table 9.3, the annual river discharges of N would be sufficient to account for 10% of the annual algal production on the continental shelf or 2% of that in the open ocean. This is clearly not an insignificant nutrient import to the ocean's primary production coming from the continents.

**Ocean circulation**. Compare the productivity of the open ocean at $\sim 57$ g C m$^{-2}$ yr$^{-1}$ with the quadrupled production rate of upwelling zones,

**TABLE 9.8** Biomass of ecosystems of the main biomes each with distinct vegetative structure. Metric ton ha$^{-1}$ ($= 10^2$ g m$^{-2}$).

| Characteristics | Tundra | Taiga | Temperate | Steppe | Deserts | Tropical | Savannas |
|---|---|---|---|---|---|---|---|
| Biomass t ha$^{-1}$ | 5–28 | 100–300 | 370–400 | 10–25 | 4.3–12.5 | 410–500 | 27–67 |
| Foliage % | 11%–15% | 6%–8% | 1% | 15%–18% | 3%–14% | 3%–8% | 11%–12% |
| Wood % | 6%–15% | 70%–72% | 75%–76% | 0%–10%% | 10%–40% | 74%–77% | 47%–82% |
| Roots % | 70%–83% | 22%–23% | 24%–26% | 82%–85% | 57%–87% | 20%–28% | 6%–42% |
| NPP (t ha$^{-1}$ yr$^{-1}$) | 1–2.5 | 4.5–8.5 | 9–13 | 4.2–11.2 | 1.2–9.5 | 24.5–32.5 | 7.3–12 |
| Increment (t ha$^{-1}$ yr$^{-1}$) | ≤0.1 | 1–3 | 2.5–4 | —— | ≤0.1 | 3.5–7.5 | ≤0.5 |

Source: Rodin et al., (1975).

$\sim 225$ g C m$^{-2}$ yr$^{-1}$, and with the demonstrated importance of nutrient loading, along with the cooler, more oxygenated water, it becomes clear why upwelling zones are so highly productive. Upwelling zones are a consequence of ocean circulation, an understanding of which is important because of not only its importance to marine biological productivity, but also because of its role in affecting the ocean's ability to sequester atmospheric $CO_2$. The ocean water mass is also a global heat sink, and circulation brings cold waters to the surface in upwelling zones. The importance of the ocean's role as a heat and carbon sink will become clearer in Chapter's 10 and 11.

The ocean is a very complex system with currents traveling around the planet. Ocean currents travel at both the surface of the ocean, and also deep within the ocean basin. Currents are influenced by factors such as wind, the rotation of the Earth, differences in the water's salt content, temperature and density, and even the shape of the ocean floor. Because ocean currents distribute nutrients necessary for photosynthesis, and because the currents are constantly circulating massive amounts of water, ocean currents have a significant impact upon marine productivity. Ocean currents are also a major factor affecting global heat exchange and influence global climate.

Surface ocean currents are strongly influenced by winds, which flow across the ocean surface, pushing the water in large circular ocean currents called ocean gyres (Fig. 9.4). Ocean gyres are present in every ocean and move water from the poles to the equator and back again. Water warms at the equator and cools at the poles. Because ocean water temperatures exchange with the atmosphere, the cold and warm waters circulated by the gyres also influence the climate of nearby landmasses. The Coriolis Effect, created by the rotation of the Earth, causes the gyres to rotate clockwise in the Northern Hemisphere and counterclockwise in the Southern Hemisphere.

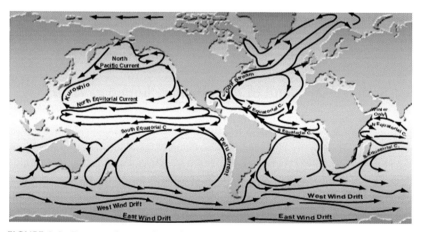

**FIGURE 9.4** Patterns of ocean circulation. *Image credit: Office of Naval Research: Oceanography*, 2019.

Downwelling occurs when wind-blown surface waters converge, pushing the surface water downwards. These are regions of low primary productivity because nutrients are consumed and are not continuously resupplied by the cold, nutrient-rich water from below the surface. Currents occur in areas where cold, denser water sinks. Downwelling currents bring dissolved $CO_2$ down to the deep ocean. Once there, the $CO_2$ moves into slow-moving deep ocean currents staying there for hundreds of years.

Along wind-blown coastlines are areas of upwelling, such as occurs along the Pacific coast off South America. Colder, nutrient-rich waters rise from the ocean floor to replace wind-blown surface waters. The ocean's bottom waters are rich in nutrients because of the decay of primary producers releases nutrients in the ocean depths. As deeper waters rise, they stimulate productivity in the area, and so upwelling areas are usually rich in marine life and biological productivity.

Downwelling and upwelling currents are important components of the deep ocean conveyor belt, and are important in physically transporting carbon compounds to different parts of the oceans. Approximately 25% of the total global marine fish catches come from five upwellings that occupy only 5% of the total ocean area (Jennings et al., 2001)

The largest ocean circulation pattern is thermohaline circulation, which is directly related to temperature and salinity (Fig. 9.4). The thermohaline circulation pattern is driven by changes in water temperature and salt content. The thermohaline circulation is sometimes referred to as the "global ocean conveyor belt," because it moves water throughout the world's oceans. The currents move massive quantities of water and nutrients, which affect biological productivity, and also distribute heat, which influences the global climate (Broecker, 2000).

Reciprocally, the ocean circulation pattern can be affected by the rising temperature from climate change. Computer models of average annual circulation patterns for the Atlantic and Pacific basins, including currents, heat burdens, and vertical mixing, are being developed (Semtner, 2005). As these models become validated by reconstructing past climates, they will offer promise for use in predicting future climate regimes.

## 9.7 Scaling from stand to the planetary boundary layer

One of the biggest challenges in refining the global carbon budget is that of scaling from short-term measurements on relatively homogeneous flat terrain to large topographically heterogeneous regions. Recent development of remote sensing technologies (Tucker et al., 1986) is helping to resolve the scaling issue by providing integrated production values for large landscape units, which can be ground-truth calibrated against empirical measurements on the ground.

**Net biome production**. Net biome production (NBP) is the net ecosystem carbon balance (NECB) (estimated at large temporal and spatial scales. This concept was developed to account for many of the landscape-scale $CO_2$ fluxes, seldom measured as NEE (net ecosystem exchange) at eddy covariance measurement sites, and would include landscape wide disturbances, such as fire, which remove C from the system through nonrespiratory processes. Net biome-scale production can thus be achieved as the spatial and temporal average of NECB over a heterogeneous landscape (Chapin et al., 2006):

$$NBP = \int_T \int_A NECB(x, t)dxdt/(T)(A) \tag{9.2}$$

But harvest measurements and eddy-covariance techniques for estimating NPP can be labor-intensive and necessarily applied only to small areas. Since the productivity of vegetation may vary greatly over the landscape, regional estimates of productivity by harvest methods can be prohibitively expensive. A NASA satellite, Moderate Resolution Imaging Spectroradiometer (MODIS), is now providing integrated estimates of GPP over large areas for studies of global change. MODIS replaced older satellites, such as LANDSAT and NOAA-AVHRR, which provided early estimates of global primary production (Schleisinger and Bernhardt, 2013).

Satellite measurement of GPP uses the differential absorption of light by chlorophyll and other leaf pigments. Chlorophyll preferentially absorbs light in the blue and red portions of the solar spectrum. Despite its strong absorption of red light (760 nm), chlorophyll shows little absorption of infrared light at wavelengths of 800—1200 nm. Thus, to provide an index of the underlying "greenness" of the Earth's surface, satellites measure the surface reflectance in discrete portions of the visible and infrared spectrum. Bare soil shows similar reflectance in the infrared and red wavebands, but vegetation shows an infrared/red ratio >1.0 as a result of the absorption of red light by chlorophyll. The normalized difference vegetation index (NDVI) is calculated as:

$$NDVI = NIR - VIS/NIR + VIS, \tag{9.3}$$

where:

NIR is reflectance in the near-infrared, and
VIS is reflectance in the visible red wavebands.

This index minimizes the effects of variations in background reflectance and emphasizes variations in the data that occur because of the density of green vegetation. NDVI allows global mapping of a greenness index for the Earth's land surface. Such satellite measurements of greenness can provide estimates of NPP, assuming that greenness is directly related to leaf area since, the leaf area index, LAI, has long been known to be a good predictor of NPP.

The MODIS satellite provides an estimate of GPP using the relationship:

$$GPP = \varepsilon \times NDVI \times PAR \qquad (9.4)$$

where:

$\varepsilon$ is a measured coefficient expressing the efficiency of conversion of sunlight energy into plant growth, and
PAR is a measure of photosynthetically active radiation.

The satellite's measurements of NDVI are coupled to independent measurements of surface climate conditions that affect $\varepsilon$. Remote sensing of biomass is more difficult than for LAI and NPP. Synthetic aperture radar (SAR) is used to measure vegetation biomass based on the absorption of microwave radiation by the water held in woody biomass. LiDAR has been used to estimate forest height, which can be correlated to biomass. Aircraft-based and satellite LiDAR systems also can estimate forest biomass from measurements of its height.

**The scientific challenge.** In order to analyze the global carbon cycle, realistic dynamic models, which scale from local to regional carbon fluxes, are being developed. Such models are needed to quantitatively predict how human alterations of the global carbon cycle (i.e., fossil fuel combustion and forest clearing) will affect atmospheric $CO_2$ concentrations in the future. International agreements aimed at minimizing rise in atmospheric $CO_2$ levels will need to depend on such model projections. The scientific challenge is how to obtain be most accurate empirical measures of these flux values. One key question is to determine how well eddy covariance data determined at the stand-level will represent these regional-scale processes. Eddy-flux measurements will have to deal with variability in landscape ecosystem types, management practices, and land-use patterns. Satellite sensing may overcome some of these limitations. But a remaining challenge will be obtaining accurate measures, and partitioning thereof, for photosynthetic and respiratory fluxes from soils and vegetation. Scientific ingenuity will be needed to integrate contrasting methodologies to obtain the most accurate scaling from the stand to the planetary boundary layer.

## 9.8 Recommended reading

Bylinsky, M., Mann, K.H., 1973. An analysis of factors governing productivity in lakes and reservoirs. Limnology & Oceanography 18 (1), 1−14. https://doi.org/10.4319/lo.1973.18.1.0001.

Houghton, R.A., Skole, D.L., 1990. Carbon, pp. 393−408. In: Turner, B.L., Clark, W.C., Kates, R.W., Richards, T.F., Mathews, T.T., Meyer, W.B. (Eds.), The Earth as Transformed by Human Action. Cambridge Univ. Press, Cambridge. https://www.amazon.com/Earth-Transformed-Human-Action-Biosphere/dp/0521446309.

Lieth,, H., 1975. Primary production of the major vegetation units of the world, pp. 203−215. In: Lieth, H., Whittaker, R.H. (Eds.), Primary Productivity of the Biosphere. Springer Verlag, Berlin-Heidelberg-New York, 339 pp. https://www.springer.com/gp/book/9783642809156.

Reichle, D.E., Franklin, J.F., Goodall, D.W. (Eds.), 1975. Productivity of World Ecosystems. The National Acadamies Press, Washington, DC, p. 166 pp. http://nap.edu/20114/.

Whittaker, R.H., Likens, G.E., 1973. Carbon in the biota, pp. 281–302. In: Woodwell, G.M., Pecan, E.V. (Eds.), Carbon and the Biosphere. National Technical Information Service, Springfield, Virginia, 392 pp. https://www.biodiversitylibrary.org/bibliography/4036#/summary/.

Woodwell, G.M., Whittaker, R.H., 1968. Primary production in terrestrial communities. American Zoologist 8, 19–30.

Chapter 10

# The global carbon cycle and the biosphere

Energy flow leads to cyclic processes. "Geochemical processes opperate only because of a flow of energy from a higher to a lower potential or intensity; hence energy is no less important than matter in the geochemical cycle" (Mason, 1958). This is clearly so also for the biogeochemical cycle. While solar energy is necessary for the maintenance of life on Earth, it is equally as

The Global Carbon Cycle and Climate Change. https://doi.org/10.1016/B978-0-12-820244-9.00010-X
**183**

important that there be a sink for the outflow of thermal energy. Otherwise, the planet would continuously heat up and life would soon cease to exist. Thus, the global ecological system, the biosphere, can be defined as that part of the surface of the planet which is ordered by the flow of energy, facilitated through the photosynthetic process. All biological processes depend upon the absorption of solar photons and the transfer of heat to celestial sinks (Morowitz, 1968). This can be seen in the trophic dynamic organization of the ecosystem discussed in the preceeding chapter. Now it will become evident again in the coupled flow of carbon and energy through the biosphere and the global cycle of carbon.

The basic elements of the global carbon cycle have been known, and their fundamental components quantified, for well over half a century. Summarizing the global carbon cycle here in Sections 10.5 and 10.6 is an opportunity to review some of the pioneering studies in this field as well as recent advances in our knowledge. A landmark representation of the global carbon cycle was made by Bolin et al. (1979). The most comprehensive treatment of the global carbon cycle is contained in the landmark publication, *The Carbon Cycle*, edited by Wigley and Schimel (republished 2005) summarizing the state of knowledge in 1993, and updated in 2000 with extractions from the IPCC report *Climate Change, 1994* (IPCC, 1994) and the *IPCC Second Assessment Report, Climate Change 1995: The Science of Climate Change (IPCC, 1995)*. Knowledge of the carbon cycle continues to rapidly advance, and new information is being continuously documented (IPCC, 2007b; IPCC, 2013; IPCC, 2014; IPCC, 2018). All contain extensive references for the serious student and researcher. The fluxes of anthropogenic $CO_2$ are being estimated more accurately. Improved estimates of the ocean carbon burden and associated carbonate system are also being made. On land, there is a better understanding of the atmospheric buildup of $CO_2$ since 1750 resulting from land-use change, and how the terrestrial biosphere responds to changing climate. Globally, refinement has been made in identifying major $CO_2$ fluxes. Let us begin with an overview of the natural carbon cycle.

## 10.1 The components of the global carbon cycle

**The atmosphere**. Industrialization and changes in the landscape have disturbed natural geochemical cycles of carbon and other elements significantly for several centuries. Projection of future trends in atmospheric $CO_2$ and global carbon flows, under human disturbance, requires understanding of the natural exchanges between biotic and other pools (Fig. 10.1). Direct measurement of the concentration of $CO_2$ in the atmospheric pool is possible only under current conditions. However, analyses of air trapped in glacier ice have helped refine estimates of atmospheric content in the several centuries before 1800 to within a range of 260−285 ppm, or about $5.50 \times 10^{17}$ to $6.10 \times 10^{17}$ g C total in the atmosphere. Every year the biosphere exchanges

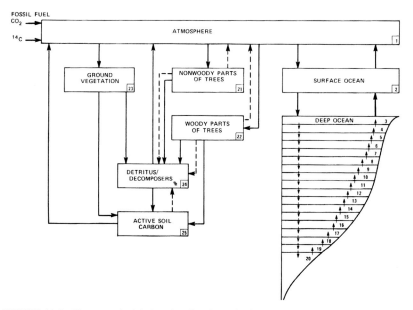

**FIGURE 10.1**  The natural global cycle of carbon showing the major reservoirs (pools) and pathways (fluxes) of carbon flow in the biosphere, as illustrated in the structure of an early, multidimensional box model. *Image credit: Reichle et al., 1985, after Emanuel et al., 1984.*

about 100 times as much $CO_2$ as the small annual atmospheric increase or the annual input to subfossil sediments. The carbon cycle absorbs about 1/2 the anthropogenic emissions into the ocean and land surface; the remainder accumulates in the atmosphere. A CO2 molecule released to the atmosphere has a residence time of 3.5 years before dissolving in the ocean.

**The land**. The draw-down of atmospheric $CO_2$ by the photosynthesis (GPP) of green plants is partly offset by total plant respiration ($R_A$) which recycles $CO_2$ back to the atmosphere (Fig. 10.1). Loss of $R_A$ from GPP leaves more or less half of the carbon assimilated as net primary production (NPP). Additional respiration by decomposers and consumers results in at least 90% of the net carbon fixed (NPP) being released back to the atmosphere as $CO_2$ Over 90% of this heterotrophic respiration is microbial. Natural fires caused mostly by lightning, or burning by early humans, hasten the return of most of the remaining $\sim 10\%$ of the organic carbon in NPP to the atmosphere. Most organic residues are rapidly recycled to $CO_2$ or $CH_4$ by bacteria and fungi. Consumers hasten the remineralization of organic residues, and otherwise stimulate growth in grazing, and metabolize the carbon fixed by plants.

The relative magnitude of the gaseous exchange of the biosphere with the atmosphere is illustrated by the intraannual (seasonal) atmospheric $CO_2$ fluctuations evident in the interannual increase curve of atmospheric $CO_2$

(Fig. 10.2). Atmospheric $CO_2$ fluctuates up and down on a seasonal basis because of the changing balance in $CO_2$ emission and uptake between photosynthesis and respiration by green plants. This is most evident in the Northern Hemisphere, because of its much larger mass of land and plants. In the spring as temperatures become warm, snow melts, and soil moisture is replenished, the photosynthetic activity of plants across the Northern Hemisphere increases. The increase in photosynthesis draws down the $CO_2$ in the atmosphere. In the autumn, as the transition to winter begins, a reversal in the biospheric $CO_2$ flux begins; photosynthesis decreases and respiration producing $CO_2$ increases; the net flux of $CO_2$ to the atmosphere becomes positive, and global atmospheric $CO_2$ concentrations increase.

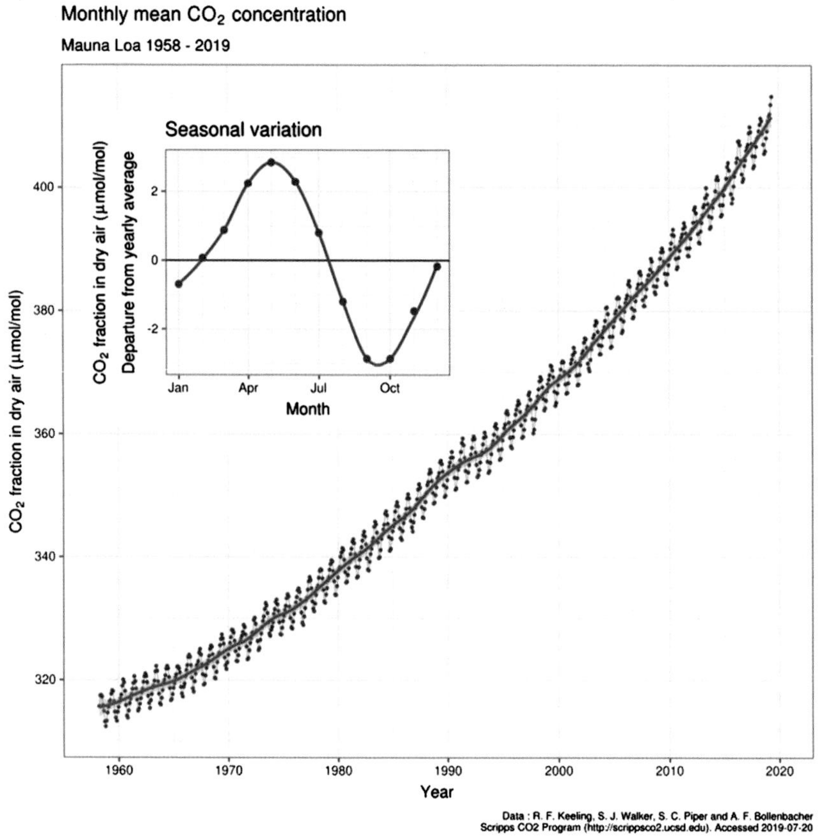

Monthly mean $CO_2$ concentration
Mauna Loa 1958 - 2019

Data : R. F. Keeling, S. J. Walker, S. C. Piper and A. F. Bollenbacher
Scripps CO2 Program (http://scrippsco2.ucsd.edu). Accessed 2019-07-20

**FIGURE 10.2** Interannual fluctuations in atmospheric CO2 concentrations reveal the "breathing" of the biosphere across the seasons of the year. *Image credit: by Delone—own work. Data from Dr. Peter Tans. NOAA/ESRL and Ralph Keeling, Scripps Institution of Oceanography. Courtesy, Wikipedia Contributors, 2019.*

A recent report (Koch et al., 2018) illustrates the influence of the biosphere in controlling the levels of atmospheric $CO_2$. When Europeans first colonized the Americas, 90% of the indigenous population was wiped out by disease and conflict. Nature then reclaimed the agricultural plots and fire-managed lands abandoned by these peoples, resulting in a $7.4 \times 10^{15}$ g $CO_2$ increase in the atmosphere. These events may have contributed to the "Little Ice Age" along with the increased volcanic activity of the period—yet another example of the biospheric connection with the atmosphere through reduced photosynthesis and less $CO_2$ draw-down from the atmosphere.

**The ocean**. Within the ocean, organic biomass and especially productivity vary so widely that using weighted averages for the world can sometimes be misleading. Dense concentrations of plant life (beds of seaweeds and other algae) and animals (coral reefs) mostly occupy small shallow areas. Other coastal and upwelling waters that have sufficient nitrogen and phosphorus are responsible for much of the ocean biomass, production, and $CO_2$ release (Borges et al., 2006), within an area comprising perhaps 10% of the total ocean area.

The oceans, particularly the areas of open ocean due to their vast size, have been the most important sink for atmospheric $CO_2$. A small phytoplankton biomass, with high productivity, and a carbonate water chemistry, result in significant uptake of $CO_2$ – the products of which precipitate to the oceanic carbon sink. Knowledge of oceanic processes that determine $CO_2$ uptake have come mainly from extensive ocean surveys that have elucidated ocean circulation and atmospheric-ocean exchange patterns, and the chemical and biochemical equilibria, needed to estimate $CO_2$ uptake.

The surface ocean layer, heated by the sun and agitated by winds, is relatively well mixed and, on average, near equilibrium with atmospheric $CO_2$. Under much of the warm ocean, the large amount of energy required to overturn the thermocline (warm layers over cold) makes the intermediate waters relatively stagnant. Salinity gradients may further stabilize this thermal layering. The cold deep ocean below 1000 m is, thus, relatively isolated from surface waters and air. When surface waters are sufficiently cold and/or saline, surface waters sink (i.e., mix) to various depths. A complex worldwide circulation pattern involves descending waters mostly in polar regions (e.g., Greenland and Antarctic seas and the North Pacific), slow gyration at all levels in major basins (Broecker and Peng, 1982; 1984), and upwelling of deep water elsewhere.

**The geosphere**. Most of the world's carbon is stored inertly in the Earth's lithosphere. Much of the carbon stored in the Earth's mantle was stored there when the Earth formed. Some of it has been deposited in the form of organic carbon from the biosphere. Of the carbon stored in the geosphere, about 80% is limestone and its derivatives, which originate from the sedimentation of calcium carbonate in the shells of marine organisms. The remaining 20% is stored as kerogens formed through the sedimentation and burial of terrestrial organisms under high heat and pressure. Organic carbon stored in the geosphere can remain there for millions of years. Carbon can leave the geosphere in several ways. Carbon dioxide is released during the metamorphosis of carbonate rocks when they are subducted into the Earth's mantle. This carbon dioxide can be released into the atmosphere and ocean through volcanoes and hotspots. It can also be removed by humans through the extraction of kerogens in the form of fossil fuels. After extraction, fossil fuels are burned to release energy, thus emitting the carbon they store into the atmosphere (Wikipedia, 2018).

**Volcanic activity.** The solid Earth contains a huge quantity of carbon, far more than is present in the atmosphere or oceans. Some of this carbon is released from the rocks in the form of carbon dioxide, through vents at volcanoes and hot springs. Carbon dioxide from volcanic eruptions is a natural flux which is periodic, generally unpredictable, and which can contribute large amounts of $CO_2$ to the atmosphere. The volcanic flux published in reports ranges from 1.8 to $8.6 \times 10^{13}$ of carbon as $CO_2$ emitted, on average, per year (Mörner and Etiope, 2002; Kerrick, 2001).

**Carbon reservoirs in the global cycle.** The global carbon cycle has four major subcomponents, or carbon pools. The contemporary values for these pools are (Trabalka, 1985):

- the *atmosphere*, $0.720 \times 10^{18}$ g C,
- the *oceans,* which include dissolved inorganic carbon, living and nonliving marine biota ($3.86 \times 10^{19}$ g C),
- the *terrestrial biosphere,* into which are usually grouped freshwater ecosystems and nonliving organic material such as soil carbon ($1.7 \times 10^{18}$ g C), and
- the *sediments:* ocean sediments ($10^{23}$ g C), plus carbonaceous rocks ($\sim 10^{24}$ g C), and fossil fuels ($3.7 \times 10^{18}$ g C).

The estimated values of key parameters in the global carbon cycle, and their associated uncertainties, are given in Table 10.1.

## 10.2 Carbon cycle regulators

Regulation of the rates of flux in the global carbon cycle occurs in enumerable biogeochemical processes, but overall through major feedbacks in the biogeochemical cycle of carbon.

**The weathering $CO_2$ thermostat.** The removal of $CO_2$ from the atmosphere, and the long-term geological cycle of carbon, is referred to as the "weathering

**TABLE 10.1** Values, and uncertainties of parameters, in the global carbon cycle (Solomon et al., 1985).

| Parameter | Value | Uncertainty range |
|---|---|---|
| Total fossil fuel release (1860–1982) | $170 \times 10^{15}$ g C | $150–190 \times 10^{15}$ g C |
| Fossil fuel release (1982) | $5 \times 10^{15}$ g C | $4.5–5.5 \times 10^{15}$ g C |
| Recoverable fossil fuel resources | $3.93 \times 10^{18}$ g C | 3.68 $-4.25 \times 10^{18}$ g C |
| Total carbon in the atmosphere | $720 \times 10^{18}$ g C | $715–725 \times 10^{15}$ g C |
| $CO_2$ conc. in the troposphere (1982) | 341 ppm | 339–343 ppm |
| Airborne fraction fossil fuels (1958–82) | 0.58 | 0.50–0.80 |
| Annual Atmospheric $CO_2$ cycle | 7 ppm | 0–15 ppm |
| Annual increase in atmospheric $CO_2$ | 1.5 ppm | 1.3–1.7 ppm |
| Historic atmospheric $CO_2$ concentration | 280 ppm | 260–285 ppm |
| Surface area of ocean | $3.61 \times 10^{14}$ m$^2$ | – |
| Carbon in ocean surface layer (0–75m) | $630 \times 10^{15}$ g C | $590–720 \times 10^{15}$ g C |
| Carbon in intermediate and deep ocean | $38 \times 10^{18}$ g C | $32–42 \times 10^{18}$ g C |
| Carbon in ocean sediments | $10^{23}$ g C | ? |
| Marine particulate carbon flux | $4 \times 10^{15}$ g C | $2–6 \times 10^{15}$ g C |
| Gross annual ocean $CO_2$ uptake | $107 \times 10^{15}$ g C | $106–108 \times 10^{15}$ g C |
| Net ocean $CO_2$ uptake | $2.4 \times 10^{15}$ g C | $1.5–3.3 \times 10^{15}$ g C |
| River export of carbon to oceans | $0.7 \times 10^{15}$ g C | $0.3–1.3 \times 10^{15}$ g C |
| Historical terrestrial biomass | $900 \times 10^{15}$ g C | 700 $-1100 \times 10^{15}$ g C |
| Contemporary biomass | $560 \times 10^{15}$ g C | $420–660 \times 10^{15}$ g C |
| Historic soil carbon | $1.7 \times 10^{18}$ g C | $1.7–3.0 \times 10^{18}$ g C |
| Contemporary soil carbon | $1.5 \times 10^{18}$ g C | $1.2–1.8 \times 10^{18}$ g C |
| Net C flux from terr. biosphere since 1800 | $120 \times 10^{15}$ g C | $90–180 \times 10^{15}$ g C |
| Gross ann. terrestrial plant $CO_2$ uptake | $120 \times 10^{15}$ g C | $90–120 \times 10^{15}$ g C |
| Net terrestrial primary production | $60 \times 10^{15}$ g C | $45–62 \times 10^{15}$ g C |
| Ann. tropical forest clearing (1970–80)[a] | $7.5 \times 10^{6}$ ha yr$^{-1}$ | $3.2–30 \times 10^{6}$ ha yr$^{-1}$ |
| Ann. net carbon flux from land clearing | $1.3 \times 10^{15}$ g C | $0.0–2.6 \times 10^{15}$ g C |

[a]calculated by author from fractional values given and using Table 9.3 land area.

$CO_2$ thermostat" (Archer, 2010). Chemical weathering acts as a sink for carbon, whereby carbon is recruited from atmospheric $CO_2$ and eventually ends up in the chemical form of calcium carbonates in ocean sediments:

$$CaSiO_3 + CO_2 \leftrightarrow CaCO_3 + SiO_2$$

The weathering of $CO_2$ acts as a regulator (i.e., a thermostat) in controlling atmospheric $CO_2$ concentrations. The mechanism underlying the thermostat concept is the effect of the Earth's temperature on the hydrologic cycle. As the temperature of the Earth rises, the pace of the hydrologic cycle increases, since the higher vapor pressure of water in a warmer atmosphere generates more rain as it rises in the atmosphere and cools. Runoff of rain water from the continents accelerates chemical weathering of the CaO component of igneous rocks. As the Earth's temperature rises, the "thermostat" will withdraw more $CO_2$ from the atmosphere. The weathering thermostat operates over a time frame of 100,000 years (Archer, 2009), so it will take over a hundred thousand years for the thermostat to reset Earth's climate to its natural, preindustrial carbon cycle balance – if additional anthropogenic emissions ceased today.

**The oxygen homeostat.** Although the global cycles of carbon and oxygen occur independently, they are, in a small but very critical way, interconnected to each other. Homeostasis is the property of a system by which a variable is actively regulated to remain at a nearly constant level. Each of these variables is controlled by separate homeostatic regulators (in this case photosynthesis and respiration) which together maintain a quasi-equilibrium of $CO_2$ and $O_2$ in the atmosphere necessary for life. Plant photosynthesis is the major determinant of the amount of oxygen present in the atmosphere. The oxygen-carbon cycle can be summarized as, plants taking in carbon dioxide and releasing oxygen during photosynthesis. This released oxygen is taken up by animals and decomposer organisms who in turn release carbon dioxide back to the atmosphere through respiration. The continual recycling of carbon and oxygen with the atmosphere will continue as long as photosynthesis and respiration in the biosphere occur. To describe this phenomenon, Archer (2010) has coined the term "the oxygen homeostat." Within bounds optimal for most life, temperature raises respiration rate (a biochemical process) more rapidly than does photosynthesis (both a biochemical and a biophysical process). Thus, as the Earth's temperature increases, the "homeostat" counterbalances the "thermostat" by producing more $CO_2$. The oxygen homeostat operates over a time frame of 2 million years (Archer, 2010).

**The ocean calcium carbonate pH-stat.** The pH, or acidity, of the ocean is a balance between the weathering input of $CaCO_3 + H_2O \rightarrow Ca(OH)_2 + CO_2$ from sedimentary rocks on land, with the $OH^-$ ions from $Ca(OH)_2 \rightarrow Ca^{2+} + 2OH^-$ acting as a base, and the solution of atmospheric $CO_2$ in surface ocean waters, $CO_2 + H_2O \leftarrow \rightarrow H_2CO_3 \leftarrow \rightarrow HCO_3^- + H^+$, with $H^+$ ions acting as the acid. Balancing of the oceans $CaCO_3$ budget controls the ocean's pH. This mechanism is the $CaCO_3$ pH-stat, and it operates over a time frame of several thousand years.

## 10.3 Units of measure for the global scale

Discussion of the global carbon cycle occurs on a spatial scale and magnitude that requires new units of measurement to characterize it. Table 10.2 summarizes the most frequently used measures.

**TABLE 10.2** Units of measure for the global carbon cycle.

| Quantity | Symbol[a] | Value |
|---|---|---|
| 1 megagram | Mg | $10^6$ g |
| 1 metric ton (=tonne) | Mt | $10^6$ g |
| 1 gigagram | Gg | $10^9$ g |
| 1 teragram | Tg | $10^{12}$ g |
| 1 petagram | Pg | $10^{15}$ g |
| 1 giga ton | Gt | $10^{15}$ g |
| 1 ppm $CO_2$ in atmosphere | ppm | 2.13 Gt C |
| 1 ppm $CO_2$ in atmosphere | ppm | $2.12 \times 10^{15}$ g C |
| 1 giga ton C | Gt C | 0.47 ppm $CO_2$ |
| 1 Gt C | Gt C | 3.664 Gt $CO_2$ |
| Mass Earth | M | $5.976 \times 10^{24}$ kg |
| Mass atmosphere | Ma | $5.137 \times 10^{18}$ kg |
| Volume atmosphere | Va | $4.2 \times 10^{15}$ m$^3$ |
| Effective thickness atmosphere | ha | 8.2 km (27,000 ft) |
| Mass ocean | Ms | $1.384 \times 10^{21}$ kg |
| Mean ocean volume | Vs | $1.350 \times 10^{18}$ m$^3$ |
| Mean ocean depth | hs | 3730 m |
| Surface area Earth | Ae | $5.101 \times 10^{14}$ m$^2$ |
| Land area | Al | $1.481 \times 10^{14}$ m$^2$ |
| Ocean area | As | $3.620 \times 10^{14}$ m$^2$ |
| Ice sheets/glacial sea | Is | $0.14 \times 10^{14}$ m$^2$ |

[a]*Symbols generally follow reference standards used in Bolin (1981).*
Modified from Clark, 1982; O'Harra, 1990; Author.

## 10.4 History of carbon dioxide in the atmosphere

Originally, the Earth was ocean-free, without an atmosphere, and all carbon existed in meteorites and igneous rocks. Since then, many geochemical changes have occurred. Evolution of the Earth's atmosphere during the last 100 million years may be divided into three stages (Gammon et al., 1985).

**Early Earth to the last Ice Age.** During the first stage, from $10^8$ years B.P., until the end of the last Ice Age, very large and poorly understood changes in atmospheric $CO_2$ levels occurred. From an initial concentration, perhaps as high as several thousand parts per million in Cretaceous times, the atmospheric $CO_2$ levels gradually fell toward much lower values of 200–300 ppm characteristic of the glacial-interglacial periods of the past million years. Analyses of sedimentary carbon isotope records indicate that the atmospheric $CO_2$ levels varied in a regular, periodic fashion, with low values $\sim 200$ ppm during cold glacial phases, and high values $\sim 270$ ppm during warm, interglacial periods.

**The Ice Ages.** The waxing and waning of ice sheets and glaciers of the Ice Ages during the last several million years followed a rhythm of several $10^4 - 10^5$ years coinciding with variations in the Earth's orbit around the sun. Carbon dioxide concentrations in the atmosphere (measured from air bubbles trapped in glacier ice) have varied systematically with the Ice Ages. The unstable carbon cycle during this time is characterized by carbon leaving the atmosphere for the oceans during cooler glacial times, explained by the fact that cooler water can hold more dissolved $CO_2$ than warmer water. In this case, it is the same carbon cycle that has gone from stabilizing the climate of the Earth on a million-year timescale to amplifying climatic changes derived from wobbles in the Earth's orbit on a tens-of-thousands year timescale. It is still not clear what drives the carbon cycle during glacial periods, but Archer (2010) provides an excellent review of current thinking on the subject.

**The postglacial period.** The second stage is the postglacial period of generally steady atmospheric $CO_2$ concentrations (260–290 ppm), which lasted from 10,000 years B.P. until the beginning of the 19th century and the Industrial Revolution. The important difference between glacial cycles and global warming is that today's climate is driven by $CO_2$ inputs to the atmosphere, while the glacial cycles were caused by wobbles in the Earth's orbit about the sun.

**The contemporary period post 1800.** The third stage, from 1800 B.P. to the present, is one where the impact of human activity can clearly be seen. Before the industrial and agricultural revolutions, humans had relatively little impact upon the global cycle of carbon. The preindustrial atmosphere, before large fluxes of $CO_2$ from fossil fuel combustion, contained a total of $590 \times 10^{15}$ g C

($\sim$277 ppm). The observed levels of $CO_2$ in the atmosphere have been: 1880, 280 ppmv; 1957, $\sim$315 ppmv; 1994, $\sim$358 ppmv; 2018, 406 ppmv. The atmosphere of 342 ppmv $CO_2$ at the beginning of the 1990s contained $720 \times 10^{15}$ g of elemental carbon. In July 2018 the atmospheric concentration of $CO_2$ was 406 ppm (NOAA, 2018).

Carbon dioxide represents 0.04% of the atmosphere. Other major gases containing carbon in the atmosphere are methane and chlorofluorocarbons (the latter introduced solely by human activity). These are all greenhouse gases whose concentration in the atmosphere are increasing, and contributing to the rising average global surface temperature through the greenhouse effect (see Chapter 11). Atmospheric $CO_2$ is being continuously measured at various monitoring stations, and atmospheric $CO_2$ is continuously reported (IPCC, 2007b) to be continuously rising, with an increase from the rate in the 1990s of $3.2 \pm 0.1$ Gt C yr$^{-1}$ to $4.1 \pm 0.1$ Gt C yr$^{-1}$ during the period 2000−05.

The annual increase is due to the net effect of several processes that regulate global land-atmosphere and ocean-atmosphere fluxes (Table 10.3). The airborne fraction (atmospheric increase in $CO_2$ concentration/fossil fuel emissions) provides a basis for assessing short- and long-term changes in these processes. From 1959 to the present, the airborne fraction has averaged 0.55, with remarkably little variation. This means that the terrestrial biosphere and the oceans together have consistently removed 45% of fossil $CO_2$ for nearly the last 60 years, and that the recent higher rate of atmospheric $CO_2$ increase largely reflects increased fossil fuel emissions.

Modern instrumentation and telecommunications allow us to monitor and record atmospheric $CO_2$ concentrations on a regular basis using direct chemical measurements (Fraser et al., 1986), in ice cores (Oeschger and Stauffer, 1986), tree rings (Stuiver, 1986), and from old spectrographic plates (Stokes and Banard, 1986). Daily average global $CO_2$ concentrations are monitored through a system of worldwide stations (Gammon et al., 1986); calculated, integrated, average global atmospheric $CO_2$ concentrations for the Earth are reported in real-time on the internet (e.g., 409.95 ppm $CO_2$ for August, 2019 [see, https://www.co2.earth/global-co2-emissions/; and https://www.esrl.noaa.gov/gmd/ccgg/trends/monthly.html/]).

## 10.5 Uptake of carbon dioxide by the oceans

Approximately three-quarters of the Earth's surface is covered by oceans to an average depth of 3730 m. For the atmospheric $CO_2$ influx to move into the oceans, the p$CO_2$ of the surface ocean needs to be about 2 ppm lower than that of the atmosphere. Seawater p$CO_2$ varies widely by season and geography, from 250 ppm to around 500 ppm. Uptake by the deep ocean depends upon vertical mixing and ocean currents. From the ocean surface downward, the waters are well-mixed only to a depth of $\sim$75 m. The surface water temperature decreases from 30°C near the equator to $-1.5$°C at the highest latitudes.

**TABLE 10.3** Atmospheric carbon dioxide fluxes (Gt C yr$^{-1}$ or $10^{15}$ g C yr$^{-1}$). Errors represent ± standard deviation of uncertainty estimates and not interannual variability which is larger. The atmospheric increase (first line) results from fluxes to and from the atmosphere: positive fluxes are inputs to the atmosphere (emissions); negative fluxes are losses from the atmosphere (sinks); and numbers in parentheses are ranges. Note that the total sink of anthropogenic $CO_2$ is well constrained. Thus, the ocean-to-atmosphere and land-to-atmosphere fluxes are negatively correlated: if one is larger the other must be smaller to match the total sink, and vice versa.

| | 1980s | | 1990s | | 2000–05[c] |
|---|---|---|---|---|---|
| | TAR | TAR revised[a] | TAR | AR4 | AR5 |
| Atmospheric increase[b] | 3.3 ± 0.1 | 3.3 ± 0.1 | 3.2 ± 0.1 | 3.2 ± 0.1 | 4.1 ± 0.1 |
| Emissions (fossil + cement)[c] | 5.4 ± 0.3 | 5.4 ± 0.3 | 6.4 ± 0.4 | 6.4 ± 0.4 | 7.2 ± 0.3 |
| Net ocean-to-atm. flux[d] | −1.9 ± 0.6 | −1.8 ± 0.8 | −1.7 ± 0.5 | −2.2 ± 0.4 | −2.2 ± 0.5 |
| Net land-to-atm. flux[e] | −0.2 ± 0.7 | −0.3 ± 0.9 | −1.4 ± 0.7 | −1.0 ± 0.6 | −0.9 ± 0.6 |
| Partitioned as follows | | | | | |
| Land use change flux | 1.7 (0.6–2.5) | 1.4 (0.4–2.3) | 0[a] | 1.6 (0.5–2.7) | 0[a] |
| Residual terrestrial sink | −1.9 (−3.8–0.3) | −1.7 (−3.4–0.2) | 0[a] | −2.6 (−4.3–0.9) | 0[a] |

[a] TAR (Third Assessment Report) values revised according to an ocean heat content correction for ocean oxygen fluxes (Bopp et al., 2002) and using the Fourth Assessment Report (AR4) best estimate for the land use change flux.

[b] Determined from atmospheric $CO_2$ measurements (Keeling and Whorf, 2005; updated by S. Piper until 2006) at Mauna Loa (19° N) and South Pole (90° S) stations (conversion factor of 2.12 Gt C yr$^{-1}$ = 1 ppm).

[c] Fossil fuel and cement emission data are available only until 2003 (Marland et al., 2006). Mean emissions for 2004 and 2005 were extrapolated from energy use data with a trend of 0.2 Gt C yr$^{-1}$.

[d] For the 1980s, the ocean-to-atmosphere and land-to-atmosphere fluxes were estimated using atmospheric $O_2$:$N_2$ and $CO_2$ trends, as in the TAR. For the 1990s, the ocean-to-atmosphere flux alone is estimated using ocean observations and model results, giving results identical to the atmospheric $O_2$:$N_2$ method (Manning and Keeling, 2006), but with less uncertainty. The net land-to-atmosphere flux then is obtained by subtracting the ocean-to-atmosphere flux from the total sink (and its errors estimated by propagation). For 2000 to 2005, the change in ocean-to-atmosphere flux was modeled (Le Quéré et al., 2005) and added to the mean ocean-to-atmosphere flux of the 1990s. The error was estimated based on the quadratic sum of the error of the mean ocean flux during the 1990s and the root mean square of the 5-year variability from three inversions and one ocean model presented in Le Quéré et al. (2003).

[e] Balance of emissions due to land use change and a residual land sink. These two terms cannot be separated based on current observations.

Source: IPCC, 2007b.

The gross features of deep-ocean circulation are that the densest water is produced near the surface at high latitudes in winter, mostly in the North Atlantic around Antarctica, being supplied, both north and south, with new bottom water from the Indian and Pacific Oceans (Chapter 9, Figure 9.4). Surface waters exchange about one-seventh of their carbon with the atmosphere each year, but the capacity of this small volume to take up additional $CO_2$ is limited. It is the transfer of excess $CO_2$ to the deep oceans by advective water flow, and by mixing, that affects the oceans' short-term response to atmospheric $CO_2$.

The ocean contains the largest active pool of carbon near the surface of the Earth; its living plus dead organic carbon pool (Table 10.4) of $725 \times 10^{15}$ g C is $\sim 47\%$ of biosphere's organic carbon pool. The small phytomass of plankton has a rapid turnover time of $\sim 1$yr, but the deep ocean inorganic carbon pool $(38,127 \times 10^{15}$ g C) does not rapidly exchange with the

**TABLE 10.4** Estimated oceanic carbon pools.

| Pool | Volume ($10^9$ km³) | Inorganic carbon ($10^{15}$g) | | Organic carbon ($10^{15}$ g) | |
|---|---|---|---|---|---|
| Surface (0−75 m) | $0.024^a$ | $580^a$ | $(560^b-660^c)$ | 50 | $(30^a-64^d)^e$ |
| Intermediate (75−1000 m) | $0.25^a-0$ | $37,542^c$ | $(6000^a -7840^d)$ | $675^{f,g}$ | $(125^h-480^i)$ |
| Deep water (1000 m) (Polar, if separated) | $1.1^a$ | | $(25,000^d -31,800^a)$ $(5000^d-0)$ | | $(440^h-1650^i)$ |
| Total | 1.374 | 38,122 | | $725^g$ | $(\sim 600-2200)$ |

$^a$ Baes et al., 1976. Implies 6.2 year residence time if income and loss are near $93 \times 10^{15}$ g/yr.
$^b$ Björkstrom, 1979,1−2 year residence time.
$^c$ Killough and Emanuel, 1981, 142−535 year residence time.
$^d$ Munro and Olson, 1982, 5−440 year residence time (Olson et al., 1985).
$^e$ If the intermediate of $30 \times 10^{15}$ g c/yr input is used for the surface pool only, residence time of 1 to 2 years for living plus dead organics matter follows from the top line. Alternate incomes of $+-40\%$ of these estimates would make correspondingly wider ranges of estimated turnover.
$^f$ For organic matter in the remaining layers together, the middle income estimate of $4 \times 10^{15}$ g/yr implies residence times of 142 to 535 years, assuming pools of $570 \times 10^{15}$ and $2140 \times 1015$, respectively. For alternate inputs of $2 \times 10^{15}$ or $6 \times 10^{15}$ g/yr, residence times become 50% shorter or longer, respectively.
$^g$ For world average of (5 mg/L) x ($1350 \times 10^9$ km³) below a 75-m mixed layer, a pool size of $675 \times 10^{15}$g implies a residence time estimate of 170 years for carbon in combined intermediate and deep water, near the lower range of note f. Taking the middle input estimate as in f, divided by $200 \times 10^{15}$ g for intermediate waters only, implies a residence time of 50 years. If $1 \times 10^{15}$ g C or less per year reaches the deep waters, and if these total near $440 \times 10^{15}$ g of combined dissolved and particulate organic carbon, a residence time near 440 years would be implied for the world average ocean.
$^h$ Mopper and Degens, 1979 (low range, wet oxidation).
$^i$ After Gordon and Sutcliffe, 1973 (high range, dry combustion).
**Source:** Olson et al., 1985

atmosphere. At equilibrium the sea contains about 56 times as much carbon as the atmosphere. The mean oceanic turnover time is of the order of 250 years. The atmosphere contains more carbon than all of the Earth's living vegetation. The terrestrial organic carbon pool (terrestrial ecosystems, including freshwater ecosystems) contains $829 \times 10^{15}$ g C g, or ~ 53% of the biosphere's organic carbon pool (Table 9.3), and has a much more rapid average turnover time of ~ 16 years (Fig. 8.7).

The inorganic chemistry of carbon in the ocean is largely determined by bicarbonate chemistry previously discussed in Chapter 8.4. In natural waters, over geologic time, basic minerals have consumed the hydrogen ions produced in the bicarbonate reaction, and bicarbonate and carbonate have accumulated in the oceans to the extent that 90% of the inorganic carbon in the ocean is in the form of the bicarbonate ion, and the rest is in the form of the carbonate ion. Photosynthesis by phytoplankton withdraws $CO_2$ from the water. Although there is a rise in alkalinity owing to the simultaneous consumption of $H^+$ ions, this is more than overcome by the loss of dissolved carbonate to the formation of calcium carbonate. The carbon and alkalinity, thus taken out of the inorganic carbon pool, falls from the surface water as detritus. The chemistry of sea water allows the oceans to hold 10 times more $CO_2$ than if there was no buffering.

**The biological and physical carbon pumps**. The oceanic biological carbon pump is a carbon flux driven by organisms that live in the ocean. Just like the terrestrial carbon cycle, the oceanic biological carbon pump is all about photosynthesis and respiration, and the production of waste products and decomposing biomass. The biological pump plays a major role in transforming carbon compounds into new forms of carbon compounds, moving carbon throughout the ocean, and eventually depositing it as sea floor sediments.

The ocean contains many dissolved chemicals which are especially important to the ocean carbon cycle and the shell-building organisms that live in the oceans. The ocean carbonate system, often called the physical carbon pump, is linked to the biological pump and plays a very big role in transporting carbon down to deep ocean sediments, where it is stored for millions of years. When $CO_2$, and then enters into a series of reversible chemical reactions that produce bicarbonate ions ($H^+$, $CO_3^-$), hydrogen ions ($H^+$), and carbonate ($CO_3^{2-}$) ions. The carbonate ions are especially important to marine organisms, because they combine with calcium ions ($Ca^{2+}$) to form calcium carbonate ($CaCO_3$). Shell-building organisms, such as corals, oysters, lobsters, pteropods, sea urchins, and some species of plankton, use calcium carbonate to build their shells, plates, and inner skeletons.

The oceans of the world contain about 50 times the amount of carbon currently in the atmosphere in the form of $CO_2$. All but a small fraction of this carbon is dissolved in the form of the bicarbonate ion, an accumulated product of the reaction of $CO_2$ with basic substances of the Earth's crust. This carbon pool dwarfs the other rapidly exchanging pool in the terrestrial carbon cycle.

The capacity of the oceans to take up additional $CO_2$ is so great that it will be the ultimate global carbon sink for future fossil carbon released by mankind.

**The oceanic sink for carbon**. To assess the mean ocean sink, seven methods have been used (IPCC, 2007a,b): (1) partial pressure of $CO_2$ at the ocean surface and gas exchange estimates; (2) atmospheric inversions using atmospheric $CO_2$ and atmospheric transport models; (3) measurement of carbon, oxygen, nutrients, and chlorofluorocarbons (CFCs) in seawater to estimate anthropogenic $CO_2$ uptake; (4) estimates of the age distribution of water based upon CFC observations combined with atmospheric $CO_2$ history; (5) simultaneous measurements of the increase in the atmospheric $CO_2$ and the decrease in atmospheric $O_2$; (6) change in $^{13}C$ in the atmosphere or the oceans; and (7) ocean general circulation models.

Estimates from all methods are consistent in estimating the global oceanic sink for anthropogenic $CO_2$. Four of the estimates are better constrained than the others. Ocean uptake of atmospheric $CO_2$ of $-2.2 \pm 0.5$ Gt C yr$^{-1}$, centered around 1998, based on the atmospheric $O_2/N_2$ ratio, needs to be corrected for the oceanic $O_2$ changes (Manning and Keeling, 2006). The estimate of $-2.0 \pm 0.4$ Gt C yr$^{-1}$, centered around 1995 and based on CFC data, provides a constraint from observed physical transport in the ocean. The mean estimates of $-2.2 \pm 0.25$ and $-2.2 \pm 0.2$ Gt C yr$^{-1}$, centered around 1995, provide constraints based on a large number of ocean carbon measurements. These well-constrained estimates all point to a decadal mean ocean $CO_2$ sink of $-2.2 \pm 0.4$ GT C yr$^{-1}$, centered around 1996 (IPCC, 2007b).

**El Niño**. El Niño is an irregularly occurring and complex series of climatic changes affecting the equatorial Pacific region and beyond every few years, characterized by the appearance of unusually warm, nutrient-poor water off the coast of northern Peru and Ecuador, typically in late December. El Niño is the warm phase of the El Niño Southern Oscillation and is associated with a band of warm ocean water that develops in the center and east central equatorial Pacific, including off the coast of South America. The normal Pacific pattern is a warm pool in the west that drives deep atmospheric convection. Local winds cause nutrient-rich cold water to upwell along the South American coast. During El Niño conditions, warm water and atmospheric convection move eastwards. In strong El Niños, deeper thermoclines off South America result in upwelled water which is warm and nutrient poor. The cool phase is called La Niña with sea surface temperatures in the eastern Pacific below average and air pressure high in the eastern and low in the western Pacific. Both El Niño and La Niña cause global changes of both temperature and rainfall. During La Niña, sea surface temperatures across the equatorial eastern central Pacific Ocean will be 3−5°C lower than normal. Appearance of La Niña persists for at least 5 months. It has extensive effects on the weather in North America.

## 10.6 Carbon exchange between the atmosphere and terrestrial ecosystems

The relative magnitude of the major biospheric carbon pools illustrated in Fig. 10.1 is summarized in Table 10.5. Two geochemical indicators of the biosphere's important role in the global carbon cycle are:

- the annual variation in the Northern Hemisphere of $^{13}C/^{12}C$ in atmospheric $CO_2$ is consistent with the participation of Rubisco in the drawdown of atmospheric $CO_2$, and

**TABLE 10.5** Carbon in major pools of the biosphere. Contemporary estimates using Whittaker & Likens, 1973 and IPCC 2014 in parentheses. Percentages of total carbon pools (columns 2 and 3) are based upon Reiner's 1973 calculation using Bolin's 1970 values.

| Carbon pool | Carbon content Gt C ($10^{15}$ g C) | Percent of total nonsedimentary organic carbon | Percent of total nonsedimentary carbon |
|---|---|---|---|
| Atmosphere | 700 (829) | | 1.76 |
| Terrestrial plants | 450 (827) | 10.82 | 1.13 |
| Terrestrial detritus | 700 (1500) | 16.83 | 1.76 |
| Marine phytoplankton | <5 (<2) | 0.12 | 0.00012 |
| Zooplankton, fish | ≤5 (<3) | 0.12 | 0.00012 |
| Marine detritus | 3000 | 72.12 | 7.53 |
| Sea surface layers | 500 (900) | | 1.25 |
| Deep sea layers | 34,500 (37,100) | | 86.55 |
| Sediments | 20,000,000 | | |
| Coal and oil | 10,000 (<2,800) | | |
| **Totals** | **20,049,860** | **100.01** | **99.98** |

**Sources:** Sverdup, et al., 1942; Rubey 1951; Hutchison, 1954; Revelle and Suess, 1957; Morowitz, 1968; Bolin, 1970; Reiners, 1973; Whittaker and Likens, 1973; Knauer, 1993; IPCC, 2013. Based on Reiners (1975), after Bolin (1970).

- the annual variation in the Northern Hemisphere of $^{18}O/^{16}O$ in the atmospheric $CO_2$ is consistent with the exposure of the entire atmospheric $CO_2$ pool to carbonic anhydrase in leaf water, which can have a much larger effect than that due to cloud water equilibration on ocean mixing in climate models.

**Carbon in biota**. The organic carbon of terrestrial ecosystems is stored in biota and soils. Estimates for the amount of carbon in terrestrial biota worldwide range from $4.50 \times 10^{17}$ g C (Table 10.5) to $8.29 \times 10^{17}$ g C (Table 9.3). Global estimates for GPP for the world's terrestrial ecosystems range from $90 \times 10^{15}$ to $120 \times 10^{15}$ g C $yr^{-1}$, approximately twice the annual net primary production. Annual GPP includes mortality of plant parts, the exudation of soluble organics from live tissues, and the turnover of organic molecules within plants (i.e., photorespiration and nonphotorespiration). Because some of the carbon fixed in photosynthesis is derived from respiration within the plant canopy, or from that emanating from the soil, atmospheric uptake is overestimated. Although estimates of GPP are low by the amount of photorespiration, they overestimate the exchange of carbon between the atmosphere and terrestrial ecosystems. The fraction of carbon in NPP that is derived from sources other than the atmosphere is generally thought to be 30% −40% (Monteith et al., 1964). If 30%−40% of the carbon fixed in GPP is cycled within the plant canopy, the exchange between terrestrial ecosystems and the atmosphere could well be $60 \times 10^{15}$ to $80 \times 10^{15}$ g C $yr^{-1}$, or 30% less than previously thought. This is the metabolic flux, the normal "breathing" of the terrestrial biosphere. Another flux emanate from the terrestrial biosphere, is when forests are cut and burned or allowed to decay (DeFries et al., 2002).

**Carbon in soil and detritus**. Roughly 20% of living biomass is belowground, primarily in roots of primary producers. Across all ecosystem types, the total belowground carbon, living roots plus detritus, is 1.5−2 times the aboveground carbon pool, or approximately 60% of the terrestrial system's carbon content (Reiners, 1975; Houghton et al., 1985). The greatest uncertainty in estimates is for tropical forests, where the largest carbon pool exists. Estimates for the worldwide carbon pool in soil carbon and litter have been in the range of $1200 \times 10^{15}$ to $2200 \times 10^{15}$ g C (Schlesinger, 1977; Post et al., 1982).

There have been several attempts to estimate the storage of soil organic matter and litter in world ecosystems (Table 10.6). Most of these estimates fall within the range of $1200 \times 10^{15}$ to $2200 \times 10^{15}$ g C, based upon compilations by soil groups (Bohn, 1976; 1982; Buringh, 1984), by vegetation types (Schlesinger, 1977, 1984; Ajtay et al., 1979; Bolin et al., 1979), and by life zones (Post et al., 1982; Zinke et al., 1984), or based on modeling of plant production and decomposition (Meentemeyer et al., 1982). The most extensive data set is that developed by Zinke et al. (1984) which included values from

**TABLE 10.6** Carbon balance in terrestrial detritus by biome (Schlesinger, 1979).

| Ecosystem matter | Mean soil area, kg C m$^{-2}$ | World soil organic, ha x 10$^8$ | Total world surface litter, mt C x 10$^9$ | Amount in surface litter, mt C x 10$^9$ |
|---|---|---|---|---|
| Tropical forest | 10.4 | 24.5 | 255 | 3.6 |
| Temperate forest | 11.8 | 12 | 142 | 14.5 |
| Boreal forest | 14.9 | 12 | 179 | 24.0 |
| Woodland & shrub | 6.9 | 8.5 | 59 | 2.4 |
| Tropical savanna | 3.7 | 15 | 56 | 1.5 |
| Temperate grassland | 19.2 | 9 | 173 | 1.8 |
| Tundra & Alpine | 21.6 | 8 | 173 | 4.0 |
| Desert scrub | 5.6 | 18 | 101 | 0.2 |
| Extreme desert, ice | 0.1 | 24 | 3 | 0.02 |
| Cultivated | 12.7 | 14 | 178 | 0.7 |
| Swamp, marsh | 68.6 | 2 | 137 | 2.5 |
| Totals | | 147 | 1456 | 55.2 |

3583 measured soil profiles and allowed patterns of soil organic content to be mapped for world life zones (Fig. 10.3). The best estimate of the pool of soil organic carbon is probably $1500 \times 10^{15}$ g, with a $\pm 20\%$ uncertainty, much of which lies in the treatment of histosols, the peatland soils of the world (Armentano, 1979). In vegetation-based compilations, peatlands are included in categories such as tundra, boreal forests, and wetlands. These categories contain $489 \times 10^{15}$ g C, or 34% of Schlesinger's (1977) estimate of the global soil carbon pool. The content of soil organic matter in tropical soils is fairly well known; differences among estimates of the global pool in tropical soils are mainly related to differences in the area assigned to tropical regions (Schlesinger, 1986; Brown and Lugo, 1982; Post et al., 1982). The decomposition rates of detritus vary greatly among ecosystem types and across

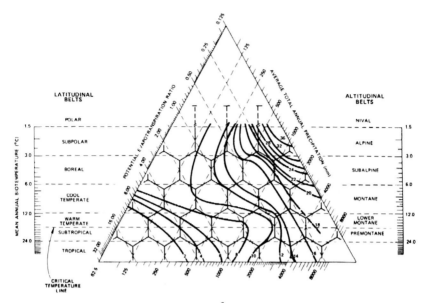

**FIGURE 10.3** Contours of soil carbon (kg C m$^{-2}$) plotted on a Holdridge (1967) life-zone chart. *Source: Zinke et al., 1984. Courtesy, Oak Ridge National Laboratory.*

climatic gradients: for forests ranging from a high of $8.0 \times 10^2$ g C m$^{-2}$ yr$^{-1}$ in mountain rain forests to $0.84-8.08 \times 10^2$ g C m$^{-2}$ yr$^{-1}$ in temperate coniferous forests; $250-670 \times 10^2$ g C m$^{-2}$ yr$^{-1}$ in grasslands; $0.60 \times 10^2$ g C m$^{-2}$ yr$^{-1}$ in tundra and $0.10 \times 10^2$ g C m$^{-2}$ yr$^{-1}$ in deserts (Ajtay et al., 1979).

## 10.7 Modeling carbon in the biosphere

**Compartment models of biosphere carbon flow.** The biosphere is here defined as the entire portion of the globe that can support life, including the atmosphere, oceans, terrestrial surface, and the belowground. The first attempt to formulate a quantitative carbon cycle model was by Craig (1957) using the natural distribution of radiocarbon to estimate the exchange time of $CO_2$ between atmosphere and the oceans. Björkstrom (1979) elegantly presented in detail the mathematical formulation of a coupled atmosphere:ocean:land carbon model. More detailed global models of land and ocean followed, as represented by the "natural" global carbon cycle compartment model by Olson et al. (1985) for the early Holocene (Fig. 10.4, Table 10.7). More fluxes have been added to represent a more detailed global carbon cycle, linked with other nutrient elements, and a dynamic model begins to emerge. This model presentation serves to illustrate the complexity of computer models when expanded in dimensions of time, ecosystem type, and landscape diversity, with

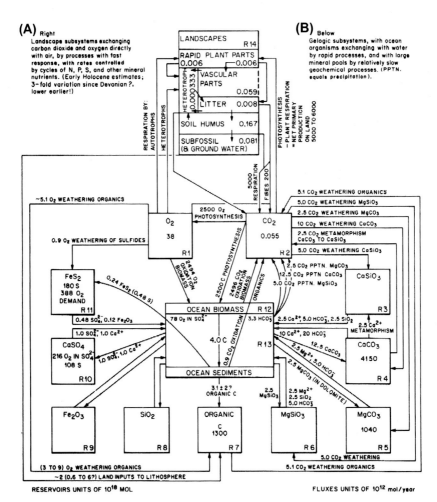

**FIGURE 10.4** A compartment model of the global carbon cycle with couplings to other elements. The model construct incorporates rapid ecological processes (A) with slow geologic processes (B) averaged over the latter portion of post-Cambrian time. (A) Landscapes is early Holocene (recent) time had approximately equal quantities of rapidly cycling (mostly photosynthetic) tissue from woody and nonwoody parts of plants. The latter probably were of negligible mass before the late Silurian Period about 400 million years ago. Estimated values and uncertainties are given in Table 10.7. (B) Summary of oceanic and lithospheric cycles. Note: 1 mol carbon dioxide = 12 g carbon. *Source: Olson et al., 1985. Courtesy U.S. Department of Energy and Oak Ridge National Laboratory.*

fluxes quantified by differential equations representing the rates of biogeochemical processes each determined, in turn, by many variables.

Such mathematical computer models have become the means by which quantified, complex, biogeochemical processes, e.g., chemical and biological processes in the ocean (Baes and Killough, 1986), using empirical data,

**TABLE 10.7** Simplified global carbon inventory and budget estimates for recent, early Holocene times. Values here are $10^{15}$ g C $yr^{-1}$.

| Major Carbon Pool | Inferred Pool of C ($10^{15}$ g C) | Annual Balance Sheet | | | | |
|---|---|---|---|---|---|---|
| | | Income ($10^{15}$ g C $yr^{-1}$) | Loss ($10^{15}$ g C $yr^{-1}$) | Difference[a] ($10^{15}$ g C $yr^{-1}$) | Turnover = Loss ÷ Pool ($yr^{-1}$) | Residence Time = Pool ÷ Loss(year) |
| Atmosphere | 600 | 220 | 220 | 0(?) | 0.333 | 3 |
| Ocean exchange | | 90 | 90 | | | |
| Landscapes | | 120 | 130(?) | | | |
| Ocean | | | | | | |
| Inorganic | 38,000 | 90 | 90 | 0(?) | ~0.00237 | 422 |
| Organic | 675 | 40 | 40 | | 0.06 | 17 |
| Recent Sedimentary Org | ~400 | 0.0048 | 0.048 | | | |
| Landscapes | | | | | | |
| Live | | | | | | |
| Heterotroph | 4(?) | 60 | 60 | 0 | 15.0 | 0.067 |
| Rapid plant | 144 | 130 | 130 | 0 | 0.9 | 1.1 |

*Continued*

**TABLE 10.7** Simplified global carbon inventory and budget estimates for recent, early Holocene times. Values here are $10^{15}$ g C yr$^{-1}$.—cont'd

| Major Carbon Pool | Inferred Pool of C ($10^{15}$ g C) | Annual Balance Sheet | | | Turnover = Loss ÷ Pool (yr$^{-1}$) | Residence Time = Pool ÷ Loss(year) |
| --- | --- | --- | --- | --- | --- | --- |
| | | Income ($10^{15}$ g C yr$^{-1}$) | Loss ($10^{15}$ g C yr$^{-1}$) | Difference[a] ($10^{15}$ g C yr$^{-1}$) | | |
| Slow plant | 652–756 | 15–25 | 15–25 | −0.1(?)−0.2 | 0.023–0.033 | 43–30 |
| Litter | 100 | 60 | 60 | 0 | 0.6 | 1.67 |
| Soil organic (1 m) | 1700–2000 | 5.2–10.1 | 5–10 | −0.2–0.1 | 0.003–0.005 | 340–200 |
| Subfossil organic | 970–3000 | 1.0–0.33 | 0.976–0.03 | 0.024–0.03 | 0.001–0.0001 | 1,000–10,000 |
| Soil carbonate | 1,100 (±100) | 0.03 | 0.029 | ~0.001(?) | 0.0000263 | 37,900 |
| Rocks[b] | | | | | | |
| Carbonates | 62,200,000 | 0.18 | 0.18 | 0 | $0.28 \times 10^{-8}$ | $345 \times 10^{6}$ |
| Organic | 10,300,000 −15,600,000[b] | 0.037–0.06 | 0.037–0.06 | 0 | $0.35–0.38 \times 10^{8}$ | $278–260 \times 10^{6}$ |

Note: Values are prior to major expansion of human populations, agriculture, and forestry, but after most postglacial forest accumulation had become well advanced. [a] The negative estimates for "difference" indicate the loss of carbon from the pool on the same line, which could be inputs of $CO_2$ to the atmosphere, or from landscapes to water bodies, or to subfossil and rock pools. Positive estimates reflect plausible but unproven precivilization storage rates, neglecting such exports as erosion. Net values are more likely to be between these bounds than not, but could have changed significantly over prehistoric intervals. [b] From Figure 10.3.

**Source:** Olson et al., 1985; Trabalka, 1985.

simulate the flows of carbon on a biospheric level. Initially, most models consisted of coupled series of internally mixed reservoirs ("boxes" or "compartments") linked by exchanges ("fluxes"), e.g., one- and two-dimensional ocean models (Björkstrom, 1986); box diffusion models (Siegenthaler and Joos, 1992; Siegenthaler and Sarmiento, 1993). Caldeira et al. (2005) provide the taxonomy of eight types of two-dimensional ocean models (surface and deep ocean as separate boxes), with references to all the models developed in each category and a comparison of model results with observed data. The next stage in development of carbon cycle simulation models were three-dimensional, general circulation models—to simulate atmospheric $CO_2$ distribution, e.g., GISS 3-D Tracer Model (Heinman et al., 1986), and for analysis of seasonal and geographical patterns of atmospheric $CO_2$ distribution (Fung, 1986). General circulation models were particularly useful for modeling $CO_2$ uptake and sinks in the oceans, e.g., GFDL (Sarmiento et al., 1992), LODYC (Orr, 1993), HAMOCC-3 (Maier-Reeimer and Hasselmann, 1987).

Compartment models of the terrestrial biosphere, which include temperature feedbacks and land use change, at different hierarchical geographic scales have evolved from the 1980s forward (Harvey, 2005). Ecosystem models of carbon dynamic in terrestrial systems have addressed carbon storage, $CO_2$ fertilization (Terrestrial Ecosystem Model, Melillo et al., 1993), temperature feedbacks (IMAGE Model, Rotmans and den Elzen, 1993), nitrogen feedback (Century Model, Schimel et al., 1994), and response to climate change (Smith and Shugart, 1993). Dynamics were incorporated into these models to predict the rates of exchange between compartments by quantifying the fluxes with time-dependent differential equations describing the physical/chemical/biological processes governing the fluxes. Globally averaged carbon cycle models have been used to project future atmospheric levels (Trabalka and Reichle, 1986), for several decades. The utility of models in distilling our knowledge of the global carbon cycle and simulating future atmospheric $CO_2$ levels from anthropogenic emissions, as well as response from proposed mitigation responses, is nowhere better seen than in the reports of the Intergovernmental Panel on Climate Change (IPCC).

The rapid development of simulation models led to model comparison exercises to evaluate robustness and accuracy (Enting et al., 1994). By 1999, international workshops were comparing estimates of NPP by a number of different computer simulation models (Kicklighter et al., 1999); the Climate and Earth System Modeling Centers around the world recently prepared their state-of-the-art models to participate in the sixth phase of the Coupled Model Intercomparison Project in 2018 at the Fall 2018 American Geophysical Union (AGU) Meeting (CMIP5, 2018). Such models are the means by which the global fluxes are obtained for global carbon cycle models, such as that depicted in Figs. 10.4 and 11.9.

In all modeling efforts, one is faced with the issue of uncertainties in model outcomes due to variations in input parameters, and because of different levels of structural complexity and how transfer processes are treated in the different models. In some cases, the uncertainties are simply expressed ± the range of the projections of the suit of models; in other cases, some sophisticated uncertain analyses are performed on model parameters (Gardner, R. H. and J. R. Trabalka, 1985); sometimes, the uncertainties are the upper and lower plausible outcomes or value of parameters (i.e., forest clearing can be no more or less than the actual geographical coverage); in other cases the mathematical function to fit observed data can be constrained by least squares regression (MathWorks, 2019).

Constraints can also come from better knowledge of biochemical processes and better measurements of parameters. For example, precise measurements of atmospheric $O_2$ concentrations can provide constraints on several important components of the global carbon cycle. Seasonal variations in physical and biotic driven concentrations of $O_2$ in seawater can be used to constrain net rates of carbon fixation in photosynthesis by marine biota. Interannual variations in atmospheric $O_2$ concentrations, mainly driven by uptake in fossil fuel combustion and exchanges with terrestrial biotas, can be used to partition the net biospheric uptake of anthropogenic emissions into oceanic and terrestrial components Keeling and Severinghaus, 2000).

**Known unknowns in the carbon cycle**. On geologic time scales, the carbon cycle acts to stabilize the gaseous composition of the atmosphere and climate, maintaining oxygen and carbon dioxide levels suitable for life on Earth (see Gaia Hypothesis, Chapter 2.4). The slowest changes in the global carbon cycle are mediated by the "weathering $CO_2$ thermostat," which provides a stabilizing factor which maintains the climate at a "set point," which itself has changed through geologic time as tectonic plates cause continents to move and mountains to rise. Throughout the past 800,000 years for which ice core data exist, there has been no discernible atmospheric $CO_2$ trend, and a rather remarkable stability, that is, until the present-day human-infusion of $CO_2$ into the atmosphere and the onset of global warming. As water temperature increases, its ability to dissolve $CO_2$ decreases. Global warming is anticipated to reduce the ocean's ability to absorb $CO_2$, leaving proportionally more in the atmosphere which would lead to even higher temperatures. This is known as positive feedback on the carbon cycle-climate system. Such feedbacks could lead to unprecedented consequences, and illustrate why we must understand the global carbon cycle and climate systems better.

Our knowledge of the global biogeochemical cycle of carbon and the biosphere's carbon metabolism is extensive, and our ability to project the consequences of increased concentrations of atmospheric $CO_2$ is profound. But there is always room for improvement. The limitations of carbon cycle models, which scale from local to regional carbon fluxes, have identified inadequacies in empirical studies that provide the data for model parameters.

Key questions which could improve resolution of models are (in no particular order of priority):

**(1)** How well do eddy-flux measurements at the stand-level represent regional-scale processes? This may be related to specific management practices (age, plantation, fertilization, etc.) or simple bias in choosing representative sites?

**(2)** How accurately do we understand the carbon dynamics of tropical forests comprising 50% of the terrestrial biosphere?

**(3)** Will an increase in global temperature lead to a larger increase in respiration (primarily microbial) relative to photosynthesis and contribute a positive feedback to atmospheric $CO_2$ levels?

**(4)** What are the effects of deforestation, reforestation, and/or conversion to croplands on carbon storage in soils?

**(5)** Given the projected increases in global mean temperatures, what will the $CO_2$ feedbacks be from oxidative metabolism of carbon stored in peatlands?

**(6)** Is the atmospheric flux to terrestrial GPP overestimated because of below canopy recycling of heterotrophic respiratory $CO_2$?

**(7)** What is the ocean's future capacity to sequester carbon from increased atmospheric $CO_2$, and at what rates?

**(8)** Will global warming alter the circulation of ocean currents, and what effect might this have on ocean-atmosphere $CO_2$ fluxes?

**(9)** To what extent will increased levels of atmospheric $CO_2$ stimulate photosynthesis in the terrestrial biosphere and increase NPP?

**(10)** What is the maintenance energy (and carbon) requirement of vegetation and has it been properly accounted for in GPP estimates?

**(11)** Should ocean circulation change, will the limiting nutrient constraints in regional oceans change and affect marine GPP?

**(12)** How will global change affect pathways and magnitudes of fluxes of carbon stored in nonreactive reservoirs?

**(13)** How does carbon interact will other rate-limiting elements in the biosphere's biogeochemistry?

**(14)** Have we missed any significant changing climate-induced feedbacks on the global carbon cycle?

"There are known knowns. These are things we know that we know. There are known unknowns. That is to say, there are things that we know we don't know. But there are also unknown unknowns. These are things we don't know we don't know." (BrainyMedia Inc., 2019). And this is always the challenge facing science.

## 10.8 Recommended reading

Archer, D., 2010. The Global Carbon Cycle. Princeton University Press, Princeton, NJ, 205 pp. https://press.princeton.edu/books/ebook/9781400837076/the-global-carbon-cycle.

Bolin, B., 1970. The carbon cycle. Scientific American 23, 124−132. https://doi.org/10.1038/scientificamerican0970-124.

Defries, R.S., Bounoua, L., Collatz, G.J., 2002. Human modification of the landscape and surface climate in the next fifty years. Global Change Biology 8 (5), 438−458. https://doi.org/10.1046/j.1365-2486.2002.00483.x.

Gates, D.M., 1971. The flow of energy in the biosphere. Scientific American 224 (3), 88−103. https://www.jstor.org/stable/24923119.

IPCC, 2014. In: Pachauri, R.K., Meyer, L.A. (Eds.), Climate Change 2014: Synthesis Report. Contribution of Working Groups I, II and III to the Fifth Assessment Report of the Intergovernmental Panel on Climate Change. Core Writing Team. IPCC, Geneva, Switzerland, 151 pp. https://www.ipcc.ch/report/ar5/syr/.

Schleisinger, W.H., 1997. Biogeochemistry: An Analysis of Global Change. Academic Press, New York, 588 pp. https://sites.duke.edu/biogeochemistry2015/files/2015/08/3rd-edition-BGC-through-Ch-2.pdf.

Trabalka, J.R. (Ed.), 1985. Atmospheric Carbon Dioxide and the Global Carbon Cycle, 316 pp. DOE/ER-0239, Office of Technical and Scientific Information, Oak Ridge, TN, U.S. Dept. of Energy DOE/ER-0239, 315 pp. https://www.osti.gov/servlets/purl/6048470/.

Wigley, T.M.I., Schimel, D.S., 2005. The Carbon Cycle. Cambridge Univ. Press, Cambridge, 292 pp. http://catdir.loc.gov/catdir/samples/cam031/00023735.pdf.

# Chapter 11

# Anthropogenic alterations to the global carbon cycle and climate change

The Global Carbon Cycle and Climate Change. https://doi.org/10.1016/B978-0-12-820244-9.00011-1

The increased energy needs due to human population growth, and the increasing industrialization, have resulted in the release of carbon from long-term storage in some of the natural reservoirs of the Earth. Conversion of large areas of forest and other natural ecosystems to agriculture to feed the growing population has been a source of additional $CO_2$. Due to the greenhouse effect, the increasing atmospheric $CO_2$ concentrations have driven up worldwide temperatures. While the possibility of global warming due to atmospheric $CO_2$ increases from fossil fuels was hypothesized as early as 1938 (Callendar, 1938), and predicted based upon scientific evidence decades ago (Broecker, 1975), it is now a reality. Rising atmospheric $CO_2$ concentrations and the resulting environmental impacts from global warming have now become very important, because they impact directly upon human society. Since global warming is a result of anthropogenic activity, we enter a complex realm of human behavior influenced by personal welfare, health, economics, social values, and political policies. Debate will continue over the factors affecting climate change and its consequences. Much is at stake. In the ensuing debate, the methods by which data are collected, analyzed, and interpreted should be known, and the underlying scientific principles made clear. These major issues should be addressed with facts, not ideology.

Reliable sources of information are readily available on the internet. The following websites, supported by government agencies, offer documented, verifiable, peer-reviewed scientific data (Table 11.1). There are many other reliable sources sponsored by professional scientific societies and national academies of

**TABLE 11.1 Internet sources of data relative to the issue of climate change.**

Intergovernmental Panel on Climate Change (IPCC)—all the data used from the scientific community in international climate change assessments: https://www.ipcc.ch/

U.S. Global Change Research Program, /https://www.globalchange.gov/nca4/

U.S. Department of Energy (DOE), Carbon dioxide Information & Analysis Center (CDIAC)—data on all aspects of the global carbon cycle and climate models:/https://cdiac.ess-dive.lbl.gov/GCP/

U.S. Department of Energy (DOE), Energy Information Administration (EIA)—energy use and emission statistics:/https://knoema.com/atlas/sources/EIA/

U.S. National Oceanographic and Atmospheric Administration (NOAA), National Centers for Environmental Information—ocean and atmospheric data on carbon and environmental parameters:/https://www.nodc.noaa.gov/

U.S. National Aeronautics and Space Administration (NASA)—remotely sensed data (land/water/atmosphere) on the biosphere:/https://climate.nasa.gov/

U.S. National Science Foundation, National Center for Atmospheric Research (NCAR)—global atmospheric data:/https://www.eol.ucar.edu/data-software/

United nNations fFood and aAgricultural Agency (FAO)—agriculture, forestry, and land use statistics:/http://www.fao.org/faostat/en/#home/

United Nations, United Nations Statistics Division - Energy Statistics—energy statistics end use:/https://unstats.un.org/unsd/energy/

U.S. Geological Service—environmental resources and climate change:/https://www.usgs.gov/

World Resources Institute—renewable resources, climate mitigation, adaptation:/https://www.wri.org/our-work/topics/climate/

The Nature Conservancy— protecting biological diversity:/https://global.nature.org/our-global-solutions/climate/

science in many countries. I will leave the objectivity of other sites offered by industry and environmental advocacy groups to the reader's judgment.

## 11.1 Changing atmospheric concentrations of $CO_2$

The evolution of $CO_2$ in the Earth's atmosphere over the past 100 million years can be divided into three stages. During the first stage from $10^8$ years BP, until the end of the last Ice Age $10^4$ years BP, very large changes in the Earth's atmosphere occurred. From an initial concentration, perhaps as high as several thousand ppm during the Cretaceous, the atmospheric $CO_2$ levels fell gradually to much lower values (200−300 ppm) characteristic of the glacial−interglacial cycles of the past few million years. The $CO_2$ levels seem to have occurred in a regular, periodic fashion, cycling at the Ice Age frequency of $\sim 10^5$ years between a low value near 200 ppm during cold, glacial periods and a higher value near 270 ppm during warm, interglacial periods.

The second stage is the postglacial period of relatively stable atmospheric $CO_2$ concentrations (260−290 ppm), which lasted from 10,000 years BP until the beginning of the 19th century Variations in concentrations were minimal and human influences inconsequential.

The third stage, from 1800 BP until the present, is one where human impact has been measured in tree rings, ice cores, and the ocean. Initially, increasing concentrations of atmospheric $CO_2$ were caused by deforestation, and later during that century by increasing dependence upon fossil fuels as the Industrial Revolution began. By the end of the 20th century, human-derived sources of atmospheric $CO_2$ inputs had come to dominate all of the natural sources of $CO_2$.

Carbon dioxide levels in the atmosphere have been monitored for many years, routine measurements beginning with those of Keeling on Mona Loa, Hawaii, in 1958 (Fig. 11.1). Mauna Loa is often used as raeference base for rising carbon dioxide levels because it is the longest, continuous series of directly measured atmospheric $CO_2$ concentrations, and fortuitously tracks very closely the average global concentration.

The reason why Mauna Loa is acceptable as a proxy for global $CO_2$ levels is because $CO_2$ mixes well throughout the atmosphere. Consequently, the trend in Mauna Loa $CO_2$ (1.64 ppm per year) is statistically indistinguishable from the

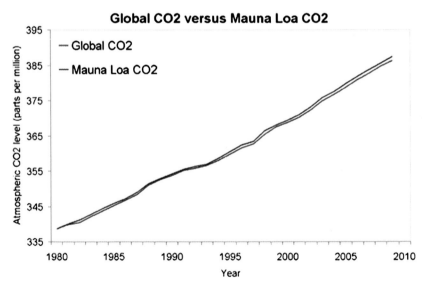

**FIGURE 11.1**   Global atmospheric $CO_2$ versus Mauna Loa $CO_2$. Measurements at Mauna Loa reflect the global average derived from many worldwide monitoring stations. *Image credit: John Cook @ Skeptical Science.*

trend in global $CO_2$ levels (1.66 ppm per year). Carbon dioxide levels are currently measured by hundreds of stations scattered across 66 countries which all report the same rising trend shown in Fig. 11.1, and these data have been shown to be continuous with ice core records of earlier atmospheric $CO_2$ levels (Fig. 11.2).

Atmospheric $CO_2$ concentrations have natural latitudinal, and seasonal variations, and global average values are the results of many measurements. Since the mid-1970s NOAA has made continuous measurements at four remote sites: Pt. Barrow, Alaska; Mauna Loa, Hawaii; American Samoa; and the South Pole. In addition to this long time series of continuous measurements, weekly whole-air flask samples have been collected at 20 cooperating global sites, with a with a major expansion to 15 sites beginning in 1979 that permitted the interpolation of the time series to create a $CO_2$ concentration surface representing a zonally averaged global concentration distribution. The global distribution of the flask sites of the NOA/GMCC (Geophysical Monitoring for Climatic Change) network permits a precise value for the global average atmospheric $CO_2$ concentration to be determined. A comprehensive review of methods and procedures is given by Gammon et al. (1985).

Satellite data (Fig. 11.3) from the Atmospheric Infrared Sounder (AIRS) on the NASA Aqua spacecraft are consistent with surface measurements and present a fuller picture of global $CO_2$ concentrations (NASA, 2008). Other, independent, collaborating measurements of historic atmospheric $CO_2$ values come from ice cores, tree rings, and ocean carbonate chemistry (see Section

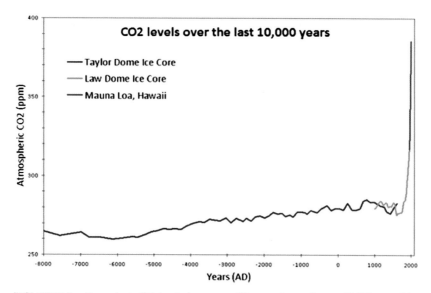

**FIGURE 11.2** Atmospheric $CO_2$ levels (parts per million, ppm) over the past 10,000 years. Blue line from Taylor Dome, Antarctica ice cores (Source: NOAA). Green line from Law Dome, Antarctica ice cores (Source: CDIAC). Red line from direct atmospheric measurements at Mauna Loa, Hawaii. *Image credit: John Cook @ Skeptical Science.*

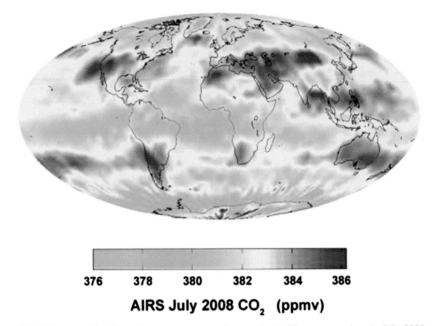

**AIRS July 2008 CO$_2$ (ppmv)**

**FIGURE 11.3** Global satellite measurements of atmospheric CO$_2$ concentrations in July 2008 from the NASA Atmospheric Infrared Sounder (AIRS) on the Aqua satellite. *Image credit: NASA, 2008.*

11.3). Online monitors now provide (https://www.esrl.noaa.gov/gmd/ccgg/trends) continuous reporting of atmospheric CO$_2$ levels, which in September 2018 stood at 408 ppm. Atmospheric levels above 400 ppm have not happened in nearly a million years.

## 11.2 The greenhouse effect

The greenhouse effect is the mechanism which maintains the temperature of our Earth. The Earth's atmosphere acts like the glass panes on a greenhouse by trapping some of the radiation received from the sun. Incident radiation energy penetrates the atmosphere and some is reflected back from the Earth's surface to the atmosphere in the form of infrared (IR) radiation. The Earth receives a large amount of radiant energy from the sun and about 30% of this energy is reflected. The longer wave IR radiation does not penetrate back through the atmosphere as efficiently as did the incident radiation. The reflected heat is absorbed by water vapor and carbon dioxide thereby warming the Earth. This process, which is known as the greenhouse effect (Fig. 11.4), occurs naturally and maintains the temperature of the Earth, making it suitable for life. The greenhouse effect is the main reason for the temperature difference between the Earth and the moon even though they are both situated nearly at the same

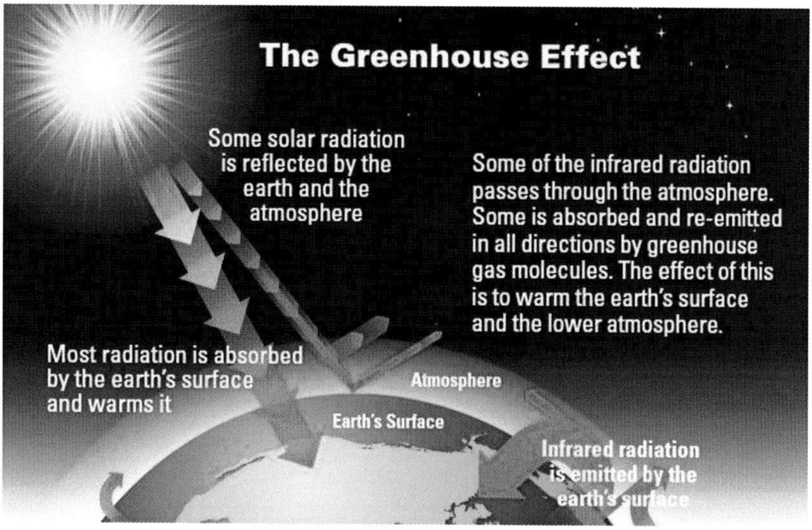

**FIGURE 11.4**    The Greenhouse Effect. *Image credit: U.S. Environmental Protection Agency, 2012.*

distance from the sun. The average temperature of the moon is around $-15°C$, while that of the Earth is around $+15°C$. This difference is due to the presence of an atmosphere on the Earth composed of gases such as oxygen, nitrogen, carbon dioxide, and water vapor. Water vapor is mainly responsible for this effect and is not a serious environmental concern since the amount of water in the atmosphere is a constant. But, the increasing levels of other greenhouse gases lead to increases in the world's temperature. Anthropogenic activities result in increased atmospheric levels of carbon dioxide ($CO_2$), methane ($CH_4$), ozone ($O_3$), sulfur dioxide ($SO_2$), chlorofluorocarbon (CFC), and nitrogen dioxide ($NO_2$), collectively known as greenhouse gases, causing the Earth to grow warmer and warmer. The consequences of the greenhouse effect are shown in the continuous records of global temperature and atmospheric $CO_2$ (Fig. 11.5).

## 11.3  Climate change

Climate, the biosphere, and the carbon cycle are interrelated in a number of ways. Terrestrial ecosystems are sensitive to changes in climate, having adapted to climatic, atmospheric, and soil conditions, such as temperature and precipitation, over long periods of time. They have limited ability to respond quickly to dramatic changes in these variables. Several excellent texts have appeared which discuss the physical basis of climate and climate change (Dressler, 2012; Farmer and Cook, 2013; Fletcher, 2013).

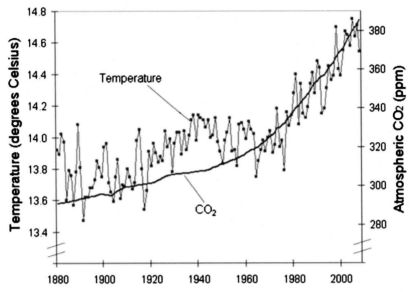

**FIGURE 11.5** Comparison of global temperature and atmospheric $CO_2$ concentrations from 1880 to 2010, with temperature deviations from historic norms. *Sources: NOAA/ERSL and NASA GISS, 2014.*

**Weather and climate** are two terms which are often confused. Weather is that which you experience when you go outside; it is the condition of the atmosphere at a particular place over a short period of time. Climate, on the other hand, refers to the weather patterns of a geographic location over long periods of time that are based upon statistical data. There are many elements that comprise weather and climate, notably temperature, atmospheric pressure, wind, solar irradiance, humidity, and precipitation. The climate of a locality is affected by its latitude, topography, and altitude as well as nearby water bodies and their currents. Climates are classified according to the average and the ranges of different variables, most commonly temperature and precipitation, which have the greatest influence upon the biosphere, and are the main variables analyzed by models projecting future climate changes. Some of the parameters considered are:

**Temperature**. Beyond the effects of temperature upon ecosystem metabolism through photosynthesis and respiration, and the resulting carbon flux, temperature controls the geographic distribution of ecosystems. The temperature influence upon biological systems can be classified according to the empirical effects of temperature upon vegetation defined as plant hardiness, or vegetation climate zones. The Köppen climate classification, which uses average monthly values of temperature and precipitation, was originally designed to identify the climates associated with certain biomes is an example. A 3°C change in mean annual temperature corresponds to a shift in isotherms of approximately 300−400 km in latitude (in the temperate zone) or 500 m in elevation. However, biological systems are influenced by more than simple average values.

While broad patterns of vegetative distribution (biomes) follow mean annual temperatures, more subtle temperature variables are also important such as: (1) The ability of tissues to withstand freezing temperatures which limits the latitudinal and altitudinal distribution of tropical vegetation, e.g., palms (Arecaceae), bromeliads (Bromeliaceae), avocado (*Persea*), and mango (*Mangifera*); (2) Length of the growing season determines biological productivity and time for reproduction, most evident in determining the latitudes where many agricultural crops may be grown, e.g., corn (*Zea mays*), wheat (*Triticum,* sp.); (3) The dates of last spring frost affect fruiting success in citrus (Rutaceae), peach relatives (*Prunus*), oak (*Quercus* sp.), and many other trees and plants; (4) Freezing ice storms can limit the northern distribution of some trees without the ability of tree limbs to droop and relieve the weight of ice on their limbs, e.g., loblolly (*Pinus taeda*) versus white pine (*Pinus strobus*).

In general, heat sums (degree days) determine the distribution of annual plants, while perennial vegetation is affected by both the heat sum of the growing season and the annual absolute minimum temperature. However, in northern climates, photoperiod control is critical for plants to avoid precocious development in the highly variable weather of early spring. The development of cool-season plants is limited by seeds triggered to germinate between 13 and 21°C, while warm-season species typically germinate between 15 and 29°C.

**Precipitation**. Water relations have an obvious impact on the success of terrestrial communities, demonstrated by the effect of drought on productivity during 2000−09 which reduced NPP by 0.55 Pg ($10^{15}$ g) carbon globally. Precipitation in the hydrologic cycle is the means by which water is returned to terrestrial environments to replenish soil moisture and deep aquifers. Soil moisture is necessary to plants for the uptake of nutrients and to maintain plant water balance necessary for photosynthesis. Not only is the annual precipitation amount a critical factor, along with temperature, it also determines the vegetative composition and distribution of the biomes. Yet, even the more-subtle pattern of precipitation during the growing season and throughout the year can be equally as important, e.g., (1) the seasonal soil moisture content must match with the plant's phenology and be adequate for plants' needs; (2) seedling germination and establishment requires sufficient soil moisture; (3) sharp variations in soil moisture content can affect nitrogen mineralization and population of microbial symbionts; and (4) moister habitats also favor trees with fleshy fruits, with advantages of dispersal by animal vectors, over wind, for seed dispersal.

## 11.4 Greenhouse gases

**Carbon dioxide**. Carbon, as an element, forms more compounds than any other element in the Periodic Table of elements, and is probably the most versatile of all the chemical elements. $CO_2$, consisting of one carbon and two oxygen atoms, is an important trace gas in the Earth's atmosphere. It is an integral part of the carbon cycle, a biogeochemical cycle in which carbon is exchanged between the Earth's oceans, soil, rocks and the biosphere. Carbon dioxide is a colorless, odorless, incombustible gas formed during respiration by plants and animals; obtained from coal, coke, or natural gas by combustion; generated from carbohydrates by fermentation; and by reaction of acid with limestone or other carbonates.

Worldwide carbon emissions from fossil fuel combustion and cement production have been steadily rising from $9.73 \times 10^{15}$ g C yr$^{-1}$ in 2010 (Fig. 11.6) to $9.90 \times 10^{15}$ g C yr$^{-1}$ in 2015 (Boden et al., 2017).

The increase in atmospheric emissions of $CO_2$ as reported in the IPCC AR-4 has continued (Le Quéré et al., 2009). The average growth rate of the atmospheric $CO_2$ concentrations from 2000 to 2008 was observed at 1.93 ppm yr$^{-1}$ (4.1 Pg C yr$^{-1}$), compared to 1.60 and 1.46 ppm yr$^{-1}$ in the 1980s and 1990s. Growth rates in 2008 and 2009 were at 1.79 and 1.86 ppm yr$^{-1}$, respectively, while in 2007 the growth rate was 2.13 ppm yr$^{-1}$. The relative decline in 2008 and 2009 can be attributed partly to the impact of the financial crisis on fossil fuel use and associated emissions. The growth rate in the period 2000–2008 was 21% higher than that in the period 1980–2000 (Dolman et al., 2010). The International Energy Agency Energy reported that carbon emissions grew by 1.7% in 2018 (3.1% in America) to an historic high of 33 billion tonnes ($8.9 \times 10^{15}$ g C). (The Economist, 2019). And, atmospheric carbon dioxide levels just hit 415 parts per million for the first time ever. There is now more $CO_2$ in the atmosphere than 3 million years ago (Harvey, 2019).

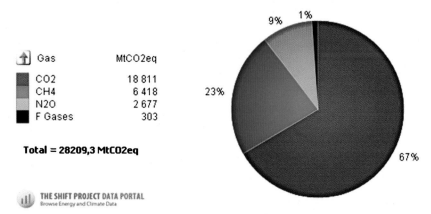

**World GHG Emissions from All Sectors in 1990 (MtCO2eq)**

| Gas | MtCO2eq |
| --- | --- |
| CO2 | 18 811 |
| CH4 | 6 418 |
| N2O | 2 677 |
| F Gases | 303 |

**Total = 28209,3 MtCO2eq**

THE SHIFT PROJECT DATA PORTAL
Browse Energy and Climate Data

**FIGURE 11.6** An estimate in 1990 of worldwide greenhouse gas emissions. Values are $10^{12}$ g CO$_2$-eq. *Source: The Shift Project, 2010, Courtesy, Creative Commons.*

**Methane.** Methane is a gas that occurs abundantly in nature and as a product of certain human activities. It is a hydrocarbon, $CH_4$, that is a product of biological decomposition of organic matter and of the carbonization of coal. Methane is ~25 times more potent than $CO_2$ over a 100-year period as a heat-trapping gas from infrared reradiation, trapping up to 72 times more radiation than $CO_2$ over 20-year period, but having a half-life (residence in the atmosphere) of only 7 years. The global pool of methane is $4.96 \times 10^{15}$ g $CH_4$; total annual methane emissions are about $535 \times 10^{12}$ g $CH_4$ $yr^{-1}$; and the net current increase in methane in the atmosphere is ~$30 \times 10^{12}$ g $CH_4$ (Schlesinger, 1997). The anthropogenic flux of methane is ~$2\times$ the natural flux (Table 11.2). The dominant natural source of methane to the atmosphere is from the anaerobic decomposition (see Chapter 5.3 Redox potential) in wetlands at $110 \times 10^{12}$ g $CH_4$ $yr^{-1}$ (Matthews and Fung, 1987). Termites and

**TABLE 11.2** Methane sources and sinks, both natural and anthropogenic (Schlesinger, 1997; after Prather et al., 1995). Units are $10^{12}$ g $CH_4$ $yr^{-1}$.

| Sources | Range | Likely |
|---|---|---|
| **Natural** | | |
| Wetlands | | |
| Tropics | 30–80 | 65 |
| Northern latitudes | 20–60 | 40 |
| Other | 5–15 | 10 |
| Termites | 10–50 | 10 |
| Ocean | 5–50 | 20 |
| Freshwater | 1–25 | 5 |
| Geological | 5–15 | 10 |
| TOTAL natural | | 160 |
| **Anthropogenic** | | |
| Fossil-fuel related | | |
| Coal mines | 15–45 | 30 |
| Natural gas | 25–50 | 40 |
| Petroleum industry | 5–30 | 15 |
| Coal combustion | 5–30 | 15 |
| Waste management systems | | |
| Landfills | 20–70 | 40 |
| Animal waste | 20–30 | 25 |

*Continued*

**TABLE 11.2** Methane sources and sinks, both natural and anthropogenic (Schlesinger, 1997; after Prather et al., 1995). Units are $10^{12}$ g $CH_4$ yr$^{-1}$.—cont'd

| Sources | Range | Likely |
|---|---|---|
| Domestic sewage treatment | 15–80 | 25 |
| Enteric fermentation (cattle) | 65–100 | 85 |
| Biomass burning | 20–80 | 40 |
| Rice paddies | 20–100 | 60 |
| TOTAL anthropogenic | | 375 |
| TOTAL all sources | | 535 |
| **Sinks** | | |
| Reaction with water | 350–560 | 445 |
| Removal in stratosphere | 25–55 | 40 |
| Removal by soils | 15–45 | 30 |
| TOTAL Sinks | | 515 |
| **Atmospheric increase** | 30–35 | 30 |

grazing animals maintain anaerobic populations of microbes in their digestive tract for fermentative digestion. Grazing animals contribute $85 \times 10^{12}$ g $CH_4$ yr$^{-1}$; humans contribute $1 \times 10^{12}$ g $CH_4$ yr$^{-1}$. The worldwide burning of forests and savannas contribute from $4 \times 10^{13}$ g $CH_4$ yr$^{-1}$ (Prather et al., 1995) to $\geq 9 \times 10^{14}$ g $CH_4$ yr$^{-1}$ (Levine et al., 1993). The mining of coal and natural gas contributes between 13% and 20% of the total methane emissions to the atmosphere.

**Other greenhouse gases**. Nitrous oxide ($N_2O$) is emitted during agricultural and industrial activities, as well as during the combustion of fossil fuels and solid waste, and has a greenhouse potency 300× that of $CO_2$ having an atmospheric residence time of 120 years. Fluorinated gases such as hydrofluorocarbons, perfluorocarbons, sulfur hexafluoride, and nitrogen trifluoride are powerful synthetic greenhouse gases that are emitted from a variety of industrial processes. Fluorinated gases are sometimes used as substitutes for stratospheric ozone-depleting substances (e.g., chlorofluorocarbons, hydrochlorofluorocarbons, and halons). These gases are typically emitted in smaller quantities, but they are potent greenhouse gases and have high global warming potential. Collectively these gases, notably $N_2O$ (not to be confused with nitrogen dioxide, $NO_2$) and hydrocarbons are emitted to the atmosphere in quantities to contribute $\sim 25\%$ of the greenhouse warming forcing equal to that of methane, with carbon dioxide contributing 50%. Water vapor also acts as a greenhouse gas, but it is a natural and stable component of the atmosphere. It is important not to focus solely on $CO_2$ and forget these other greenhouse gases.

## 11.5 Anthropogenic contributions to atmospheric $CO_2$

Over the past 150 years, atmospheric $CO_2$ concentrations have risen from $\sim 270$ ppm to over 410 ppm today. This increase is attributed largely to human activity, i.e., the combustion of fossil fuels, landscape clearing, and cement manufacture. How do we know this with certainty? One way we know is from historical records, e.g., FAO land use and UN fossil fuel combustion records (Marland et al., 2006). Historically, fossil fuel combustion and land clearing have contributed over 500 billion metric tons of carbon to the atmosphere, enough to have raised atmospheric levels to $\sim 500$ ppm, if some $CO_2$ had not been sequestered by the biosphere. While the oceans and terrestrial biosphere have had the capacity to absorb some of this excess $CO_2$ that mankind has produced, it is primarily because we produce $CO_2$ faster than it can be absorbed by the biosphere that the atmospheric levels have increased.

**Combustion of fossil fuels.** By using the isotopic ratio of $^{12}C/^{13}C$ and the $^{14}CO_2$ signature of fossil fuels, we can identify the burning of carbon-based fossil fuels, like coal, oil and natural gas, as being primarily responsible for the contemporary atmospheric $CO_2$ increase. Carbon has a unique footprint which allows scientists to determine whether the burning of fossil fuel contributed to the atmospheric carbon dioxide increases over the past 150 years. Carbon has three isotopes which are carbon-12, carbon-13, and carbon-14. Carbon in the atmosphere consists mainly of $^{12}C$ ($\sim 99\%$) and $^{13}C$ ($\sim 1\%$). A small amount of the carbon atom is in the form of the radioactive isotope $^{14}C$ ($10^{-7}\%$). In the upper atmosphere cosmic rays from the sun react with nitrogen in the atmosphere to produce $^{14}C$. Carbon-14 is unstable and over time is converted by radioactive decay back to nitrogen (5700 year half-life). After 60,000 years there is no $^{14}C$ remaining in the original sample, because it has been completely converted back to nitrogen.

Fossil fuel reservoirs are composed of coal, oil, or natural gas and over time these reservoirs have been buried deep in the ocean floor or underground. The carbon atoms found in both the atmosphere and initially in fossil fuels contain all three carbon isotopes ($^{12}C$, $^{13}C$, and $^{14}C$). After 60,000 years fossil fuel contains only radioactive $^{12}C$ (all of the $^{14}C$ has been converted to nitrogen), but the atmosphere still maintains a mixture of the three isotopes. Since it takes millions of years to create fossil fuel, the carbon dioxide that is released into the atmosphere from the burning of fossil fuel would contain no $^{14}C$. If the burning of carbon-based fossil releases carbon dioxide into the atmosphere, the amount of $^{14}C$ isotope found in atmospheric carbon dioxide should decrease over time. Measurements of the isotopic composition of atmospheric carbon dioxide do indeed demonstrate a steady decline of $^{14}C$.

Furthermore, over time the amount of $^{13}$C ($^{13}$C is a stable isotope) found in atmospheric carbon dioxide also has decreased. This is because fossil fuels also contain a much lower amount of $^{13}$C than does the atmosphere. Molecular kinetics affects $^{13}$C/$^{12}$C fractionation in photosynthesis due to a combination of: the difference in diffusion between air and stomata in plant leaves by 4.4%; the difference in kinetic constants for the reaction of $^{12}$CO$_2$ and $^{13}$CO$_2$ with ribulose biphosphate carboxylase-oxygenase (Rubisco) by 2.0%; and the thermodynamic difference in distribution between air/liquid phases by 0.4%. Because of this isotopic fractionation, the net result is that fossil fuels have a $^{13}$C/$^{12}$C ratio about 2% lower than that of the atmosphere.

Geochemists have developed time series of the historic concentrations of $^{13}$C and $^{14}$C in the atmosphere. One way they have done this has been to analyze values in tree rings. Trees lay down carbon compounds in the growth rings of their wood which can be radio-dated and give a signature of atmospheric levels at that time. Dating can go back thousands of years in ancient trees (Stuiver, 1986; Freyer, 1986) or in wooden artifacts in tombs. Gas bubbles trapped in layers of glaciers (Oeschger and Stauffer, 1986) yield similar results. Because the age of rings, or ice layers, is known exactly, a time series can be constructed of atmospheric $^{13}$C/$^{12}$C ratios through time. At no time in the past 100,000 years have the $^{13}$C/$^{12}$C ratios in the atmosphere been as low as they are today. Furthermore, the $^{13}$C/$^{12}$C ratios began to drop dramatically around 1850 AD—just at the time when the industrial revolution began and when CO$_2$ levels in the atmosphere began to increase. Just what would be expected if fossil fuels were the source of the increased atmospheric CO$_2$ (Fig. 11.7).

Other independent evidence comes from ocean chemistry. The absorption of CO$_2$ by the ocean can be traced by measurements of the $^{13}$C/$^{12}$C ratios in surface sea water which is also declining. Measurements of $^{13}$C/$^{12}$C ratios in coral and sponges, where their carbonate shells reflect historic ocean carbonate chemistry, similar to tree rings, show a similar time series as that for ice bubbles and tree rings, dropping about 0.15% since 1850.

Since fossil fuels are derived from ancient photosynthesis, and the $^{14}$C with a half-life of 5730 ± 40 years has decayed out, the fossil fuel contribution to rising atmospheric CO$_2$ content also can be calculated from the dilution of the atmospheric $^{14}$C content. The $^{14}$C content of the atmosphere has decreased approximately 0.034% yr$^{-1}$ during recent decades (Keeling et al., 1989); this 0.03% is 5× greater decrease than that observed in the atmosphere over the past 150 years (Stuiver et al., 1984). Clearly, the atmosphere's carbon isotopic composition is changing, and this change matches the isotope fingerprint of coal, oil, and natural gas. This demonstrates that the burning of fossil fuel is partly responsible for the current atmospheric carbon increase.

In 2012 the combustion of petroleum was responsible for 33% of the total CO$_2$ emissions generated by burning fossil fuels and producing cement; coal's share was 43%. Coal generates the most CO$_2$ emissions of any fossil fuel and is the world's dominant energy source.

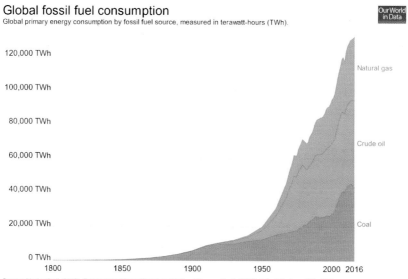

FIGURE 11.7   Fossil fuel consumption by the world. *Image credit: Smirl, 2017. Courtesy, OurWorldinData.*

In addition to the $CO_2$ emissions from the combustion of fossil fuels used to support a broad spectrum of industrial, transportation, and agricultural activities (Fig. 11.8), there are other notable sources of anthropogenic $CO_2$, notably land-use clearing and cement production. The U.S. Department of

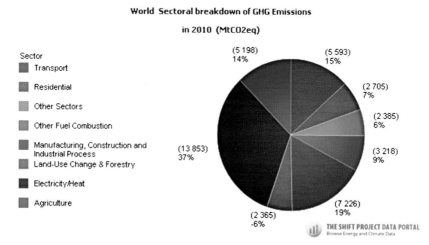

FIGURE 11.8   Breakdown of annual worldwide greenhouse gas emissions by industrial sector in 2010. Values are $10^{12}$ g $CO_2$-eq. *Source: The Shift Project, 2010. Courtesy, Creative Commons.*

Energy's Carbon Dioxide Information and Analysis Center (CDIAC) provides data on other $CO_2$ sources and trends in emissions.

**Land use and deforestation.** When agriculture replaces forests, the carbon stored in the trees and the organic matter in the soils are oxidized. Estimating these historical changes in the carbon content of terrestrial ecosystems has been an important part of establishing the Earth's carbon budget (Houghton, 1986). The average share of total $CO_2$ emissions caused by human activities during 2003−12 that were associated with deforestation and other land-use change was 8%. In fact, a new scientific publication reveals that the world loses 50 soccer fields worth of forest every minute, every day (CDIAC, 2013). Approximately 160 billion cumulative metric tons ($160 \times 10^{15}$ g C) of carbon were emitted into the atmosphere by land-use change (e.g., deforestation) between 1870 and 2013. Land use and climate changes also have significant effects on ecosystem services, which are essential and vital to human well-being (DeFries et al., 2002).

It is obvious that the metabolic flux of $CO_2$ between terrestrial ecosystems affects the seasonal levels of atmospheric $CO_2$; but, changes in the biomass of terrestrial ecosystems from changing land use can also have significant impact on atmospheric $CO_2$ levels. Some 129 million hectares of forest—an area almost equivalent in size to South Africa— have been lost since 1990 (FAO, 2015). Deforestation in tropical Africa and South America had the highest net annual loss of forests in the 2010−2015 time period, with 2.8 and 2 million hectares lost, respectively. But tropical deforestation has been to some limited extent balanced by regrowth elsewhere. The southeast United States has been a sink of about $0.07 \times 10^{15}$ g C $yr^{-1}$ during the 1960s (Delcourt and Harris, 1980). Globally, afforestation may sequester $0.7 \times 10^{15}$ g C $yr^{-1}$. Forests cover $4.1 \times 10^9$ ha of the Earth's land area. Globally, forest vegetation and soils contain $\sim 1146$ Pg ($10^{15}$ g) of carbon: 37% in low latitude forests, 14% in mid-latitudes, and 49% in high latitudes. Over two-thirds of the carbon in forest ecosystems is contained in soils and associated peat deposits. In the 1990s, deforestation in the low latitudes emitted $1.6 \pm 0.4 \times 10^{15}$ g C $yr^{-1}$; whereas forest expansion and growth in mid- and high-latitude forests sequestered $0.7 \pm 0.2 \times 10^{15}$ g C $yr^{-1}$ for a net flux to the atmosphere of $0.9 \pm 0.4 \times 10^{15}$ g C $yr^{-1}$ (Dixon et al., 1994). While concern about forest clearing has focused on the tropics, boreal forests are often overlooked, and could be a huge source of carbon if disturbed. About 300 billion tons ($3 \times 10^{17}$ g C) of carbon reside in the soils, plants, and wetlands of the boreal forest biome, equal to three decades of fossil fuel emissions (Suh, 2019).

**Cement production.** $CO_2$ emissions from cement production are 2.4% of global $CO_2$ emissions from industrial and energy sources (Marland et al., 1989). About 309 billion cumulative tons ($309 \times 10^{15}$ g C) of carbon were emitted into the atmosphere by burning fossil fuel and producing cement between 1870 and 2013. Approximately 7.7 billion tons ($7.7 \times 10^{15}$ g C) of carbon are now being emitted annually into the atmosphere by burning fossil fuels, plus that from producing cement of about 0.18 Gt. This is equivalent to the carbon emissions associated with more than 10,000 coal-fired power plants.

**Food production**. It is estimated that 14% or $1.40 \times 10^9$ g C yr$^{-1}$ (Fig. 11.7) of the anthropogenic $CO_2$ emissions to the atmosphere are contributed by the human food production system (and much higher if the herbicide and fertilizer production, harvest, and transportation to market are all included— showing the huge fossil energy requirement to support farming (Pimemtel, 2009) and prompting scientist David Pimentel to state in his classic net energy analyses of agricultural production that "potatoes are made of oil." Meat production has been growing by 3% per year, while raising livestock takes 80% of agricultural land and produces only 18% of the world's calories. FAO calculates that cattle are the world's fifth largest source of $CH_4$; in a vegan world in 2050, greenhouse gas emissions from agriculture would be 70% lower (*The Economist*, Oct. 13, 2018).

**Fire**. Fire is an important contributor of $CO_2$ to the atmosphere, both as a natural feature of the terrestrial landscape, and as result of human activities (e.g., land clearing and historic land management by indigenous peoples. Burning of residues from land use clearing, wildfires, burning of forests and grasslands, and burning of peat deposits are one of the most important disturbances on a global scale, and the second largest contributor of atmospheric $CO_2$, other trace gases, and aerosols. From 1997 to 2011, between $301-377 \times 10^6$ ha (mean $= 348$ ha) burned globally (Gigilio et al., 2013). Between 1997 and 2016, fire contributed 2.2 PgC yr$^{-1}$ to the atmosphere $(2.2 \times 10^{15}$ g C yr$^{-1})$, nearly one-fourth of the quantity from the combustion of fossil fuels (van de Werf et al., 2017). The magnitude of this atmospheric $CO_2$ source term is misleading, however, since a part of it $(5.4 \times 10^{13}$ g C yr$^{-1})$ is a component of the natural carbon cycle (Fig. 10.4), and subsequent $CO_2$ uptake in rapid vegetative regrowth counterbalances the fire emissions of $CO_2$. But fire also remains as an anthropogenic flux $(2.15 \times 10^{15}$ g C yr$^{-1}$ implied as the difference between Olson et al.'s (1985) natural flux and van de Werf at al.'s (2017) contemporary flux); it is a negative feedback from drought-induced climate warming (Li et al., 2007), and it is not dealt with well in global models of the carbon cycle.

**Firewood**. The $CO_2$ emissions from the burning of firewood is inconsequential in terms of the atmospheric flux of carbon. Use of wood on a commercial scale as a substitute for fossil carbon-based fuels has a negligible effect on the global carbon budget because: (1) is it almost carbon neutral, excepting the energy costs for harvesting and transportation which are not inconsequential, and (2) the land area required to produce a sustained yield of biomass to heat a three bedroom home would require 8 acres of coppice to produce 8 metric tons ($10^6$ g) per year (equivalent to $2.3 \times 10^6$ ha yr$^{-1}$). Land area required is economically competitive with other uses such as agricultural food production, unless marginal lands are used or agricultural waste byproducts are used or other biomass production schemes (e.g., algaculture) are used. Combustion of wood also produces 1.9 g $CO_2$ for each 1 g wood burned; 0.2 g $CO_2$-eq $NO_2$ per g wood; and 0.07 g $CO_2$-eq $CH_4$ per g wood (Crawford, 2019).

**Overall.** There was a 58% increase in $CO_2$ emissions from burning fossil fuels and producing cement between 1990 and 2012. An approximate cumulative 550 billion tons of carbon ($5.50 \times 10^{17}$ g C) were emitted into the atmosphere by all human activities between 1870 and 2013—meaning that fossil fuels and cement have contributed more than two-thirds of all net atmospheric carbon emissions (CDIAC, Global Carbon Project, 2013).

## 11.6 Where are the $CO_2$ emissions being generated?

In 2015 the world's carbon emissions were $9.90 \times 10^{15}$ g C worldwide (CDIAC, Global Carbon Project, 2013). This value continuously changes, according to country, population, and year, so that quantitative values at any point in time serve to provide relative comparisons among sectors and nations. Summarizing $CO_2$ emissions in different categories (e.g., total, per capita, by sector) serve to provide insight into the sources of anthropogenic $CO_2$ emissions and possible solutions to controlling or lowering emissions.

The Global Carbon Project is conducted in partnership between the International Geosphere-Biosphere Program, the International Human Dimensions Program on Global Environmental Change, the World Climate Research Program, and Diversitas, and all datasets and modeling output are described in the peer-reviewed literature. The Global Carbon Project's (CDIAC) 2013 report found that at the precise time emissions reductions are needed most, carbon dioxide ($CO_2$) emissions from burning fossil fuels and producing cement have reached their highest level in human history. Some of the headline numbers from the CDIAC summary include:

- The per capita carbon emissions for 2012 in India were $0.5 \times 10^6$ g C. India was responsible for 6% of global fossil fuel $CO_2$ emissions in 2012, and its emissions increased 7.7% compared to 2011.
- Per capita carbon emissions were $1.9 \times 10^6$ g C for 2012 in China. China was responsible for 27% of global fossil fuel $CO_2$ emissions in 2012, and emissions increased 5.9% compared to 2011.
- The per capita carbon emissions for 2012 in the European Union were $1.9 \times 10^6$ g C. The EU was responsible for 10% of global fossil fuel $CO_2$ emissions in 2012, and emissions decreased 1.3% compared to 2011.
- Per capita carbon $CO_2$ emissions for 2012 in the United States were $4.4 \times 10^6$ g C. The United States was responsible for 14% of global fossil fuel $CO_2$ emissions in 2012, and emissions decreased 3.7% compared to 2011.

- Growth rates for major emitter countries in 2012 were 5.9% (China), $-3.7\%$ (USA), $-1.3\%$ (EU), and 7.7% (India). The 2012 carbon dioxide emissions breakdown is coal (43%), oil (33%), gas (18%), cement (5.3%), and gas flaring (0.6%). According to data from the Global Carbon Project, $CO_2$ emissions were on track to rise by 2.7% to 37.1 billion metric tons (37.1 $\times$ $10^{15}$ g C) in 2018.

Now for the first time, the $CO_2$ emissions of 50,000 power plants worldwide, the globe's most concentrated source of greenhouse gases, have been compiled into a massive new database, called CARMA—Carbon Monitoring for Action (Center for Global Development, 2007).

## 11.7 Carbon cycle model projections of future atmospheres

The global carbon cycle is constantly changing as flux and pools change, many through direct or indirect human influence. Independent measurements of carbon pools and fluxes are continuously being updated, and new pathways and feedbacks identified and quantified. The most reliable source of documented, validated, and peer-reviewed scientific information about the changing global carbon cycle can be found in the periodic reports of the Intergovernmental Panel on Climate Change (IPCC). The 5th Assessment Report (AR5) in 2018 of the IPCC is the fifth in a series of such reports.

Assessing the impact of anthropogenic $CO_2$ emissions on the global carbon cycle uses the reference base of the natural global carbon cycle, discussed in detail in Chapter 10 and diagrammed in Figs. 10.1 and 10.3. Having established the structure and dynamics of the natural carbon cycle, and the role of the biosphere, we will now concentrate on what new has been learned about the global carbon cycle and how is it affected by human activities and anthropogenic $CO_2$ and other GHG emissions (Fig. 11.9).

**Contemporary global carbon cycle**. Referred to by some as the Anthropocene, the contemporary era's carbon cycle is dominated by human influences. Anthropogenic fluxes and the global carbon cycle are constantly changing. The IPCC representation of the global carbon cycle at the beginning of the 21st century in Fig. 11.9 is from IPCC (2013) AR5 WG1 Fig. 6 and builds upon IPCC AR4 WG1 Fig. 7 (IPCC, 2007b) Fig. 11.9 is the best summary of the modern-day global carbon cycle, validated by inputs and reviews by the international scientific community, and incorporating the most recent knowledge of pool and fluxes. Annual fluxes and exchanges are in Gt C $yr^{-1}$, equivalent to petagrams C $yr^{-1}$ (Pg C $yr^{-1}$ or $10^{15}$ g C $yr^{-1}$). Quantities of carbon in reservoirs are given in Gt C. Preindustrial "natural" fluxes are shown in black, and the additional anthropogenic fluxes, averaged for the period 2000—2009, are shown in red. The red numbers in the reservoirs show the cumulative changes in carbon from 1750 due to anthropogenic activities; a

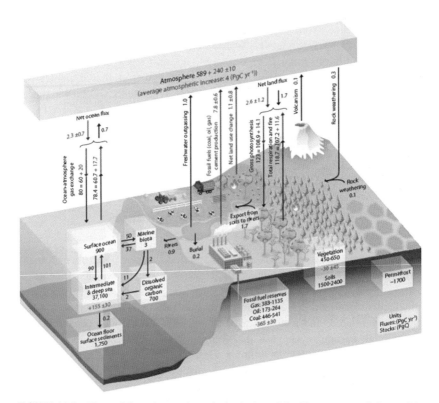

**FIGURE 11.9** The world's carbon cycle at the beginning of the 21st century as influenced by human activities, showing how carbon atoms flow between various reservoirs in the Earth system. Reservoir sizes are in Gt ($10^{15}$ g) C; fluxes are in Gt C $yr^{-1}$. The red numbers and arrows show the additional fluxes and reservoir changes caused by humans, such as the burning of fossil fuels and land use changes, averaged over 2000–2009. *Image credit: Figure 6.1 from Ciais, P., C. Sabine, G. Bala, L. Bopp, V. Brovkin, J. Canadell, A. Chhabra, R. DeFries, J. Galloway, M. Heimann, C. Jones, C. Le Quere, R. B. Myneni, S. Piao and P. Thornton, 2013: Source: IPCC, AR5 WG1, 2013.*

positive change indicates that the reservoir has gained carbon. What has changed in the modern-day global carbon cycle, and what does current scientific research tell us about these changes?

*Carbon in the atmosphere.* The preindustrial carbon content in the atmosphere, the size of the atmospheric carbon reservoir, historically had been about 589 Gt C; that value has increased by 240 Gt and now totals 829 Gt C. Annual increases of atmospheric $CO_2$ between 2000 and 2008 were 1.93 ppm $yr^{-1}$, or 4.1 Gt C $yr^{-1}$. Exchange fluxes with the atmosphere include the:

*Anthropogenic carbon flux.* The most significant difference between the pre-1750 "natural" and the contemporary carbon cycles is the increased $CO_2$ emissions from human activity. The IPCC estimate of fossil-fuel carbon emissions in 1990 was 6.4 Gt C $yr^{-1}$. In 2007, IPCC reported the increase in anthropogenic (fossil fuel + land use) $CO_2$ emissions to then be over 8.0 Gt C $yr^{-1}$ (8.0 Pg C or

$8.0 \times 10^{15}$ g C yr$^{-1}$), putting the atmospheric emission trajectory for $CO_2$ at the high end of the average of the family of high emission scenarios developed by IPCC in 2000 (Dolman et al., 2010). One year later in 2008, anthropogenic emissions were estimated by (Le Quéré et al., 2009) at $8.7 \pm 0.5$ Gt C yr$^{-1}$. This was a 36% greater emission rate by fossil fuels over that of 1990.

Fossil fuel emissions of carbon used by IPCC in Fig. 11.9 are 7.8 Gt yr$^{-1}$; land use change contributed another net $+1.1$ Gt C yr$^{-1}$ to the atmosphere (net is the difference between loss due to clearing and uptake due to regrowth). The total anthropogenic flux to the atmosphere in Fig. 11.9 is 8.9 Gt C yr$^{-1}$, including cement production of $\sim$0.18 Gt C yr$^{-1}$.

Current carbon emissions from the burning of fossil fuels are now over 2 Gt yr$^{-1}$ higher than the values used by IPCC in Fig. 11.9. The current total anthropogenic emission (fossil fuels, cement manufacture, and land clearing) is over $10 \times 10^{15}$ g C yr$^{-1}$; in 2018 energy-related emissions alone rose by 1.7%, or $151 \times 10^{12}$ g C, to $8.94 \times 10^{15}$ g C yr$^{-1}$ (IEA, 2019). (Note: land use emission estimates have proportionally higher uncertainty compared to fossil fuel emissions). As quickly as the global carbon cycle is updated with validated numbers it is out of date, because of the increasing levels of fossil fuel emissions! Fig.11.9 serves as an illustration of the contemporatry global carbon cycle, with values for natural and anthropogenic fluxes of carbon and the resulting reservoir changes, and shows the significance of the various components in the global biogeochemical cycle of carbon.

*Land clearing of vegetation.* Estimates of carbn fluxes land use change and vegetation clearing vary across recent decades and by different geographic regions. Figure 11.9 uses a value of $1.1 \pm 0.8$ Gt C as the annual land use flux to the atmosphere. [Note previous discussion of land clearing in section 11.5.].

*Natural weathering.* Natural weathering from soil formation and organic detrital leaching via ground water discharge by rivers to oceans eventually yields a net natural flux of 1 GT yr$^{-1}$ to the atmosphere.

*Volcanism.* Highly variable, volcanic activity has a long-term annual average of a net 0.1 Gt C yr$^{-1}$ flux to the atmosphere.

*Ocean exchange.* The net ocean flux of about 1.6 GT C yr$^{-1}$ from the atmosphere to the ocean is a balance of $-78.4$ Gt yr$^{-1}$ release and $+80$ Gt yr$^{-1}$ uptake, illustrating how important the ocean "sink" is in assimilating new fossil emissions of $CO^2$.

*Terrestrial exchange.* Similar to the ocean, terrestrial exchange is the result of a near balance between a $-118.7$ Gt C yr$^{-1}$ respiration flux to the atmosphere and a $+123.0$ Gt C yr$^{-1}$ uptake resulting from GPP, for a net annual uptake of 4.3 Gt C yr$^{-1}$ terrestrial uptake in the biogeochemical cycle.

*Reservoir changes.* The net terrestrial vegetation, soil, and detritus pool of 2500 Gt C lost about 30 $\pm$45 Gt C between 1750 and 2018. The intermediate and deep ocean pool of 37,100 Gt C gained 155 Gt C, and the atmospheric pool of 589 Gt C increased by 240 $\pm$10 Gt C.

*Overall.* some of the additional anthropogenic carbon is taken up by the land and the ocean (about 4.9 Gt C yr$^{-1}$), while the remainder is left in the atmosphere ($\sim$4 Gt C yr$^{-1}$), resulting in the rising atmospheric concentrations of $CO_2$. Gross fluxes generally have uncertainties of more than $\pm20\%$ but fractional amounts have been retained to achieve overall balance when including estimates in fractions of Gt C yr$^{-1}$ for riverine transport, weathering, and deep ocean burial.

**Cycle dynamics.** It is important to understand the dynamics of global carbon cycle when interpreting changing levels of $CO_2$ in the atmosphere. Considerable misunderstanding exists about the relative strength of the anthropogenic and natural fluxes of $CO_2$ with the atmosphere. Implication by some that the natural flux of $CO_2$ from the land and the ocean with the atmosphere vastly exceeds in magnitude the fossil fuel flux are both intentionally misleading and intellectually dishonest. It is the net flux, not the gross flux which is important! Refer to Fig. 11.9. The 7.8 Gt C yr$^{-1}$ fossil fuel + cement flux shown is a one-way, or **net** flux, to the atmosphere. It is not comparable to the total flux of the land plus the ocean to the atmosphere of 197.1 Gt C yr$^{-1}$(118.7 + 78.4 Gt C yr$^{-1}$), but rather should be compared with the **net** biospheric exchange of only +5.9 Gt C yr$^{-1}$. When compared to the smaller net biotic flux of 5.9 GT C yr$^{-1}$, it can be seen why the 7.8 Gt C yr$^{-1}$ fossil fuel net flux is a significant perturbation to the natural biogeochemical cycle of carbon. The biosphere is simultaneously withdrawing carbon from the atmosphere through photosynthesis and contributing through respiration. The greater $CO_2$ flux to the atmosphere from fossil fuels is a major perturbation to the global carbon cycle.

**Airborne fraction.** The airborne fraction refers to that amount of anthropogenic-derived $CO_2$ emissions which remains in the atmosphere after absorption by the biosphere. The airborne fraction is calculated using the rate of human $CO_2$ emissions and the changes in atmospheric $CO_2$ concentration. The global increase in atmospheric $CO_2$ has been directly measured since 1959 and can be calculated from ice cores for earlier periods. $CO_2$ emissions come primarily from fossil fuel combustion, with smaller contributions from land-use changes and cement production. Fossil fuel combustion data come from international energy statistics. Carbon dioxide emissions from land-use changes come from FAO and satellite data, fire records, and carbon models of biospheric metabolism (see Chapter 10).

Currently, the best estimates are that between 40% and 46% of $CO_2$ emissions stay in the atmosphere (depending on the time periods over which values are calculated) with the remainder being absorbed by sinks in the global carbon cycle (Fig. 11.9). A key question that will determine future atmospheric levels is whether the airborne fraction is changing. If the atmospheric fraction remains constant, or decreases with time, then the biosphere is capable of responding to increased releases by absorbing more $CO_2$. If the biosphere's capacity is fixed, then the atmospheric fraction should be increasing with time.

While both the land and the ocean continue to absorb carbon at an average rate of 4.8 Pg C yr$^{-1}$ (Fig. 11.7, IPCC, 2007b). Le Quéré et al. (2009) provide evidence for a long-term (50-year) increase in the airborne fraction of $CO_2$ emissions. This implies a decline in the efficiency of the $CO_2$ sinks of the land and/or ocean in absorbing anthropogenic emissions. The IPCC's conclusion about the airborne fraction was *"There is yet no statistically significant trend in the $CO_2$ growth rate since 1958 …. This 'airborne fraction' has shown little variation over this period"* (IPCC AR4, 2007). Since 1850 the airborne fraction has remained relatively constant at 43%, based upon direct $CO_2$ measurements from Mauna Loa and the South Pole, and with $CO_2$ data derived from Antarctic ice cores (Knorr, 2009). When $CO_2$ emissions were low, the amount of $CO_2$ absorbed by natural carbon sinks was correspondingly low. As human $CO_2$ emissions sharply increased in the 20th century, the amount absorbed by nature increased correspondingly. The long-term trend in the airborne fraction since 1850 was $0.7 \pm 1.4\%$ per decade, and this was not statistically significant. But, another recent study (Le Quéré et al., 2009) showed that from 1959 to 2008 an average of 43% of each year's $CO_2$ emissions remain in the atmosphere, although there was much year-to-year variability. The noise in the airborne fraction parameter was reduced by removing the variability associated with El Niño Southern Oscillation (ENSO) and volcanic activity. They found that the airborne fraction increased by $3 \pm 2\%$ per decade. This is a slightly increasing trend although only barely statistically significant. But, Hansen and Sato (2016) now report that the airborne fraction has been declining since 2000 (Fig. 11.10). Currently, the

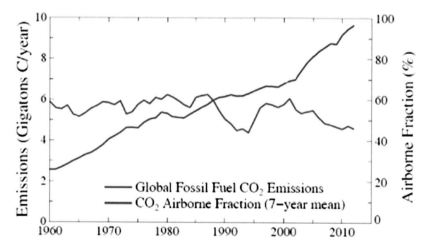

**FIGURE 11.10**  The airborne $CO_2$ fraction showing global carbon dioxide emissions (as gigatons of carbon without oxygen molecular weight added) from 1960 through 2012, and the amount of emitted $CO_2$ that has remained in the atmosphere. *Source: Hansen et al., 2013. Courtesy, Creative Commons.*

amount of carbon from $CO_2$ remaining in the atmosphere is in the range of 45% of the human emission— or around 5 gigatons.

**Methane**. After a decade of no growth in $CH_4$ concentration in the atmosphere, in 2007 and 2008 the concentration of methane in the atmosphere increased again. This is likely due to contributions from Northern Hemisphere wetlands (Rigby et al., 2008).

**Carbon dioxide uptake by terrestrial vegetation**. Contradictory results occur when one compares the regional scale estimates of biospheric $CO_2$ uptake, calculated from empirical measurements in forests and grasslands, with estimates inferred from atmospheric $CO_2$ measurements by Eddy covariance techniques. Process-based ecosystem models and ensembles of atmospheric inversions will be needed to resolve these differences.

It has been found that old-growth forests can continue to accumulate carbon, in contrast to the long-standing view that they were carbon neutral (Luyssaert et al., 2008). The significance of this finding is that it applies to over 30% of the global forest area that is still unmanaged primary forest. Half of the primary forests (600 million ha) are located in the boreal and temperate regions of the Northern Hemisphere. These forests alone sequester about $1.3 \pm 0.5$ Pg C $yr^{-1}$. These findings suggest that 15% of the global forest area, previously considered to be noncontributing to net uptake, provides at least 10% of the global net ecosystem productivity, NEP (Dolman et al., 2010).

**Terrestrial ecosystem fluxes**. Estimates of carbon fluxes from terrestrial ecosystems are still too sparse, and the ecosystems too heterogeneous, to allow accurate estimate of the net land flux (IPCC, 2007b). Large-scale biomass inventories have been limited to forests with commercial value, and inadequately address tropical forests. Direct flux measurements by eddy covariance exist only at point locations, are generally short-term measurements, and require extrapolation to obtain global estimates. Thus, two methods have been used by the International Panel on Climate Change (IPCC) to estimate the net global land-atmosphere flux: (1) deducing the value as a residual between the fossil fuel and cement emissions and the sum of ocean uptake and atmospheric increase, or (2) inferring the land-atmosphere flux simultaneously with the ocean sink by inverse analysis or mass balance computations using atmospheric $CO_2$ data, with terrestrial and marine processes distinguished using $O_2/N_2$ and/or $^{13}C$ measurements.

Method 2 was used in the IPCC's Third Assessment Report (TAR), based upon $O_2/N_2$ data (IPCC, 2007b), and corrections have now been made to the results to account for the effects of thermal $O_2$ fluxes by the ocean (IPCC, 2014; Le Quéré et al., 2003), resulting in a net land flux estimate of $-0.3 \pm 0.9$ Gt C $yr^{-1}$ during the 1980s. For the 1990s and afterward, Method 1 was adopted for estimating the ocean sink and the land-atmosphere flux. Unlike in the TAR, Method 1 is preferred for the 1990s and thereafter (i.e., estimating first ocean uptake, and then deducing the land net flux), because the ocean uptake is now more robustly determined by oceanographic research than by the atmospheric $O_2$ trends. The land-atmosphere flux has changed through time, from a small sink in the 1980s of $-0.3 \pm 0.9$ Gt C yr$^{-1}$ to a large sink during the 1990s of $-1.0 \pm 0.6$ Gt C yr$^{-1}$, and thereafter returned to an intermediate value of $0.9 \pm 0.6$ Gt C yr$^{-1}$ during the early 2000s.

**Land-use change and deforestation**. Because of human-induced land-use changes, the carbon pool in terrestrial ecosystems, and hence also its $CO_2$ flux, are not constant. Paramount among these changes is tropical deforestation. Although deforested lands are often replaced with some form of photosynthesizing agriculture, huge quantities of carbon stored in woody biomass are released to the atmosphere. For example, estimates for worldwide tropical forest clearing in 1982 (Houghton et al., 1983) ranged from $3.22 \times 10^6$ to $17.1 \times 10^6$ ha yr$^{-1}$, yielding a flux of carbon to the atmosphere from $1.2 \times 10^{15}$ to $4.2 \times 10^{15}$ g C yr$^{-1}$, of which the lower estimate would be roughly equivalent to 20% of that from the combustion of fossil fuels— clearly a very significant flux in the global carbon cycle. The rate of tropical forest clearing continues unabated. The Nature Conservancy reported in 2015 (TNC, 2015) that $2.3 \times 10^6$ km$^2$ ($2.3 \times 10^8$ ha) of tropical rainforest were cleared between 2000 and 2012— an average of $1.9 \times 10^7$ ha yr$^{-1}$. The global forest report of FAO (FAO, 2005) estimated that $\sim 1.0 \times 10^7$ ha yr$^{-1}$ of tropical rainforests were cleared between 1990 and 2005. About $3.3 \times 1015$ g C yr$^{-1}$ or about 6.5% of annual NPP is utilized in the direct harvest of food by humans. Additionally, 25%−40% is lost through wildfires, settlement by humans, and pollution (Vitousek et al., 1986).

**Residual land sink**. Forest regrowth and/or $CO_2$ fertilization of temperate and subtropical ecosystems was 3.5 gigatons C yr$^{-1}$ (one-half the fossil fuel flux) during 1992−93 (Ciais et al., 1995.) This carbon sink could also be due to climatic differences affecting respiration/photosynthesis balances. Furthermore, this may not be just a recent phenomenon, but may have been occurring for decades. When considering the issue of land-use change, global deforestation dominates over forest regrowth, and the net uptake of $CO_2$ by the terrestrial biosphere means that there must be an uptake by terrestrial ecosystems somewhere not yet identified. This missing land sink is referred to as the "residual land sink." Identifying the residual land sink depends upon the value of the land-use change flux, which has associated with it very large errors of uncertainty. Using the high value of the land-use source term of Houghton (2003), the residual land sink would be $-2.3$ ($-4.0$ to $-0.3$) and $-3.2$ ($-4.5$ to $-1.9$) Gt C yr$^{-1}$, respectively, for the 1980s and 1990s. DeFries et al. (2002) estimated a residual land sink of $-1.7$ ($-3.4$ to $0.2$) and $-2.6$ ($-4.3$ to $-0.9$) Gt C yr$^{-1}$ for the decades of the 1980s and 1990s, respectively. Although providing different estimates for the sink term, both sources agree that deforestation emissions were 0.2−0.3 Gt C yr$^{-1}$ higher in the 1990s than in the 1980s. To compensate for that increase, and to match the larger land-atmosphere uptake during the 1990s, the residual land sink must have increased by 1 Gt C $yr^{-1}$ between the 1980s and 1990s. But where is it?

Even though there are large areas of tropical forest deforestation and degradation, there remain large areas of undisturbed wilderness with minimal human impact. The net change in carbon biomass in these forests continues to be a large uncertainty. Mature, old-growth, tropical forests contain vast stores of carbon in biomass, about 46% of global C in biomass, and account for a large part of global terrestrial NPP. Recent studies of the carbon balance of mature tropical forests (Phillips et al., 1998; Baker et al., 2004) report accumulation of carbon at a rate of $0.7 \pm 0.2 \times 10^6$ g C $ha^{-1}$ $yr^{-1}$, equivalent to a net carbon uptake by global neotropical biomass of $0.6 \pm 0.3$ Gt C $yr^{-1}$. An intriguing possibility is that rising $CO_2$ levels could stimulate carbon uptake by accelerating photosynthesis. Atmospheric $CO_2$ has been rising at $\sim 1.5$ ppm (0.4%) $yr^{-1}$. If photosynthesis of tropical forests responded positively by +0.25%, or incrementally 2.5% over 10 years, it would be consistent with the reported rates of living biomass increase of 5%. Much research has yet to be done, but such changes in the carbon balance of tropical forests could have a significant effect on the global carbon budget— and could be at least a part of the missing residual land sink.

Due to the intricacies of ocean circulation, chemistry, and biology, credible estimates of $CO_2$ uptake on a global scale can only come from models. The relatively simple, one-dimensional (depth) box models have not taken up enough anthropogenic $CO_2$ to accurately account for known releases. Models of intermediate complexity, initial two-dimensional formulations, have taken up a little more $CO_2$. This provided impetus for development for more elaborate, three-dimensional models, which treat chemical and biological processes in a three-dimensional representation of ocean circulation. Such models have been better validated with isotopic tracers and incorporate important feedback effects of $CO_2$ uptake. Future challenges will be to incorporate enhanced carbon storage on the continental shelves ($0.2 \times 10^{15}$ to $0.8 \times 10^{15}$ g C $yr^{-1}$) stimulated by nutrients from river outflows, carbonate mineral reactions in sediments, changes in alkalinity, oceanic circulation, and upwelling. Uncertainties about the adequacy of the treatment of oceans in global carbon cycle models are responsible for a large portion of the uncertainties in predictions of how the biosphere will respond to both atmospheric $CO_2$ increases and climate changes over the period of the next 100–200 years. Less uncertainty in empirical data, better spatial and temporal resolution, and more tracer data will improve the utility of models.

**Climate effects on the $CO_2$-vegetation exchange.** The NOAA advanced Carbon Tracker assimilation scheme found that terrestrial uptake of $CO_2$ in the United States fell to 0.32 Pg C $yr^{-1}$ during the extensive drought of 2002, compared to an average uptake of 0.62 Pg C $yr^{-1}$ over the period 2001–2005

(Peters et al., 2007). This was also suggested by Ciais et al. (2005) who found a reduction in European uptake in the dry and hot summer of 2003 that offset four previous years. The response of carbon uptake and release to climate perturbations is nonlinear, and there is large potential for extreme climate events to significantly disturb the average annual sink behavior. Fire, as discussed earlier, is one of the main features lacking in global models, even though human deforestation and seasonal droughts are governed by El Niño. The strong nonlinear relation between droughts and fires, versus carbon emissions and deforestation, highlights a climate-carbon feedback that may lead to higher $CO_2$ concentrations if droughts become more frequent in the future (Li et al., 2007).

**$CO_2$ fertilization of photosynthetic fixation**. Since the initial IPCC Assessment Report, the possibility of a carbon cycle feedback has received much attention and modeling groups have been eager to incorporate this feedback into their models, but much uncertainty still exists as to the potential for both $CO_2$ fertilization on land, increased ocean uptake in higher $CO_2$ environments, and with carbon/nitrogen interactions. A vegetation model suggests that, while increased atmospheric temperature will diminish terrestrial net primary productivity, this reduction will be offset by photosynthetic stimulation from the associated increase in atmospheric $CO_2$ (Woodward and Smith, 2005). Empirical results show that a significant fertilization effect of over 10% upon photosynthesis is sustained only for a short period of time due to loss of nitrogen fertility (Norby, 2000). The significance of the $CO_2$ fertilization effect remains unresolved (see Section 11.9).

**Ocean $CO_2$ fluxes**. There is good agreement between inversions and other $CO_2$ flux estimates. This congruity is similar to progress made in narrowing the uncertainties between models and observations for oceanic sources and sinks for the decade of the 1990s and the early 2000s. Differences at the regional level are generally less than 0.1 Pg C yr$^{-1}$ (Dolman et al., 2010).

**Permafrost regions**. New estimates of the C stored deep in the permafrost have more than doubled the previous high-latitude inventory estimates of carbon stocks (Tarnocai et al., 2009). If validated, this would put the permafrost carbon mass at an equivalent of twice the atmospheric C pool. Thawing of permafrost, predicted to occur with global warming will expose organic C to microbial decomposition. Schuur et al., (2009) concluded that over decadal timescales the losses of $CO_2$ greatly exceed increased plant carbon uptake, and that this could make permafrost in a warmer world a large biosphere carbon source of the order of 1 Pg C yr$^{-1}$. The Arctic is now warming twice aa fast as the rest of the planet, and wildfires, unprecedented in the past 10,000 years, rage as temperatures peak at 8-10 °C warmer that average. Many Siberian and Akaskan fires are burning deeper into caron-dense peat soils nromally water-logged (The Economist,

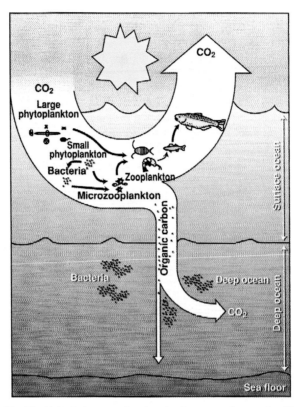

**FIGURE 11.11** The biological pump of carbon in the ocean. *Source: Reichle et al., 1999. Courtesy, Oak Ridge National Laboratory.*

2019). Peat fires produce more $CO_2$ and $CH_4$ than surface litter, illistrating yet another negative feedback in the global carbon cycle.

**The biological pump**. Illustrated in Fig. 11.11, this is a term used to describe the process whereby the ocean's biological removal of carbon from the atmosphere ultimately sequesters carbon in seafloor sediments. It is the part of the oceanic carbon cycle, responsible for the cycling of organic carbon molecules formed by phytoplankton during photosynthesis, as well as the cycling of calcium carbonate ($CaCO_3$) incorporated into the shells of organisms such as plankton and mollusks (the carbonate pump). In the "physical carbon pump," carbon compounds can be transported to different parts of the ocean in downwelling and upwelling currents.

**Future atmospheric $CO_2$ levels**. The gaseous envelope that surrounds the Earth, the atmosphere, consists almost entirely of nitrogen (78.1% by volume) and oxygen (20.9%), with a number of trace gases, such as argon (0.93%), helium, and radiating active greenhouse gases such as carbon dioxide

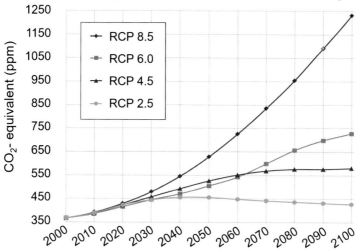

**FIGURE 11.12** Future atmospheric $CO_2$ levels as projected for the four RCP emission scenarios (IPCC SRES Report, 2007). All forcing agents' atmospheric $CO_2$-equivalent concentrations (in parts-per-million-by-volume (ppmv)) according to four RCPs. Image credit: Ilinri, Wikimedia contributors, 2019.

(0.035%) and ozone. In addition, the atmosphere contains water vapor which acts as a greenhouse gas, whose amounts are highly variable but typically around 1% by volume. The atmosphere also contains clouds and aerosols.

The IPCC, in its fifth Assessment Report (AR5) in 2014, adopted Representative Concentration Pathways (RCPs). An RCP is a projected greenhouse gas concentration, not an emissions trajectory. The AR5 supersedes Special Report on Emissions Scenarios (SRES) projections published in 2000. Four RCP pathways were selected for climate modeling and research, which describe different climate futures (Fig. 11.12), all of which are considered possible depending on the extent to which greenhouse gases are emitted in the years to come.

The four RCPs, namely RCP2.6, RCP4.5, RCP6, and RCP8.5, are labeled after the possible range of resulting radiative forcing values (Fig. 11.13) in the year 2100 ($+2.6$, $+4.5$, $+6.0$, and $+8.5$ W/m$^2$) relative to preindustrial values (Moss et al., 2008; Weyant et al., 2009). The RCPs are consistent with a wide range of possible changes in future anthropogenic (i.e., human) greenhouse gas (GHG) emissions, and are intended to represent their atmospheric concentrations. RCP 2.6 assumes that global annual GHG emissions (measured in $CO_2$-equivalents) peak between 2010 and 20, with emissions declining

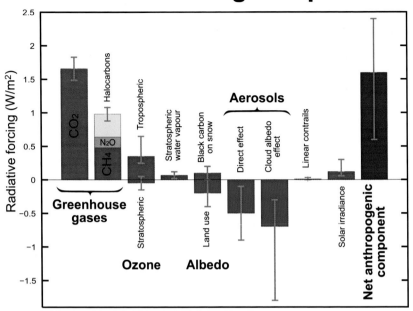

**FIGURE 11.13** Radiative-forcing components used by the IPCC in 2007 in the calculation of climate outcomes from four different representative concentration pathways (RCPs) dependent upon possible future levels of greenhouse gas emissions. Image credit: Leland Mcinnes, Wikipedia, 2019.

substantially thereafter. Emissions in RCP 4.5 peak around 2040, then decline. In RCP 6, emissions peak around 2080, then decline. In RCP8.5, emissions continue to rise throughout the 21st century (Fig. 11.12).

Over the past decade, the annual rates of atmospheric $CO_2$ increase have remained in the range of $\sim 2.2$ ppm yr$^{-1}$, driven by human carbon emissions of over nine billion metric tons ($10^{15}$ g) annually (Boden et al., 2017). In geological terms, this is an extremely rapid rate of increase. Dominated by fossil fuel burning, this massive discharge of carbon into the atmosphere is filling up the major pools in the world's carbon cycle. The oceans are becoming saturated with carbon—as seen in its rising rate of acidification. The oceans are warming, steadily losing their ability to keep a higher fraction of greenhouse gases stored in solution. The terrestrial vegetative drawdown of carbon from the atmosphere is being limited due to the combined effects deforestation and wildfires. The massive carbon store in Arctic peat may be beginning to vent higher volumes of greenhouse gases back into the atmosphere. And, meanwhile, the atmospheric concentration of $CO_2$ is now surpassing 410 ppm.

## 11.8 Climate changes and climate model projections for the future

In 2013, the Intergovernmental Panel on Climate Change (IPCC) Fifth Assessment Report concluded that, "It is extremely likely that human influence has been the dominant cause of the observed warming since the mid-20th century." The largest human influence has been the emission of greenhouse gases such as carbon dioxide, methane, and nitrous oxide. In view of the dominant role of human activity in causing it, the phenomenon is sometimes called "anthropogenic global warming" or "anthropogenic climate change" (IPCC, 2013). Climate model projections summarized in the IPCC report indicated that during the 21st century, the global surface temperature is likely to rise a further 0.3−1.7°C (0.5−3.1°F) to 2.6−4.8°C (4.7−8.6°F) depending on the rate of greenhouse gas emissions (IPCC, 2013). These findings have been recognized by the national science academies of the major industrialized nations, and are not disputed by any scientific body of national or international standing (Kirby, 2001; NASA, 2019).

**Climate models**. A climate model is a representation of the physical, chemical, and biological processes that affect the climate system. Although scientists modeling climate try to include as many processes as possible, simplifications of the real-world climate system are necessary because of limitations in both knowledge of and data on the climate system, and the available computer power necessary to run the models. Results from models can also vary due to different greenhouse gas inputs (Ehhalt et al., 2001; Meehl et al., 2005), feedbacks (Wigley et al., 2005), and the model's climate sensitivity. Climate sensitivity is the equilibrium temperature change in response to changes of the radiative forcing. Slow climate feedbacks, especially changes of ice sheet size and atmospheric $CO_2$, amplify the total Earth system sensitivity by an amount that depends on the timescale considered. The effects of clouds on the radiative balance are especially difficult to predict. Improving these models' representation of clouds is an important goal in current research. Another model improvement is better representation of the global carbon cycle.

The Fifth Assessment Report (AR5) of the United Nations Intergovernmental Panel on Climate Change (IPCC), the fifth in a series of such reports, was issued in 2014. SRES calculations of global warming (Table 11.3) under different scenarios differ in assumptions of world population growth and economic and political integration among nations; only the A1 scenarios consider choices of energy generation. Under the SRES A2 emissions scenario no actions are taken to reduce emissions and that regionally divided economic development occurs (Fig. 11.11).

An important and often misunderstood principle is that climate models do not assume that the climate will warm due to increasing levels of greenhouse gases. Rather, the models predict how greenhouse gases will interact with radiative transfer and other physical processes. Warming or cooling is thus a result and not an assumption of the models.

The accuracy of models is tested by examining their ability to simulate present-day and past climates. Climate models produce a good match to observations of global temperature changes over the last century, but they do not simulate all aspects of climate precisely. Arctic shrinkage has been faster than that predicted. Precipitation has actually increased proportionally to atmospheric humidity, and hence significantly faster than global climate models predict. Since 1990, sea level has risen considerably faster than models have predicted. The 2017 United States National Climate Assessment notes that "climate models may still be under-estimating or missing relevant feedback processes. Models can also be used to help investigate the causes of recent climate changes by comparing model predictions with observed changes due to natural and human causes. The models do not attribute the global warming that occurred in the decades between 1910 and 1945 to either natural variation or human effects, but they do indicate that the warming since 1970 is strongly affected by anthropogenic greenhouse gas emissions.

**The future carbon cycle**. Over periods of hundreds of thousands to millions of years the carbon cycle has regulated the Earth's climate. Even with plate tectonics shifting the Earth's continental photosynthetic geometry and with the venting gases from the core, the carbon cycle has performed as a negative feedback stabilizing atmospheric $CO_2$ levels. During the more recent period of glacial cycles, the carbon cycle behaved differently, acting as positive feedback and amplifying climate. Now in contemporary times, on a timescale of decades, the carbon cycle is again acting as a negative feedback and stabilizing climate by absorbing increased $CO_2$ emissions out of the atmosphere. The future carbon cycle will be driven by changes in atmospheric $CO_2$ directly, not by temperature changes resulting from changes in radiant energy as occurred during the Ice Ages. How will the carbon cycle behave in the future under simultaneous conditions of temperature change from climate warming and increased carbon into the carbon cycle from $CO_2$ emissions? We have already examined future prospects for $CO_2$ emissions in Section 11.5. Next, Section 11.9 will examine the projected impacts of climate change, which include important feedbacks that will impact the carbon cycle.

**Measurements of changing temperature**. There can be no doubt that mean global temperatures have been increasing. The U.S. National Oceanographic and Atmospheric Administration (NOAA) has recorded average temperature of the atmosphere to have increased by 0.83°C (1.5°F) since 1950 (Fig. 11.5). The U.S. National Aeronautics and Space Administration (NASA) recently documented the 10 hottest years globally—all but one falling within the last decade (Fig. 11.14). The Earth experienced its hottest June in records going back to 1880 — averaging 60.6°F, 1.7°F higher than the 20th Century average. Europe, Russia, Africa, Asia, and South America also experienced the hottest month in history (NOAA July 18, 2019, reported by Associated Press).

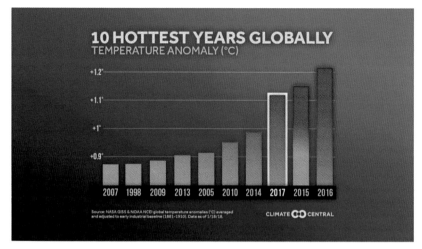

**FIGURE 11.14**    The 10 hottest years globally. *Source: NASA, 2019.*

The historic Paris Climate Agreement generated a request of the Intergovernmental Panel on Climate Change (IPCC) to prepare a Special Report (IPCC, 2018) on 1.5°C increase above preindustrial temperatures. Scientists and government representatives worked to assure that the summary for policymakers (SPM) accurately conveyed evidence presented in the report. The total increase between the average temperature of the 1850–1900 period and the 2003–2012 period is 0.78 (0.72–0.85) °C, according to the SPM for the IPCC fifth assessment report (AR5). The 2017 NOAA reported average global temperature across land and ocean surface areas was 0.84°C (1.51°F) above the 20th-century average of 13.9°C (57.0°F). The IPCC's recent reports show policymakers, business leaders, and energy system planners that under a carbon-intensive trajectory the world will, in about 30 years, entirely "burn through," i.e., exhaust the carbon budget remaining to stay below 1.5°C. The Global Carbon Project's aim in 2013 (CDIAC, 2013) was that the nations' collective carbon dioxide emissions put the world's temperature on track to rise 3.2–5.4°C above preindustrial levels by 2100.

## 11.9 The effects of climate change

**Distribution of ecosystems**. Changes in precipitation and temperature will lead to largescale adjustments in the distribution of vegetation and global NPP (Emanuel et al., 1985). Particularly significant could be changes in the distribution vegetation in the Northern Hemisphere — a slight shift of forests northward could increase rates of carbon storage, but agricultural regions could also shift northward across political boundaries (VEMAP, 1995), raising the specter of international conflicts.

**TABLE 11.3** Warming increases (°C) projected by the radiative forcing functions resulting from different assumptions of GHG emission scenarios (Fig. 11.12; *Table from: Wikimedia contributors, 2019*).

| Forcing function scenario | 2046–2065 mean and likely range°C | 2081–2100 mean and likely range°C |
|---|---|---|
| RCP2.6 | 1.0 (0.4–1.6) | 1.0 (0.3–1.7) |
| RCP4.5 | 1.4 (0.9–2.0) | 1.8 (1.1–2.6) |
| RCP6.0 | 1.3 (0.8–1.8) | 2.2 (1.4–3.1) |
| RCP8.5 | 2.0 (1.4–2.6) | 3.7 (2.6–4.8) |

Source: IPCC, 2013, Summary for Policymakers. In: Climate Change 2013.

**Temperature change and net primary productivity**. Changes in the vegetative biomass, or rates of productivity due to climatic changes or pollution, can also upset the biospheric $CO_2$ flux to the atmosphere. Model estimates (Table 11.4) of terrestrial global NPP (Lieth, 1975; Melillo et al., 1993), using temperature-light-moisture algorithms suggest potential global terrestrial NPP to be between 53.2 and $63.0 \times 10^{15}$ g C $yr^{-1}$. A similar approach using AVHRR arrives at $48 \times 10^{15}$ g C $yr^{-1}$ (Potter et al., 1993), the lower value possibly due to human disturbance.

**$CO_2$ fertilization**. The issue that rising atmospheric $CO_2$ levels could act as a "fertilizer" and stimulate photosynthesis in the biosphere, thereby reducing the airborne fraction of anthropogenic $CO_2$ emissions, remains inconclusive. There have been predictions that future $CO_2$ fertilization could cause future NPP to rise by $\sim 13\%$–$60 \times 10^{15}$ g C $yr^{-1}$ (Melillo et al., 1993). However, long-term field experiments have demonstrated that projections of stimulated terrestrial plant production from rising $CO_2$ concentrations remain tenuous. Coupled climate-carbon cycle models are sensitive to this possible, so-called

**TABLE 11.4** Estimated global NPP by terrestrial ecosystems.

| Net primary production | Source |
|---|---|
| $48.0 \times 10^{15}$ g C $yr^{-1}$ | Potter, 1993 |
| $48.6 \times 10^{15}$ g C $yr^{-1}$ | Whittaker and Likens, 1973 |
| $53.2 \times 10^{15}$ g C $yr^{-1}$ | Melillo et al., 1993 |
| $63.0 \times 10^{15}$ g C $yr^{-1}$ | Lieth, 1975 |

$CO_2$ fertilization-effect feedback on atmospheric $CO_2$, but model projections are uncertain because of the expectation that nutrient limitations through the nitrogen (N) cycle will reduce this $CO_2$ fertilization effect. Whether N limitation would cause a reduced stimulation of net primary productivity (NPP) under elevated atmospheric $CO_2$ concentrations was assessed over an 11-year study in a free-air $CO_2$ enrichment (FACE) experiment in a deciduous forest in Tennessee, USA (Norby et al., 2010). FACE technology is an important step forward in measuring the integrated response of an intact forest ecosystem (Norby et al., 1999). During the first 6 years of the experiment, NPP was significantly enhanced in forest plots exposed to 550 ppm $CO_2$ compared with NPP in plots under ambient $CO_2$ levels, and this was a consistent and sustained response. However, the enhancement of NPP under elevated $CO_2$ declined from 24% in 2001–2003 to 9% by 2008 due to nitrogen limitations. Global analyses that assume a sustained $CO_2$ fertilization effect are not supported by this FACE experiment. The amplitude of the atmospheric $CO_2$ cycle at Mauna Loa (Keeling et al., 1989) rose from a near constant 5.2 ppmv (peak to trough) during 1958 to the mid-1970s, rising to 5.8 ppmv for most of the 1980s and rising still, providing weak evidence for $CO_2$ fertilization (Schimel et al., 2005).

**Temperature changes—soil carbon**. The soil carbon pool will change as a result of global warming with increasing temperature stimulating microbial decomposition — and acting as a positive feedback to atmospheric $CO_2$ levels— nowhere more pronounced than in the permafrost regions where water levels become lower due to an increase in meltwater evaporation in warmer temperature regimes. However, in some situations increased temperature could stimulate tundra vegetation and result in net storage of $CO_2$ (Oechel et al., 1994). Since peatland and boreal soils contain 24% of the total carbon in soils of the world, if these soils are drained and warmed decomposition will increase and a massive negative of $CO_2$ to the atmosphere will occur.

**Precipitation changes**. Records for both wet and dry weather are being set around the world as evidence from study of rainfall data from 50,000 weather stations around the world in the Global Precipitation Climatology Center in Germany (Lehmann et al., 2018), with severe consequences for agricultural production and food security. Empirical data are supporting climate model projections. Wet regions are becoming wetter, and dry areas drier. The number of months with record-high rainfall increased in central and eastern United States by more than 25% between 1980 and 2013. Severe rainfall events and flooding are occurring more often in North America, northern Europe, and northern Asia. Hurricane losses in 2018 were $51 billion in the United States, over the long-term average of $34 billion. The number of dry months in sub-Saharan Africa increased by nearly 50%, one-third of which would not have occurred without the influence of climate change.

**Hydrological cycle**. Global warming is resulting in a melting of the polar ice caps with a resulting rise in sea water and coastal flooding. During the past 100 years (1880−1980) sea level has risen ∼ 80 mm, in some years averaging 1−2 mm $yr^{-1}$ (Gornitz et al., 1982).

Most climate models predict a more humid world to result from global warming in which water will be enhanced through evaporation, transpiration, precipitation, and surface runoff. Also expected will be grater variability in rainfall. Associated with these phenomena will be an increase in cloudiness. Most of the anticipated temperature increase will be confined to higher latitudes, and large areas of temperate North America and Asia may experience a reduction in soil moisture. Other areas of the Earth such as North Africa and the Middle East will experience extreme drought, while still others will receive increased rainfall. These climate changes could result in large-scale changes and shifts in terrestrial vegetation and net primary production (Emanuel et al., 1985). Water is the main factor, along with temperature, in controlling NPP in the terrestrial biosphere. Greater rainfall will mean increased runoff and river discharges to the oceans.

Scientists and Experts from Universities and Institutions in the Great Lakes Region (Environmental Law and Policy Center, 2019), in an assessment of the impacts of climate change on the Great Lakes, detail how climate change could affect and threaten public health, fish and wildlife, water quality, and the regional economy.

**Temperature change effects on agriculture**. Changes in temperature and precipitation will lead to large-scale shifts in distribution of terrestrial vegetation and global net primary production. Particularly sensitive will be agricultural regions where climate extremes will affect length of growing seasons and sustainability of crops (USGCRP, 2009; 2018a,b).

Higher $CO_2$ levels can affect crop yields. Some laboratory experiments suggest that elevated $CO_2$ levels can increase plant growth, although rising $CO_2$ also reduces the nutritional value of most food crops. Other factors, such as changing temperatures, ozone, and water and nutrient constraints, may counteract these potential increases in yield. For example, in 2010 and 2012, high nighttime temperatures affected corn yields across the U.S. Corn Belt (Fig. 11.15), and premature budding due to a warm winter caused $220 million in losses of Michigan cherries in 2012 (USGCRP, 2014). Each degree Celsius increase in global mean temperature, on average, reduces global yields of wheat by 6 percent, rice by 3.2 percent, maize by 7.4 percent, and soybean by 3.1 percent (Zaho et al., 2017).

**Melting of polar ice caps**. Over the past three decades of global warming, the oldest and thickest ice in the Arctic has declined by a stunning 95% according to NOAAA's annual Arctic Report Card (Osborne et al., 2018). The total volume of ice in the Arctic sea ice system in September is estimated to have declined by 78% between 1979 and 2012 according to the Pan-Arctic Ice

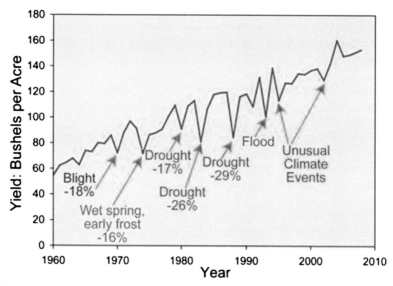

**FIGURE 11.15** Despite technological improvements that increase corn yields, extreme weather events have caused significant yield reductions in some years. *Source: U.S. EPA, January 19, 2017 (web site withdrawn), after USGCRP, 2009.*

Modeling and Assimilation Group (PIOMAS, 2018). Few locations have been monitored more closely than Greenland. Reporting from Helhein, the Associate Press (Borenstein, 2019) states that a NASA satellite found Greenland's ice sheet to have lost about 255 billion metric tons of ice annually between 2003 and 2016. The Helheim glacier has shrunk six miles since 2005. In the summer of 2019 alone, 440 billion tons of ice will melt off Greenland's giant ice sheet — enough to flood the state of Pennsylvania a foot deep. This phenomenon is occurring across the polar regions. Melting polar ice sheets are the drivers of sea level rise world-wide.

**Thawing of the Arctic permafrost**. Huge quantities of organic carbon reside in the frozen sediments of Arctic permafrost, some reaching to depths of 1500 m. An estimated 1672 billion metric tons (1400—1800 Pg soil carbon) carbon are locked up in the permafrost region of the Northern Hemisphere, compared to about 780 billion tons of carbon in the atmosphere (Schuur et al., 2008; Tarnocai et al., 2009). As global warming becomes sufficient to melt layers of the permafrost, releasing methane and $CO_2$ from the exposed detritus to the atmosphere in what modelers of the global carbon cycle would call a negative feedback. Recent research shows that permafrost off-gassing from arctic lakes could double by 2100, from the annual average of $25 \times 10^9$ g $CH_4$ released in 2006 (Anthony et al., 2018). Existing models currently attribute ∼20% of the permafrost carbon feedback to the atmosphere in this century to methane, with the rest due to carbon dioxide from terrestrial soils. The

$0.8 \times 10^{15}$ g $CO_2$ emissions from the arctic could reach 1.1 billion tons $CO_2$-eq per year in the future if permafrost continues to thaw (Schuur et al., 2008). If these new data on Arctic thermokarst lakes are included, methane becomes the dominant Arctic carbon source, responsible for 70%–80% of the warming this century attributed to permafrost carbon. Adding thermokarst methane to the models makes the feedback's effect similar to that of terrestrial land-use change, which is the second-largest contributor of human-made warming after fossil-fuel emissions.

Generalizations of the carbon budgets of high attitude ecosystems with future climate change are still premature; few comprehensive studies have been completed and the dynamics of a warming in tundra environments are complicated. A first attempt to develop a carbon budget for a subarctic catchment in northern Sweden using a total carbon ($CO_2$ and $CH_4$ fluxes) vegetation model (Tang et al., 2015) illustrates the dilemma. Using a summary of current research and new measurements, their model shows that with climate warming there is a general increase in the carbon sink of the birch forest and tundra heath ecosystems, many on drier sites, but with greater $CH_4$ emissions for the entire catchment which included drier, permafrost sites and also wetter, nonpermafrost sites. This result was due to increased growth ("densification") of birch forests and encroachment of birch into tundra heath. But, being both simultaneously a carbon sink and a methane source illustrates the complexity of carbon metabolism in high latitudes and the difficulty of simple conclusions about climate change effects on their carbon budgets. Recently, biogeochemical models have been developed that simulate microbial $CO_2$ and $CH_4$ production under the low-temperature, anaerobic, moisture and redox potential conditions of Arctic soils (Zheng. et al., 2019). The vast geographical area occupied by Arctic and subarctic ecosystems and their huge carbon reservoirs in litter, soils, and peat make better understanding of the carbon cycle, and especially methane, in these ecosystems a high priority.

**Methane hydrates**. Methane hydrate is a form of water ice that contains a large amount of methane within its crystal structure. Potentially large deposits have been found under sediments on the ocean floors of the Earth. On the sea floor the icelike solid fuel is formed, and only remains stable, at high pressure and low temperature. Methane hydrates are fragile and global warming of the oceans could affect their stability. This is evident in the history of the Earth where climate changes in the past may have led to the destabilization of methane hydrates and the release of methane. It is likely that between 1000 and 5000 Gtons carbon are present in methane hydrates, equivalent to 100–500 times the annual emissions from the combustion of coal, oil, and gas. Also, methane hydrates may be tapped as a new source of energy by humans unable to divest from their appetite for fossil fuels.

**Ocean acidification**. As temperatures rise, massive coral bleaching events and infectious disease outbreaks are becoming more frequent. Additionally, carbon dioxide absorbed into the ocean from the atmosphere has already begun to reduce calcification rates in reef-building and reef-associated organisms by altering seawater chemistry through decreases in pH, by a process termed ocean acidification (NOAA, 2018). Ocean acidification involves a shift toward pH-neutral conditions rather than a transition to acidic conditions (pH < 7). An estimated 30%−40% of the carbon dioxide from human activity released into the atmosphere dissolves into oceans, rivers, and lakes. To achieve chemical equilibrium, some of it reacts with the water to form carbonic acid. Some of the resulting carbonic acid molecules dissociate into a bicarbonate ion and a hydrogen ion, thus increasing ocean acidity ($H^+$ ion concentration). Between 1751 and 1996, surface ocean pH is estimated to have decreased from approximately 8.25 to 8.14 (Wikipedia, 2019). In the starkest warning yet of the threat to ocean health, the International Program on the State of the Ocean (IPSO, 2013) warns that the oceans are becoming more acidic at the fastest rate in 300 million years, due to carbon dioxide emissions from burning fossil fuels. This acidification is unprecedented in the Earth's known history, leading to changes in marine ecosystems.

**Ocean warming**. Unlike the atmosphere which is heated by solar energy from the bottom (i.e., the Earth's surface), the oceans are heated from the surface. Because warm water is less dense and "floats" on the surface, it provides stability to the ocean water column and prevents exchanges with deeper layers. Surface waters have a mean temperature of 18°C; cold deep waters, which contain ∼95% of the oceans volume, have a mean temperature of 3°C.

Atmospheric winds create surface turbulence and mixing; they also form currents, e.g., Gulf Stream in the Atlantic. The Trade Winds drive currents east to west in the Atlantic and Pacific oceans along the equator where these currents collide with continents and then proceed either north or south along the continents to the northern and southern latitudes. The cyclic pattern of circulation in each of the major oceans is called a gyre. This global circulation of surface currents transfers heat from the tropic to polar regions. More than half of the net excess solar energy received in the tropics is transferred to the poles in this manner, the remainder being transferred directly to the atmosphere.

**Sea level rise**. During the next century sea level is projected to rise (Table 11.5) between 0.24 and 0.82 m (24−82 cm or 9.5−32.3 inches) (IPCC, 2014), having already risen around 19 cm (7.6 in.) in the 20th Century. Between 1880 and 2013 the global average sea level rose about 22 cm (8.8 in.), while just between 1993 and 2016 the global averagre sea level rose 8.5 cm (3.4 in.) (The Economist, August 17, 2019). About half of sea level rise is due to the melting of glaciers and polar ice caps; the remaining rise is due to thermal expansion of water due to warming. Coastal impacts will be varied geographically, depending upon shoreline topography and local tidal patters. Extreme events, such as storm surges, would be triggered by regional combinations of moon phases and storm patterns. For low-lying coasts and tidal

**TABLE 11.5** Future sea level rise (in meters) projected from different radiative forcing function scenarios from assumptions of different GHG emissions.

| Forcing Function Scenario | 2046 − 2065 Mean and likely range (meters) | 2081- 2100 Mean and likely range (meters) |
| --- | --- | --- |
| RCP2.6 | 0.24 (0.17−0.32) | 0.40 (0.26−0.55) |
| RCP4.5 | 0.26 (0.19−0.33) | 0.47 (0.32−0.63) |
| RCP6.0 | 0.25 (0.18−0.32) | 0.48 (0.33−0.63) |
| RCP8.5 | 0.30 (0.22−0.38) | 0.63 (0.45−0.82) |

Source: IPCC, 2013, Summary for Policymakers. In: Climate Change 2013.

marshes, even several centimeter rises can invite devastating storm surges, erosion, and salt water intrusion. If the Paris Agreement's preferred target to keep warming below 1.5°C relative to pre-industrial levels is met, sea levels will rise another 50 cm (20 in.). Failure to control emissions will result in an additional sea level rise of another 30−40 cm (12−16 in.) (The Economist August 17, 2019). Even if atmospheric $CO_2$ composition were fixed today, global mean temperature and sea level rise would continue for decades due to oceanic thermal inertia. The economic costs of flooding in coastal cities around the world are staggering, as will be the impacts to human populations living in low-lying areas.

**Animal effects**. Heat waves, which are projected to increase under climate change, could directly threaten livestock. In 2011, exposure to high temperature events caused over $1 billion in heat-related losses to U.S. agricultural producers USGCRP, 2014). Climate change may increase the prevalence of parasites and diseases that affect livestock. Warmer temperatures also have been related to increased fire frequencies and insect outbreaks in forests (Portland State University, 2018); particularly sensitive are mountain forest ecosystems (National Park Service, 2016). Of 143 breeding bird species evaluated in the Massachusetts Audubon's State of the Birds report (Mass Audubon, 2017), 43% (61 species) are highly vulnerable to the effects of climate change predicted to occur by 2050. Of the other species, 15% (22 species) are likely vulnerable. The unusually cold winter of 2017−2018 on the southeastern U.S. coast depressed the shrimp fishery in 2018, and associated colder water impact on the offshore invertebrate food base is thought to have been the cause of reduced nesting by loggerhead turtles (Kincaid, 2018). New examples of climate change impacts appear continuously and frequently. And should you not believe that global warming is occurring, then explain the recent appearances of opossums in New England states, alligators in North Carolina, and armadillos in Tennessee.

**Fisheries**. The ranges of many fish and shellfish species may change. In waters off the northeastern United States, several economically important species have shifted northward since the late 1960s. American lobster, red hake, and black sea bass have already moved northward by an average of 119 miles (NOAA, 2014). Winter warming in the Arctic is contributing to salmon diseases in the Bering Sea and resulting in reductions in the Yukon Chinook Salmon (USGCRP, 2014). Warmer temperatures have caused disease outbreaks in coral, eelgrass, and abalone (USGCRP, 2014). The U.S. NOAA has identified the potential impacts of climate change upon fisheries.

In short, climate change can affect living marine resources (LMRs) via changes in:

- Genotype (natural selection, selective breeding),
- Vital rates (reproductive rate, emigration, immigration),
- Physiology rates (growth, consumption, respiration, metabolism, thermal tolerance),
- Susceptibility to disease, and
- Trophic interactions.

These changes can result in a variety of subsequent changes to:

- Mortality,
- Productivity,
- Species distribution,
- Nutritional value of prey,
- Movement of migratory species, and
- Habitat structure and location.

And those changes can in turn impact other parameters, such as:

- Species relative abundance,
- Community composition and predator-prey overlaps,
- Food web structure,
- Energy and matter fluxes, invasive species, and
- Life histories.

**Disease vectors and human health**. Climate change threatens human health and wellbeing in many ways, including impacts from changes in food production, water supplies, increased extreme weather events, wildfire, decreased air quality, threats to mental health, and illnesses transmitted by food, water, and disease-carriers. Climatic conditions strongly affect waterborne diseases and diseases transmitted through insects, snails and other cold-blooded animals. Changes in climate are likely to lengthen the seasons of transmission for important vector-borne diseases and to alter their geographic ranges. For example, climate change is projected to widen significantly the area of China where the snail-borne disease schistosomiasis occurs. Malaria is strongly influenced by climate. Transmitted by *Anopheles* mosquitoes, malaria kills over 400,000 people every year—mainly African children under 5 years of age. The *Aedes* mosquito, vector of dengue, is also highly sensitive to climate conditions, and studies suggest that climate change is likely to continue to increase exposure to dengue. Global warming is enabling the movement of diseases such as Zika virus and yellow fever to higher latitudes. The World Health Organization expects climate change to cause approximately 250,000 additional deaths per year between 2030 and 2050; 38,000 deaths due to heat exposure in elderly people, 48,000 due to diarrhea, 60,000 due to malaria, and 95,000 due to childhood undernutrition (WHO, 2018). Good estimates of the economic impacts of workplace lost time caused by climate change stress are not available.

Although climate change may affect the growth rate of the global economy, negative impacts will be substantially greater in poorer, hotter, and lower-lying countries (Toi, 2018). The economic impact of climate change on health is striking. "The direct costs from damage to health due to climate change could reach $2 - $4 billion per year by 2030" (World Bank @ Climate Reality Project). The cost of air-pollution-related disease on lost economic output is $1.7 trillion annually in countries of the Organization for Economic Co-operation and Development (OECD )— with $1.4 trillion of that from China and another $500 billion from India (OECD @ Climate Reality Project).

How do all these impacts of climatic change relate to ecological energetics? A few years ago, many predictions of the ecological impacts discussed above would have been viewed as the prognostications of environmental extremists. Now today they are reality. One might ask, why is the measurement of the detailed bioenergetic relationships of species, food chains, and trophic levels important now that there are new, less tedious, remote-sensing technologies which allow direct measurement of ecosystem properties? And now we understand much better how ecological health is related to human health.

Besides the "bottoms up," traditional measurements being validations of the accuracy of remotely sensed values, their primary importance is that they allow prediction, and understanding, of the responses of species and ecosystems. Predictions are made possible through incorporation into the models of algorithms representing functional ecological relationships and responses to environmental variables. This allows the models to "predict" changes in the biosphere that we can now only extract after the fact from monitoring using remote sensing. We need to continuously improve our predictive scientific capability, develop more effective communication of science in the face of dogma, and expect an educated public's better understanding and acceptance of scientific information.

## 11.10 Recommended Reading

Archer, D., 2010. The Global Carbon Cycle. Princeton Univ. Press, Princeton, 216 pp. https://press. princeton.edu/books/ebook/9781400837076/the-global-carbon-cycle.

Broecker, W.S., 1975. Climatic change: are we on the brink of a pronounced global warming? Science 189 (4201), 460−463. https://www.jstor.org/stable/1740491.

Farmer, G., Cook, J., 2013. Climate Change Science: A Modern Synthesis. In: Physical Climate, vol. 1. Springer Verlag, Berlin-Heidelberg-New York, 564 pp. https://www.springer.com/gp/book/9789400757561.

Gornitz, V., Lebedeff, S., Hansen, J., 1982. Global sea level rise in the past century. Science 215, 1611−1614. https://doi.org/10.1126/science.215.4540.1611.

NASA, 2018. Scientific Consensus: Earth's Climate Is Warming. Climate Change: Vital Signs of the Planet. NASA. /https://climate.nasa.gov/scientific-consensus/.

Schuur, E., Vogel, J., Crummer, K., Lee, H., Sickman, J., Osterkamp, T., 2009. The effect of permafrost thaw on old carbon release and net carbon exchange from tundra. Nature 459 (7246), 556−559. https://www.nature.com/articles/nature08031.

Scientists and Experts from Universities and Institutions in the Great Lakes Region, 2019. An Assessment of the Impacts of Climate Change on the Great Lakes. Environmental Law and Policy Center. http://elpc.org/wp-content/uploads/2019/03/Great-Lakes-Climate-Change-Report.pdf.

Toi, R.S.J., 2018. The economic impacts of climate change. Rev. Environ. Economics and Policy 12 (1), 4−25. https://doi.org/10.1093/recp/rex027/.

Trabalka, J.R., Reichle, E. (Eds.), 1986. The Changing Carbon Cycle: A Global Analysis. Springer Verlag, New York, 592 pp. https://link.springer.com/book/10.1007/978-1-4757-1915-4.

USGCRP, 2014. Ch. 6: agriculture. Climate change impacts in the United States, pp. 150−174. In: Hatfield, J., Takle, G., Grotjahn, R., Holden, P., Izaurralde, R.C., Mader, T., Marshall, E., Liverman, D. (Eds.), The Third National Climate Assessment. In: Melillo, J.M., Richmond, T.C., Yohe, G.W. (Eds.), U.S. Global Change Research Program,p 150−174. https://nca2009.globalchange.gov/Search for other works by this author on.

# Chapter 12

# Carbon, climate change, and public policy

The global warming since preindustrial time has been barely 1°C—small compared to weather fluctuations, yet seasonal mean temperature anomalies in most land areas are now large enough to be noticeable, as are the environmental consequences. No one has been more persistent and erudite in issuing

The Global Carbon Cycle and Climate Change. https://doi.org/10.1016/B978-0-12-820244-9.00012-3

the warning of climate change than Jim Hansen, former director of the NASA Goddard Institute for Space Studies and more recently of Columbia University's Earth Institute.

*"Global warming over the past several decades is now large enough that regional climate change is emerging above the noise of natural variability, especially in the summer at middle latitudes and year-round at low latitudes. Despite the small magnitude of warming relative to weather fluctuations, effects of the warming already have notable social and economic impacts. Global warming of 2°C relative to preindustrial would shift the 'bell curve' defining temperature anomalies a factor of three larger than observed changes since the middle of the 20th century, with highly deleterious consequences. There is striking incongruity between the global distribution of nations principally responsible for fossil fuel $CO_2$ emissions, known to be the main cause of climate change, and the regions suffering the greatest consequences from the warming, a fact with substantial implications for global energy and climate policies."* *(Hansen and Sato, 2016).*

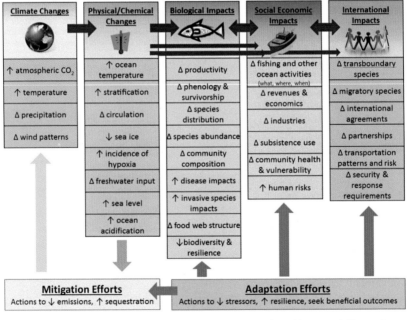

**FIGURE 12.1** Schematic diagram illustrating current and/or projected impacts of climate changes on major components of marine and coastal ecosystems. *Source: Link et al., 2015. Courtesy, NOAA.*

## 12.1 What are the potential consequences of inaction?

**Impacts on environmental systems**. There are many potential consequences of climate change, depending upon the extent of temperature change and the time period over which this change might take place. Climate change involves many factors, those of paramount importance to the biosphere being increasing temperature and altered precipitation regimes (Fig. 12.1). These lead to physical and chemical changes in environmental parameters, with resulting impacts upon natural biological and human systems, all with social consequences at local, regional, and global scales.

Scientists say they are now watching the unfolding of massive worldwide coral bleaching events, spanning the globe from Hawaii to the Indian Ocean. And they fear that as a result of warmer sea temperatures, the ultimate result could be the loss of more than 12,000 square kilometers, or over 4500 square miles, of coral—with particularly strong impacts in Hawaii and other U.S. tropical regions. This is being brought on by a combination of global warming, a very strong El Niño event, and the so-called warm "blob" in the Pacific Ocean, say the researchers, part of a consortium including the National Oceanic and Atmospheric Administration (Mooney, 2015). There is much at risk. Marine and coastal fisheries support over $200 billion in economic activity and 1.8 million jobs in the United States alone each year.

Many worry about the future of the coastal salt marshes, the nursery of the southern U.S. coastal fisheries, and those of other countries worldwide. The barrier islands are the buffer for the impact of coastal storms; behind them lie the salt marshes. As rising sea levels and storms erode the barrier islands, the coastline is pushed inland. But the coast is now becoming increasingly developed with communities and their associated revetments and sea walls. As communities harden their defenses against the encroaching ocean, there is less room for the marsh ecosystems to migrate inland (SC etv, 2019).

Charleston, South Carolina, is an example of the impacts of sea level rise and increased frequency of storm events along the southeastern U.S. Atlantic coast. Mean tide sea level has risen 18″ in the last 100 years; it is now rising faster than it has been and is projected to rise another 1.5−2.5 feet in the next 50 years. Rising sea levels, high tides, winds, and rain from tropical storms, when combined, can cause serious problems. The energy in warmer ocean waters supports stronger storms, and warmer air holds more moisture. In October 2015, tropical storm Joaquin dropped 26.88 inches of rain over a 4-day period in Mt. Pleasant, SC, just north of Charleston. In 1970 Charleston had flooding 4 times, a not unusual event in the low country; but, in 2017, it flooded 34 times (SC etv, 2019)! NOAA (2017) recorded 50 flood days in 2016. Tide gauges of the U.S. National Oceanic and Atmospheric Administration (NOAA) have been measuring water levels for over a century along the U.S. coastline (NOAA, 2017). The number of days with high tide flooding in 2016 was above the local flood-frequency trend at the majority of the 28

## Past and future ocean heat content changes

Annual observational OHC changes are consistent with each other and consistent with the ensemble means of the CMIP5 models for historical simulations pre-2005 and projections from 2005–2017, giving confidence in future projections to 2100 (RCP2.6 and RCP8.5) (see the supplementary materials). The mean projected OHC changes and their 90% confidence intervals between 2081 and 2100 are shown in bars at the right. The inset depicts the detailed OHC changes after January 1990, using the monthly OHC changes updated to September 2018 [Cheng et al. (2)], along with the other annual observed values superposed.

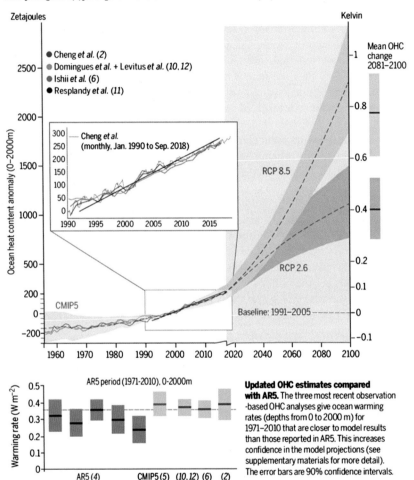

FIGURE 12.2 Past and future ocean heat content changes (OHC). Annual observational OHC changes are consistent with each other and consistent with the ensemble means of the CMIP5 models (Taylor et al., 2012) for historical simulations pre-2005 and projections from 2005 to 2017, giving confidence in future projections to 2100 (RCP2.6 and RCP8.5) (see the supplementary materials). The mean projected OHC changes and their 90% confidence intervals between 2081 and 2100 are shown in bars at the right. The inset depicts the detailed OHC changes after January 1990, using the monthly OHC changes updated to September 2018 (Cheng et al., 2017), along with the other annual observed values superposed. *Image credit: Cheng et al., 2019, Courtesy,* Science *Magazine.*

locations examined (more than half of the trends are accelerating in time). Three all-time records for annual-flood days were either tied (Key West, FL) or broken (Charleston, SC; Savannah, GA).

Coastal habitats help defend coastal communities from storms and inundation, and provide the foundation for tourism and recreation-based economies in many coastal communities (Link et al., 2015). The city of Boston expects sea levels to rise at least 1.5 feet higher than they were in 2000 and 3 feet higher by 2070. The U.S. National Oceanic and Atmospheric Administration released study estimates that cities along the U.S. East Coast could see even higher levels than anticipated, up to 8.2 feet by 2100, a rise from previous predictions of 6.6 feet (NOAA, 2017). Sea levels, already up by 16−21 cm in the past 150 years, could rise another meter in this century; coupled with crop drought, fire, floods, and storms, total damages could reduce U.S. GDP by one-tenth this century (USGCRP: Fourth National Climate Assessment, 2017; The Economist, Dec., 2018).

About 90% of the additional energy accumulated in the atmosphere as a consequence of greenhouse gases, and the resulting greenhouse effect, winds up being stored in the oceans. A steady rise in ocean temperature down to 2000 m and been occurring since the 1950s, with a record high predicted for 2018 (*The Economist*, Sept. 22, 2018). The U.S. National Center for Atmospheric Research reports that 2018 will be the warmest year on record in the oceans—as 2017 was and 2016 before that, 40%−50% warmer than predicted in the last IPCC AR5 report. Ocean heat content (OHC) has been steadily increasing for decades. Scientific analyses (Cheng et al., 2019) show that during 2015 and 2016, the heat stored in the upper 2,000 meters of the world ocean reached a new 57-year record high (Fig. 12.2). This heat storage amounted to an increase of $30.4 \times 10^{22}$ Joules (J) since 1960, equal to a heating rate of 0.33 Watts per square meter (W m$^{-2}$) averaged over Earth's entire surface—0.61 W m$^{-2}$ after 1992 (Cheng et al., 2019).

It is the warming sea water that has been driving the increased frequency and intensity of storms in the Atlantic and Pacific oceans. Tropical storms derive their strength from the energy contained in warm ocean waters. The number of category 4 and category 5 hurricanes world wide have nearly doubled from the 1970s to the 2000s, with tropical cyclone duration and strongest wind speeds having increased nearly 50% over the past 50 years (Than, 2019).

The scientific literature anticipates similar future impacts of climate change on terrestrial environments. Climate change will, in general, make higher-latitude regions of the Earth more tropical, and growing seasons will become longer. The wheat belt will shift to more northern latitudes. From 30% up to 60% of the Amazon rainforest could be converted into a type of dry savanna according to Brazil's National Space Research Institute (WRI, 2013). Extreme drought conditions are projected by mid-century over some of the most populated areas on Earth—southern Europe, south-east Asia, Brazil, the

U.S. Southwest, and large parts of Australia and Africa (Union of Concerned Scientists, 2011). Impacts are expected on terrestrial animal populations through habitat loss and fragmentation, physiological sensitivities, alterations in the timing of species life cycles, and indirect effects on species interactions (Friggens et al., 2018). With a rise in global temperature of $2°C$, ecosystems covering between a twelfth and a fifth of the Earth's land mass can be expected to undergo transformation to another type, i.e., savanna to desert (*The Economist*, Oct. 13, 2018).

Since the increase of temperature has more significant impact on autotrophic and heterotrophic respiration than it does on GPP, the NEP will decline affecting the global carbon cycle (Grace and Zhang, 2006). And in an extensive review of potential climate change impacts in the U.S. Intermountain Region, the U.S. Forest Service has projected effects of climate change on hydrology, water resources, soils, native fish, forest vegetation, nonforest vegetation, ecological disturbances, terrestrial animals, outdoor recreation, infrastructure, cultural resources, and ecosystem services (Halofsky et al., 2018).

Without global warming, only half of the loss of western U.S. forests to fire between 1984 and 2015 would have occurred (Abatzoglou and Williams, 2016). Global warming induces huge quantities of $CO_2$ releases through thawing of arctic permafrost or outgassing from ocean waters from altered ocean circulation patterns. Accelerating the rate of injection of $CO_2$ into the atmosphere, whatever and how many the sources, tends to increase $CO_2$ absorption into the ocean faster. In the short term, doubling atmospheric $CO_2$ halves the $CO_3^{2-}$ concentration in seawater, halving the ability of the seawater to absorb more $CO_2$. These heretofore unexperienced feedbacks upon the global carbon cycle could have serious and unforeseen consequences.

**Impact on the carbon-climate system.** The potential for strong, multiple feedbacks to the global carbon cycle is high. Melting polar ice caps and glaciers alter albedo, methane hydrate, and peat carbon reservoirs sensitive to warming, possible ocean warming altering currents and $CO_2$ fluxes, $CO_2$/temperature/moisture stimulation of global photosynthesis, and human activity injecting multiple greenhouse gases in huge quantities are some of the feedbacks which could quickly reset the global carbon cycle and possibly destabilize it. For instance, if temperature increases affect ocean circulation, it could result in a positive feedback of $+100-200$ ppmv in atmospheric $CO_2$ levels and even more global warming (Schimel et al., 2005).

If climate change modifies ocean circulation through a combination of direst warming and seawater dilution by melting polar ice caps and glaciers, it is very possible that resulting changes in the ocean circulation pattern could affect atmospheric $CO_2$ concentrations through four basic mechanisms (Keir, 2005):

- circulation induced change in calcium carbonate ($CaCO_3$) production would change $CaCO_3^{2-}$ in the deep ocean,
- circulation induced alterations in the ratio of organic C to $CaCO_3$ production could occur,
- change in upper ocean water mixing or thermocline circulation could alter the vertical gradient of $CO_3^{2-}$, and
- in high latitude areas of deep water formation, changing thermocline may interact with air-surface water exchange and biological production to alter atmospheric $CO_2$ levels through chemical solubility and the biological pump.

If indeed it will take the order of 100,000 years for the carbon cycle's weathering $CO_2$ thermostat to reset climate to something near the pre-$CO_2$ atmospheric excursion of contemporary industrial times (see Chapter 10.2, Carbon Cycle Regulators), are there any records in the Earth's history which might provide insight into the consequences of such a rapid rise in atmospheric $CO_2$ content? The Paleocene-Eocene Thermal Maximum (PETM), also called the Initial Eocene Thermal Maximum (IETM), was an interval of maximum temperature lasting approximately 100,000 years during the late Paleocene and early Eocene epochs of around 55 million years ago (Encyclopedia Britannica, 2018). This interval had the highest global temperatures of the Cenozoic Era, extending from 65 million years ago to the present. The underlying causes of this temperature excursion are unclear, and some authorities associate the PETM with the sudden release of frozen methane hydrates from deep ocean sediments caused by massive volcanic eruptions. The onset of the PETM was rapid, occurring within a few thousand years. The ecological consequences were huge, with widespread extinctions in both marine and terrestrial species and whole ecosystems. Sea surface and continental air temperatures increased by more than $5°C$ ($9°F$) during the PETM (Archer, 2005; Schimel et al., 2005), with sea surface temperatures in the high-latitude Arctic possibly being as warm as $23°C$ ($73°F$), which conditions are comparable to those in contemporary subtropical and warm-temperate seas.

## 12.2 Do we know enough?

Historically this question would have been raised by skeptics about the prospects of global warming and climate change. But, due initially to humankind's actions over past centuries that have disrupted the global carbon cycle, and now humankind's inactions over past decades by refusing to alter environmentally damaging practices, climate change is a reality beyond doubt. We have known more than enough for decades about the prospects and consequences of global warming. And our worst fears are upon us. The question now becomes, do we know enough to take corrective actions and simultaneously mitigate and adapt to climate change? Corrective action means limiting the anthropogenic releases of $CO_2$. Mitigation means minimizing the effects of changes, and adaptation means

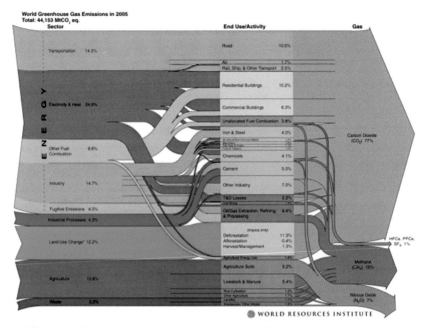

**FIGURE 12.3** Worldwide greenhouse gas emission in 2005. *Image Credit: World Resources Institute, 2005.*

adjusting to changes which we cannot ameliorate. All of these actions will require humankind to rely upon knowledge of ecological energetics, the flow of energy in the biosphere, and the global biogeochemical cycle of carbon.

What are the sources of atmospheric $CO_2$, and where might one begin to look to minimize $CO_2$ emissions to the atmosphere? Fig. 12.3 gives us a perspective of the world's energy system from which to begin. The major energy demands of society, which rely heavily upon fossil fuel sources, are transportation, primarily automotive and trucking, electrical usage by commercial and residential buildings, and the industrial base of the economy. Major fluxes come from land-use change, primarily from agriculture through soil carbon loss and methane generation through agricultural practices and animal husbandry, but also through natural gas infrastructure.

All of these many sources and pathways of carbon flux generated by human activity are the template against which we can apply human ingenuity to reduce our $CO_2$ emissions. There are unlimited opportunities, and we shall examine a few.

**TABLE 12.1** Historical timeline of milestones in establishing international climate policy.

| Date | Action |
| --- | --- |
| November 8. 1980 | The **intergovernmental panel on Climate Change**, IPCC established. |
| November 1990 | The IPCC releases the **First Assessment Report** saying "emissions resulting from human activities are substantially increasing the atmospheric concentrations of greenhouse gases" leading to calls by the IPCC and the second World Climate Conference for a global treaty. |
| December 11, 1990 | The United Nations General Assembly establishes the intergovernmental negotiating committee (INC) for a **Framework Convention on Climate Change**. |
| June 1992 | IPCC SAR 1992 serves as a basis for The United Nations Framework Convention on Climate Change, which opens for signature at the **Earth Summit** in Rio de Janeiro. |
| December 11, 1997 | The third conference of the parties achieves an historical milestone with adoption of the **Kyoto Protocol**, the world's first greenhouse gas emissions reduction treaty. |
| July 2001 | A major breakthrough is achieved at the second part of the sixth conference of the parties meeting in Bonn, with governments reaching a broad political agreement on the operational rulebook for the 1997 Kyoto Protocol. |
| March, September 2007 | IPCC 2007 synthesis Report. Contribution of Working Groups I, II and III to the **Fourth Assessment Report** of the intergovernmental panel on Climate Change. IPCC receives 2007 Nobel Peace Prize |
| December 2009 | World leaders gather for the fifteenth conference of the parties in copenhagen, Denmark, which produced the **Copenhagen Accord**. Developed countries pledge up to USD 30 billion in fast-start finance for the period 2010 −2012. |
| December 2011 | At the seventeenth conference of the parties, governments commit to a new universal climate change agreement by 2015 for the period beyond 2020, leading to the launch of the ad hoc Working Group on the Durban Platform for Enhanced Action or ADP. |
| September 27, 2013 | The UN intergovernmental panel on Climate Change (**IPCC**) releases the **Working Group 1** contribution to its **Fifth Assessment Report** (AR5), on the science of climate change. |

*Continued*

**TABLE 12.1** Historical timeline of milestones in establishing international climate policy.—cont'd

| Date | Action |
|------|--------|
| March 31, 2014 | The UN intergovernmental panel on Climate Change (**IPCC**) releases the **Working Group 2** contribution to its **Fifth Assessment Report** (AR5), on impacts, adaptation, and vulnerability. |
| December 2015 | The **Paris Agreement** for the first time brings all nations into a common cause based on their historic, current, and future responsibilities. About 195 nations agreed to combat climate change and unleash actions and investment toward a low-carbon, resilient, and sustainable future. |
| December 2–14, 2018 | At the **Katowice Climate Change conference** government representatives showcase the critically important cooperation between state and nonparty actors if the world is to meet the 1.5C goal of the Paris Agreement. |

Source: United Nations Climate Change, 2019.

## 12.3 International accords

The Intergovernmental Panel on Climate Change was established in 1988 by the World Meteorological Organization (WMO) and the United Nations Environment Program (UNEP) to assess scientific, technical, and socioeconomic information concerning climate change, its potential effects, and options for adaptation and mitigation. IPCC has been the focus in the international community on climate change, integrating scientific knowledge with governmental leadership. For its several decades of contributions (Table 12.1), the IPCC was awarded the Nobel Peace Prize in 2007.

**Setting limits on global warming**. The IPCC Special Report SR1.5 (IPCC, 2018) on Global Warming focused on the impacts of global warming of 1.5°C above preindustrial levels, and related global greenhouse gas emission pathways, in the context of strengthening the global response to the threat of climate change, sustainable development, and efforts to eradicate poverty. It is the first publication in the Intergovernmental Panel on Climate Change Sixth Assessment Report (IPCC, 2018). It presents the scientific basis for addressing the following questions:

- What are the impacts of 1.5°C global warming above preindustrial levels?
- How can emissions be brought to zero by mid-century, within the remaining carbon budget for limiting global warming to 1.5°C?

- Why is it vital to maintain the global temperature increase below 1.5°C to avoid the intensity and frequency of extreme events, on renewable resources, ecosystems, biodiversity, food security, and human populations?
- What is the feasibility of mitigation and adaptation options?
- What are the low-carbon, climate-resilient, economic development pathways in response to climate change?

**Atmospheric $CO_2$ targets and temperature increases**. The 2017 average global temperature reported by NOAA across land and ocean surface areas was 0.84°C (1.51°F) above the 20th-century average of 13.9°C (57.0°F). The atmosphere's $CO_2$ concentration has now gone to over 410 ppm. At 450 ppm mean global temperature will rise to 2°C above preindustrial levels, and the Earth will be warmer than it has been in millions of years. Based upon 1994 calculations, $CO_2$ emissions would have had to have been reduced by two-thirds of 1994 emissions to 2 Gton carbon per year to stabilize at 450 by 2200. Reduction by only one-third, to 4 Gton C $yr^{-1}$ in 1994 would have resulted in stabilization at only 650 ppmv in 2200 (Schimel et al., 2005). A detailed discussion of the various $CO_2$ stabilization profiles used by the IPCC and the Pade approximation formula and coefficients to calculate future atmospheric $CO_2$ stabilization profiles are provided by Wigley (2005).

Fossil fuel emissions in 2018 were 10 Gton carbon (World Resources Institute, 2018). The world does not seem to be taking mission reduction targets very seriously. Forecasts of future fossil fuel use, and hence $CO_2$ emissions, are important for anticipating future climate change. The trajectories of estimates of future $CO_2$ emissions are many and vary considerably due to different assumptions of population growth, economic growth, development of technologies for producing and consuming energy, and progress in decarbonizing the energy system (Edmonds, 2005). We shall examine some of these factors.

**TABLE 12.2** Cumulative $CO_2$ emissions limits from a 2011 emissions baseline necessary to limit global warming to <1.5°C and <2°C, with associated probabilities.

| Net warming probability[a] | <1.5°C | <2°C |
| --- | --- | --- |
| >66% | 400 Gt $CO_2$ | 1000 Gt $CO_2$ |
| >50% | 500–600 Gt $CO_2$ | 1300–1400 Gt $CO_2$ |
| >33% | 700–900 Gt $CO_2$ | 1500–1700 Gt $CO_2$ |

[a]probability that emission limit will keep global temperature at or below limit.
Source: IPCC, 2014.

## 12.4 Mitigation and adaptation

Can we take corrective actions and simultaneously mitigate and adapt to climate change? Corrective action means limiting the anthropogenic releases of $CO_2$. Mitigation means making climate change impacts from global warming less grievous, and taking actions to reduce greenhouse gas emissions or increase the amount of carbon dioxide absorbed and stored by natural and manmade carbon sinks. Adaptation means adjusting natural or human systems to those changes whose corrections are beyond our control, by exploiting beneficial opportunities or moderating negative effects. At the same time research will need to continue on understanding how ecological systems can adapt to climate change by:

- quantifying the factors regulating the biologically driven fluxes of geochemically important elements, especially those related to climate,
- identifying the processes that determine how organisms physiologically acclimate and genetically adapt to environmental change, and
- understanding the impact of long-term environmental changes on the stability, diversity, and function of ecosystems.

The IPCC in 2014 estimated the cumulative $CO_2$ emission reductions from 2011 forward that would be necessary to limit global warming to both $<1.5°C$ and $<2°C$ (Table 12.2). The difference between a $+1.5°C$ and a $+2°C$ world makes a huge difference. Arctic summers would be ice-free once every decade in $+2.0°C$ world, but only once in a century in a $+1.5°C$ world (IPCC, 2018). To achieve a $+1.5°C$ the world would need to eliminate 42 billion metric tons ($4.2 \times 10^{16}$) of $CO_2$ annually from emissions; renewables would have to treble their current 25% share of electricity production; and internal combustion engines, which currently power 499 out of every 500 vehicles, would have to nearly vanish (*The Economist*, Oct. 13, 2018). Progress is being made—the number of electric cars is rising, green finance is becoming available, and zero-carbon technologies are being developed. But, challenges to decarbonization will come from newly improved efficiencies of extracting oil from shale and abundant supplies of price-competitive natural gas. Nevertheless, the scale of needed change is immense.

These goals can be achieved by decarbonizing the energy system through the development of renewable energy sources such as biomass, solar, wind, and nuclear; developing low-carbon alternatives for transportation and the production of electrical power; improving the efficiency of energy use; reducing $CO_2$ emissions from fossil energy production through new "clean technologies" and carbon sequestration; and dealing with methane emissions from municipal waste management and agriculture/animal husbandry. The share of electrical production by different energy sources in the United States in 2018 was: natural gas 34%, coal 28%, nuclear $\sim 20\%$, hydro $\sim 7\%$, and nonhydro renewables $\sim 10\%$ (EIA, 2018). Figure 12.4 provides a reference

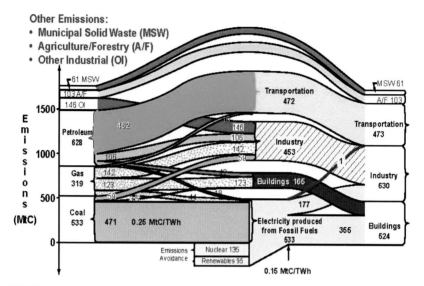

**FIGURE 12.4** Carbon flows in the energy system and sources of emissions in the United States in 1995 in millions of metric tons ($10^{12}$ g C). *Source: Energy Information Administration (1998a,b) after Reichle et al. (1999). Courtesy, Oak Ridge National Laboratory and U.S. Department of Energy.*

point as to the potential for carbon reductions in each of the energy production/use sectors.

**Greenhouse gas technologies**. Many of these of energy technologies with promise to reduce greenhouse gas emissions were summarized in a 1997 report by 11 of the U.S. Department of Energy's National Laboratory directors for the U.S. Department of Energy (National Laboratory Directors, 1997). They examined many novel concepts for energy efficiency savings with appliances, building operations, industrial processes, and transportation. They examined clean energy options through higher generating efficiencies and low-carbon fuels, more efficient electrical distribution, hydrogen and fuels cells, advanced energy storage systems, and carbon capture and sequestration from fuels before and after combustion. The U.S. DOE National Laboratories believed that annual carbon emission reductions of 25%–50% in the United States from 400 to 800 MtC ($4-8 \times 10^8$ g C $yr^{-1}$) by 2030 could be possible with an aggressive federal R&D program. New lithium-glass electrolyte batteries hold great promise for energy storage (*The Economist*, Dec. 1, 2018). Stratospheric sunscreens of $SO_2$ spewed by statellites at 20 km are now technologically possible and economically feasible, which could slow global warming by $0.3°$ C in 15 years (The Economist, 2019). Let us examine some of these technologies and where progress is possible, particularly where different sources of electrical energy produce different intensities of $CO_2$) (Figs. 12.3, 12.4, Table 12.3).

**TABLE 12.3** The lifecycle carbon intensity of electricity sources: greenhouse gas emissions per kilowatt (Moomaw et al., 2011).

| Technology | Description | 50th percentile (g $CO_2$/kWh) |
|---|---|---|
| Hydroelectric | Reservoir | 4 |
| Wind | Onshore | 12 |
| Nuclear | Various 2nd generation reactors | 16 |
| Biomass | Terrestrial and aquatic | 18 |
| Solar thermal | Parabolic trough | 22 |
| Geothermal | Hot dry rock | 45 |
| Solar photovoltaic | Polycrystalline silicon | 46 |
| Natural gas | Combined cycle, no scrubbing | 469 |
| Coal | Generators, no scrubbing | 1001 |

Source: Courtesy, IPCC.

**Decarbonization of the energy system**. The world will need to substantially decarbonize the global energy system by the middle of the 21st century and attain zero net emissions around 2070 in order to have a chance of keeping average global temperatures from rising more than 2°C. Electrical power produced by nuclear energy would be one source of carbon-free energy, and could offer substantial reductions in the $>1.5 \times 10^{15}$ g C $yr^{-1}$ produced in the United States annually from fossil-fueled power plants. This is not an unreasonable assumption considering that France produces over 70% of its electricity from nuclear power. The United States was once a world leader in nuclear energy R&D. It was a chemical engineer at the Oak Ridge National Laboratory who expanded my perspective nearly five decades ago from the carbon metabolism of forests to the atmospheric pool of $CO_2$ and the global carbon cycle, with his concerns about the potential of off-gas discharge of $^{14}CO_2$ and $^{14}CH_4$ from light-water nuclear reactors. Nuclear energy has great potential for decarbonization of the energy system, especially inherently safe, modular plants, but it will require satisfactory long-term waste disposal options, a knowledgeable and educated citizenry, and political leadership—all of which currently are in short supply.

**Renewable energy** contributed about 10% of the total power-sector U.S. electricity supply in 2010 (6.4% from hydropower, 2.4% from wind energy, 0.7% from biopower, 0.4% from geothermal energy, and 0.05% from solar energy (NREL, 2012). To evaluate the $CO_2$ emission reductions possible with renewables, compare the kilograms of $CO_2$ produced per million ($10^6$) Btu—for coal, 93.3; natural gas, 53.07; and fuel oil, 73.16. (A value of 25 kg C per $10^6$ Btu is equivalent to 1 g C per $4.22 \times 10^5$ J.)

Increasing the supply of renewable energy would replace carbon-intensive energy sources and significantly reduce U.S. global warming emissions. The Union of Concerned Scientists claim that producing electricity in the United States with 25% renewable energy would lower power plant $CO_2$ emissions by 277 million metric tons annually ($2.77 \times 10^{14}$ g $CO_2$) by the year 2025—the equivalent of the annual output from 70 typical (600 MW) new coal plants (Union of Concerned Scientists, 2009). The U.S. Department of Energy's National Renewable Energy Laboratory (NREL) explored the feasibility of generating 80% of the country's electricity from renewable sources by 2050. Although these goals were exceedingly optimistic, they found that renewable energy could reduce the electricity sector's $CO_2$ emissions by approximately 81% (NREL, 2012).

The potential of electricity generated from solar energy to reduce greenhouse $CO_2$ emissions is large. Photovoltaic cells convert solar energy directly to electricity. Solar thermal technologies are based on the concentration of solar radiation and its conversion to heat for running conventional steam-powered systems. New applications of high-efficiency multijunction solar cells make them costcompetitive with cheaper silicon cells and yield 40% efficiencies (*The Economist*, Mar. 30, 2019). Other technologies, such as solar towers and parabolic dishes, are being developed for commercial deployment. Photovoltaic battery recharging systems offer great potential in supporting the growing development of electric vehicles. Carbon dioxide reductions through improved efficiencies in housing ($-7\%$ reduction in $CO_2$) and transportation ($-10\%$ to $-15\%$ in $CO_2$) offer huge opportunities; by reaching 1970 levels by 2050. This could amount to as much as a 33.5% reduction in energy use, saving over 15 mt C per year ($10^{12}$ g C $yr^{-1}$) by U.S. households through reduction in $CO_2$ emissions by changed behavior (vehicle type, mileage driven, house size/design, etc.), in addition to low-carbon biofuels and electrification of vehicles and homes (Wei et al., 2011). While optimistic, it is possible.

**Biomass energy** is produced by growing crops or forest products that can be burned for fuel. Since, their carbon comes from the atmosphere rather than from fossil reservoirs, biofuels can in theory be carbon neutral, or close to it. The $CO_2$ generated from harvesting and transporting the biomass must be taken into consideration. If the greenhouse gases from a bioenergy plant are captured and sequestered, the whole system can be carbon negative while also making heat, electricity, and fuels. The more crops you plant, burn, and sequester, the more carbon dioxide you remove from the air. The pros of biomass energy are that it is renewable, dependable, near carbon neutral, widely available, and can be used in many forms. The cons are that it takes energy to grow and harvest bioenergy feedstocks, it is not totally clean when burned, it can lead to deforestation, it requires large acreages, and the biodiesel and ethanol products are not as energy efficient as gasoline.

**Hydropower** is likely approaching its potential in most developed countries. Hydropower is the leading renewable source for electricity generation globally, supplying 71% of all renewable electricity. Reaching 1064 GW of installed capacity in 2016, it generated 16.4% of the world's electricity from all sources. By the end of 2015 the leading hydropower generators among 150 countries were China, the United States, Brazil, Canada, India, and Russia (World Energy Council, 2019). Other than the production of concrete for the dams, this technology is virtually $CO_2$-emissions free. The technical potential for hydropower development around the world is much greater than the actual production: the percent of potential hydropower capacity that has not been developed is 71% in Europe, 75% in North America, 79% in South America, 95% in Africa, 95% in the Middle East, and 82% in Asia—Pacific (IEA, 2010). But dams replace global climate change impacts with local ecological impacts, primarily in-stream and riparian; a tradeoff preferably to be avoided.

In 2017, it was estimated that the cumulative installed wind power capacity could reach approximately 539,291 MW globally. New installations of wind power capacity reached 52,552 MW in 2017 (Statistica, 2019). The potential use of wind energy is estimated to be 7−10 etajoules (EJ) per year by 2020−25, which if accepted as a simple displacement of other energy sources represents 0.1−0.2 Gt C $yr^{-1}$ not emitted (Rosenbloom, 2007). In 2017, the electricity generated from wind turbines avoided an estimated 189 million tons of carbon. This reduction is equal to roughly 11% of the U.S. 2017 power sector emissions, or 40.3 million cars' worth of $CO_2$ emissions (AWEA, 2019).

**Improved energy efficiency**. Smart urban design, demand reduction, and improved GPP energy use intensity by industry are examples of energy efficiencies. Improved efficiency in buildings, both in their heating, ventilation, and air conditioning (HVAC) systems, and in their design offer significant energy savings (Brown and Levine, 1997). The delivery of energy services such as heating, cooling, and lighting drives consumer energy demand. The more efficiently these technologies operate, the less energy is required to deliver these energy services. If maximum efficiencies were obtained, demand for coal, oil, natural gas, and biomass would be reduced by 31%, 47%, 40%, and 40%, respectively, resulting in annual emissions savings of as much as 3.65 gigatons carbon ($3.65 \times 10^{15}$ g C $yr^{-1}$) as $CO_2$ globally (Cullen, 2018). Clearly, there are opportunities here.

**Industry and construction**. More than five billion metric tons ($10^{15}$ g) of cement—the raw material for concrete and mortar—is produced globally each year, adding another 6% to global carbon emissions. Steel, half of which goes into construction, contributes another 8%. If steel beams used in construction could be replaced by glulam (wooden beams laminated together), significant carbon savings might be achieved, assuming wood production is $CO_2$ neutral (*The Economist*, Jan. 5, 2019). In order to limit global warming to less than $2°C$, total emissions from energy use by industry alone will have to be reduced by 50%−80% by 2050 (*The Economist*, Dec. 1, 2018).

**Carbon neutral liquid fuels**. A hydrogen-based transportation system, mobile and using fuel cells or distributed from centrally located plants using coal feedstocks and onsite capture and sequestration technologies, could become alternate energy carriers for vehicles. Add to that the possibilities of synthetic fuels made from captured atmospheric $CO_2$ reacted with hydrogen, with synergies among technologies, and exciting possibilities emerge. Photo-electrochemical reduction of $CO_2$ is another feasible chemical process, whereby carbon dioxide is reduced to carbon monoxide or hydrocarbons by the energy of incident light. While electric vehicles can decarbonize ground transportation, air transportation is expected to double in the next 20 years, rising from 1.0 to $1.7 \times 10^{15}$ g $CO_2$, and it is not amenable to electrification; but $CO_2$-capture and hydrogen-based synthetic fuels could halve that number (The *Economist*, Dec. 1, 2018).

Most of these technologies are in early stages of development and deployment and the scale of actual carbon mitigation currently is in hundreds of tons of carbon annually. Remember, the IPCC's low-end estimate for the amount of carbon capture needed by 2100 is 100 gigatons ($10^{17}$ g C).

**Carbon sequestration**. Sequestration is the return of atmospheric carbon to long-term storage pools in the natural carbon cycle. Several decades ago, the U.S. Department of Energy explored the new, and then radical, approach of capturing and securely storing carbon emitted from the global energy system, i.e., carbon sequestration (Reichle et al., 1999). Most of these approaches are dependent upon development of carbon separation and capture technologies; except for those employing biospheric systems which absorb atmospheric $CO_2$ as part of the natural biogeochemical cycle of carbon. A number of technological approaches were examined, many of which are now being put into practice:

- Ocean sequestration such as direct injection, formation of $CO_2$ hydrates in the deep sea, open ocean iron fertilization,
- Carbon sequestration in terrestrial ecosystems using photosynthetic fertilization, afforestation, and soil amendments (Table 12.4). It would take 766,000,000 acres of forests to offset 16% of 2018 domestic U.S. $CO_2$ emissions,
- Sequestration of $CO_2$ in geologic formations trapping $CO_2$ in geological formations such as depleted oil and gas fields, brine formations, and coal formations, using reasonably well-understood hydrodynamics and solubility processes,

**TABLE 12.4** The potential of different terrestrial biomes to sequester carbon that might be sustained over a 25–50 year period (Reichle et al., 1999).

| Biomes | Potential C sequestration (Gt C yr$^{-1}$) |
| --- | --- |
| Agricultural lands | 0.85–0.90 |
| Biomass croplands | 0.5–0.8 |
| Grasslands | 0.5 |
| Rangelands | 1.2 |
| Forests | 1–3 |
| Urban forest and grasslands | ————[a] |
| Deserts and degraded lands | 0.8–1.3 |
| Terrestrial sediments | 0.7–1.7 |
| Boreal peatlands and other wetlands | 0.1–0.7 |
| Total | 5.65–10.1 |

[a]*No estimate available.*
Courtesy, Oak Ridge National Laboratory and the U.S. Department of Energy.

- Advanced biological processes using enzymes for biopolymer production from sugars, and microbial production of recalcitrant cellulosic compounds, and,
- Advanced chemical approaches employing chemical technologies allowing gaseous $CO_2$, or its constituent carbon, to be transformed into materials that are benign.

Now several decades later, the U.S. National Academies of Sciences, Engineering, and Medicine have issued a research agenda for negative emission and carbon sequestration technologies (NASEM, 2019), which reviews the current state-of -the-art of removing $CO_2$ emissions from the carbon cycle.

One of the most powerful tools in fighting climate change is beneath our feet. About 75%, or 1456 Gt ($1.456 \times 10^{18}$ g), of carbon in the terrestrial biosphere resides in the detritus of soil carbon, with an additional 55.2 Gt C in litter and another 560 Gt C in plants (Reichle et al., 1999). Accumulation of soil carbon is dependent primarily upon temperature and moisture control over decomposition rates, and follows a pattern opposite that of NPP, increasing in higher latitudes. Soil carbon represents the long-term NEP of terrestrial ecosystem, with woody biomass as the intermediate-term NEP sink. However, the current rate of $0.04 \times 10^{15}$ g C yr$^{-1}$ carbon storage in northern terrestrial ecosystems alone is too small to be a significant sink for human release of $CO_2$

to the atmosphere from fossil fuels (Schlesinger, 1997). But, terrestrial eco-systems as a whole have great potential (Table 12.4). Net accumulation by terrestrial biosphere during the 1990s was $\sim 0.3 \times 10^{15}$ g C yr$^{-1}$ (Reichle et al., 1999). Woodlands, prairies, algae, mangroves, wetlands, and soil withdraw carbon dioxide from the atmosphere and keeps it from going back, tipping the balance negative. Every acre of restored temperate forest can sequester 3 metric tons ($3 \times 10^6$ g) of carbon dioxide per year. In the United States, forests already offset about 13% of the country's carbon emissions. Globally, forests absorb 30% of humanity's emissions. Therefore, restoring forests can be an effective way to reduce the concentration of carbon dioxide in the air (Marland, 2005; Kumar and Nair, 2011). There is enough room globally (3.5 million square miles, roughly the size of the U.S. or three times that of India) primarily in Russia, U.S., Canada, Australia, Brazil, and China for 1 trillion more trees (on top of the 3 trillions of trees now on earth) which over decades could absorb 830 billion tons of $CO_2$ — about as much as humans have omitted in the past 25 years (Bastin, 2019). And this might be an underestimation, if one includes the potential of soil cabon reservoirs and the possibilities if replanting occurs on depleted soils.

**Adapting to climate change**. Adaptation will mean anticipating the adverse effects of climate change and taking the appropriate actions to prevent or minimize the eventual impacts. Well-planned, early adaptation that anticipates changes, conserves resources, money, and lives later (USGCRP: Forth National Climate Assessment, 2018). Adaptation to climate change has already become necessary. Examples of adaptation measures include:

- using scarce water resources more efficiently in domestic, industrial, and agricultural practices,
- adapting building codes to future climate conditions and extreme weather events,
- building natural and engineered flood defences, raising the levels of dykes, and new construction zoning codes,
- developing drought- and heat-tolerant crops, and resistance to new diseases and pests
- choosing tree species and forestry practices less vulnerable to storms and fires,
- setting aside critical habitats and land corridors to help species migrate,
- changing lifestyles to protect human health, and
- developing conservation strategies to protect the world's biodiversity.

New technologies will need to be implemented, including genetic engineering of agricultural crops and the use of more efficient irrigation systems. Desalinization plants will need to be deployed, ideally run with "carbon free" power, to supply impacted regions with domestic and agricultural water supplies. In response to rising sea levels and more frequent major storm events, coastal zone management will become a tremendous challenge. And the genetic engineering of plants and animals will become a necessity to protect gene pools and biological productivity. But, the most critical challenge will be understanding how to protect the integrity of the global biogeochemical cycle of carbon, so as to maintain a global carbon balance suitable to sustain life on Earth.

## 12.5 The economics of clean energy

**Pricing carbon**. In order to effectively limit emissions of greenhouse gases (GHGs) the cost of using carbon as an energy source needs to be high enough to motivate the necessary changes needed in economic behavior (Nordhaus, 1992). Through the market mechanism, a high carbon price raises the cost of products according to their carbon content. Raising the price of carbon achieves awareness in consumers about what goods and services are carbon intensive and which should be used more sparingly. Carbon pricing helps to educate producers about which energy sources are the most carbon intensive, and thus encourages firms to substitute low-carbon fuels. Pricing carbon provides the market with incentives for inventors and innovators to develop and introduce low-carbon products and processes that can replace the current generation's technologies.

Meanwhile, in 2019 the Clean Power Plan (CPP) issued by The U.S. Environmental Protection Agency, which would reduce $CO_2$ emissions by 19% by 2030, remains bogged down in legal challenges; and, the current administration's proposed replacement of the CPP with the Affordable Clean Energy rule would reduce $CO_2$ emissions by only 0.7%−1.5% (*The Economist*, Aug 2018). That is moving in the wrong direction. The data on decarbonizing the energy system show that if you want to avoid catastrophic climate impacts from future atmospheric $CO_2$ levels of $\sim 450$ ppm, a significant price for carbon will be more important than technology breakthroughs, although mandates, subsidies, and other government deployment programs will be needed before the $CO_2$ price becomes effective (Romm, 2008). Even with carbon pricing, aggressive energy efficiency will continue to be important.

**Carbon credits**. In the early 1980s the concept of swapping national debt with developing countries to protect natural resources was proposed as a means of protecting biological diversity (Lovejoy, 1984). The debt-for-nature swaps became a model for carbon credits. An objective of the Kyoto Protocol was to enable developed nations, which had profited from economic development based upon high-carbon GDPs, to economically assist the growing economies of developing nations, impacted by carbon emission constraints and heavily indebted to foreign creditors. A carbon credit system was devised that imposed national caps on greenhouse gas emissions of developed nations that ratified the Kyoto Protocol. These countries were aligned as Annex B countries. Each of these countries was given an allotment and corresponding number of emission allowances known as Assigned Amount Units (AAUs). Participating countries were required to reduce their emissions to well below 1990 levels and more than 5% by 2012. They could also reduce their emissions by trading in emission allowances with countries that already had surplus allowances. They could meet their targets by buying carbon credits.

Carbon credits and carbon markets are a component of national and international attempts to mitigate the growth in concentrations of greenhouse gases in the atmosphere. A carbon credit (often called a carbon offset) is a credit for greenhouse emissions reduced or removed from the atmosphere by an emission reduction project, which can be used by governments, industry, or private individuals to compensate for the emissions they generate elsewhere. Since GHG mitigation projects generate credits, this approach can be used to finance carbon reduction schemes between trading partners around the world.

One carbon credit is equal to one metric ton of carbon dioxide, or in some markets, carbon dioxide equivalent gases ($CO_2$-eq), and are bought and sold through international brokers, online retailers, and trading platforms. Businesses that find it difficult to comply with the carbon emission requirements can purchase carbon credits to offset their emissions by making finance readily available to renewable energy projects, forest protection, and reforestation projects around the world. Projects which sell carbon credits include wind, solar, geothermal, and biomass which replace fossil-fuel-powered plants. Offsetting one metric ton of carbon means that there will be one less Mt of carbon dioxide in the atmosphere than there would otherwise have been. The Kyoto Protocol provides for three mechanisms that enable countries, or operators in developed countries, to acquire greenhouse gas reduction credits:

- Under Joint Implementation (JI) a developed country with relatively high costs of domestic greenhouse reduction would set up a project in another developed country.
- Under the Clean Development Mechanism (CDM) a developed country can "sponsor" a greenhouse gas reduction project in a developing country where the cost of greenhouse gas reduction project activities is usually much lower, but the atmospheric effect is globally equivalent. The developed country would be given credits for meeting its emission reduction targets, while the developing country would receive the capital investment and clean technology or beneficial change in land use.
- Under International Emissions Trading (IET) countries can trade in the international carbon credit market to cover their shortfall in Assigned Amount Units (AAUs). Countries with surplus units can sell them to countries that are exceeding their emission targets under Annex B of the Kyoto Protocol.

These carbon projects can be created by a national government or by an operator within the country. A question has been raised over the grandfathering of allowances. Countries within the European Union's Emission Trading System (EU ETS) have granted their incumbent businesses most or all of their allowances for free. This can sometimes be perceived as a protectionist obstacle to new entrants into their markets.

**Cap and trade**. One method to reduce carbon emissions is "Cap and Trade." The Kyoto Protocol was the precursor to mandatory carbon credits. Cap and trade and carbon taxing both involve putting a price on carbon emissions. Pricing carbon involves "costing" the environmental damage from $CO_2$ emission, a highly controversial topic. The carbon tax is expensive for fossil fuel interests, but the revenues can go back to the public to offset increased power cost borne by the consumers (Whitehouse, 2018). Taxing carbon moves energy markets toward cheaper renewables. On balance, people save money. Taxing carbon moves energy markets toward cheaper renewables. The savings from avoiding climate catastrophes can be immense. If you really want to innovate, there has to be a cost to carbon pollution. Without that, there is little incentive to innovate.

The fossil fuel industry benefits from global subsidies of $5.3 trillion a year, equivalent to $10 million a minute every day, according to a startling new estimate by the International Monetary Fund (IMF) (Coady et al., 2015). The IMF says that the figure is an "extremely robust" estimate of the true cost of fossil fuels. The $5.3 \times 10^{12}$ subsidy estimated for 2015 is greater than the total health spending of all the world's governments. IMF calculated this sum as the costs not paid by polluters that are imposed on governments by the burning of coal, oil, and gas. These costs include the harm caused to local populations by air pollution, as well as to people across the globe affected by the floods, droughts and storms being driven by climate change. This subsidy warps the economy; it discourages technological innovation and behavior necessary to reduce $CO_2$ emissions.

The cap on greenhouse gas emissions that drives global warming is a firm limit on pollution. The trade part is a market for companies to buy and sell allowances that let them emit only a certain amount of carbon, as supply and demand set the price. Trading gives companies a strong incentive to save money by cutting emissions in the most cost-effective ways. The Cap and Trade System is one of the best ideas available right now to help limit emissions, but the system is not perfect as seen in comparing the pros and cons of Cap and Trade (Gaille, 2015).

**The pros of Cap and Trade are**:

- New economic resources for industries are created. Companies lower their emissions because there is a low cost for implementation. Companies with emissions credits can sell them for extra profit. This creates economic resources, because more is spent to lower emissions and the credits are a new product to be purchased for additional profits,

- A predetermined maximum level of emissions is set. However, it can be difficult to track atmospheric emissions by companies that are not regulated,
- Taxpayer resources are supplemented. The government purchases emissions credits when they are available and then sells them at a higher price to businesses when they are needed. The incomes from these purchases help to supplement the resources that taxpayers are providing to the government,
- Alternative energy resources can be funded. Research on alternative energy resources can be funded with carbon taxes in those places that work with the cap trade system,
- The average person can create change. The cap and trade system creates a new knowledge base for consumers who can choose whether or not to purchase from carbon-efficient businesses,

**While the cons of Cap and Trade are:**

- Emission credits can be given away. Credits can be sold at auction to the highest bidder. This means that a business can expand their emissions which harm a local economy which receives no economic gain in return,
- Governments can retire emissions credits. Governments also have access to credits because they also have an economic impact on society. What makes the government different from a business when an emissions credit is received is that the credit can be canceled and removed from circulation. This means taxpayer money is used to purchase something that isn't used and could potentially stagnate industrial development,
- Some credits may be artificially high in price. Many environmental agencies have discovered that they can purchase these credits and choose not to use them and hold them indefinitely, creating artificially high credit prices,
- Emissions credits are almost always cheaper than converting. For industries that use fossil fuels, the cost of converting to renewable resources can be very high. The emissions credits, offsets, and even penalties and fines for exceeding a cap limit, often are all cheaper than going through a conversion to a new source of energy. This means there is no real incentive for those industries to change their practices,
- It is relatively easy to cheat the system. Most industries do not have monitoring devices installed to determine how much output is really occurring. This makes it very easy for the average business to cheat on its emissions reports,
- Higher prices for goods and services could be created. Renewable energy resources are still relatively new and are still relatively expensive. For industries that transition into lower emissions and follow cap rules, the products that they produce will be more expensive in the future, and

- Different nations may have different standards for a maximum cap. Some nations create more emissions than others. This means a maximum cap will be defined differently in every society. Some countries may be very lenient about their emissions caps and credits. Others may be very strict.

**Carbon tax**. Another method for national governments to meet their $CO_2$ reduction goal is to tax the emissions of $CO_2$ generators. The advantages of a carbon tax are:

- Less complex, expensive, and time-consuming system to implement. This advantage is especially great when applied to markets like gasoline or home heating oil.
- Reduced risk for certain types of cheating. But under both credits and taxes, emissions must be verified.
- Reduced incentives. Companies may delay efficiency improvements prior to the establishment of the baseline, if credits are distributed past emissions.
- More centralized handling of acquired gains. It is simpler to administer, easier to ensure compliance, and better to monitor progress.
- Stabilization of the worth of carbon by government regulation rather than market fluctuations. Poor market conditions and weak investor interest have a lessened impact on taxation as opposed to carbon trading. But, when credits are grandfathered, it puts new or growing companies at a disadvantage relative to more established companies.

There are arguably some advantages of a tradable carbon credit over a carbon tax. The price may be more likely to be perceived as fair by those paying it. Investors in credits may have more control over their own costs. The flexible mechanisms of the Kyoto Protocol help to ensure that all investment goes into genuine sustainable carbon reduction schemes through an internationally agreed-upon validation process. Some proponents state that, if correctly implemented, a target level of emission reductions may somehow be achieved with more certainty, while under a tax the actual emissions might vary over time. Tradable carbon credits can provide a framework for rewarding people or companies who plant trees or otherwise meet standards recognized as "green."

A new study (Parry et al., 2018) finds that carbon taxes raise about twice as much revenue as cap and trade, and are approximately 50% better at cutting emissions. A tax of $70 on each metric ton of $CO_2$ by 2030 would raise between 1% and 2.5% of GDP in the G20 club of big economies, and allow most to approach their 2015 Paris climate agreement pledges (Parry et al., 2018). But, citizens of the State of Washington, for the second time in 2 years, have rejected a $15 per ton carbon "fee," which by 2013 would have raised one billion dollars annually (*The Economist*, Nov 2018). *The Wall Street Journal* reported that the state of Washington's tax would have reduced global emissions by only 0.02%. This illustrates how difficult it is to get started without everyone world-wide participating; local initiatives with local costs yield small global results, but you must start someplace. The IPCC has estimated that a price of well over $135 a ton would be necessary to have any effect on future climate. Energy taxes are going to be difficult to sell. Broader-scale, more universal participation might help to get things started. The need for governmental leadership is critical.

**Estimated costs of mitigation.** Projected damages and potential costs of climate change have been estimated for the United States (USGCRP, 2018), using the RCPs (representative concentration pathways) from the Intergovernmental Panel on Climate Change (IPCC AR5, 2014). The RCPs are scenarios (Table 12.5) that describe how the planet might change by year 2100 relative to preindustrial values (+2.6, +4.5, +6.0, and +8.5 W m$^{-2}$, respectively). The RCPs are defined by their total radiative forcing (cumulative measure of human emissions of greenhouse gases from all sources expressed in Watts per square meter). The high concentration pathways are dependent on assumptions of abundant fossil fuel for future production.

After beginning at slightly over 400 ppm in 2018, the four atmospheric $CO_2$ concentration pathways begin to separate. RCP2.6 peaks at just over

**TABLE 12.5** Relative concentration pathways (RCPs) with pathway descriptions and integrated assessment models used by IPCC for the year 2100. Source: IPCC, 2014.

| Scenario | Description | IA model | Reference |
|---|---|---|---|
| RCP 8.5 | Rising relative forcing 8.5 W m$^{-2}$ in 2100 | MESSAGE | Riahi et al., 2007 |
| | | | Rao and Riahi, 2006 |
| RCP6 | Stabilization without overshoot pathway | AIM | Fujino et al., 2006 |
| | To 6.0 Wm$^{-2}$ at stabilization after 2100 | | Hijioka et al., 2008 |
| RCP4.5 | Stabilization without overshoot pathway | GCAM | Smith and wigley, 2006 |
| | To 4.5 Wm$^{-2}$ at stabilization after 2100 | (MiniCAM) | Clarke et al., 2007 |
| | | | Wise et al., 2009 |
| RCP2.6 | Peak in radiative forcing at ∼ 3 Wm$^{-2}$ before 2100 and decline thereafter | IMAGE | Van Vuuren et al., 2006, 2007 |

**Source:** Courtesy, IPCC.

450 ppm $CO_2$-eq in the 2040−2050 time period and slowly decreases thereafter, remaining below 450 ppm until 2100 and beyond. RCP4.5 and RCP6 track each other until separating around year 2065; by year 2100, RCP4.5 is near 575 ppm $CO_2$-eq and RCP6 is at around 725 ppm $CO_2$-eq. RCP8.5 shows exponential growth in atmospheric $CO_2$-eq concentrations, rising rapidly and continuously until reaching 1225 ppm in year 2100. Atmospheric $CO_2$-eq concentrations resulting from RCP4.5 and RCP6 show hints of stabilizing after year 2100; not so for RCP8.5, but by that time the world will not be as we know it now.

The mid- and late-21st century (averages for 2046−2065 and 2081−2100) projections of global mean sea level rise are derived from the IPCC Fifth Assessment Report scenarios from Working Group 1 (IPCC, 2014) were shown previously in Table 11.5. The projections are relative to temperatures and sea levels in the late 20th to early 21st centuries (1986−2005 average). Temperature projections can be converted to a reference period of 1850−1900 or 1980−99 by adding 0.61°C or 0.11°C, respectively. Across all RCPs, global mean temperature is projected to rise from a minimum of 0.3°C to a maximum of 4.8°C by the late 21st century (Table 11.3). Across all RCPs, global sea level increase is projected to rise by from a minimum of 0.26 m to a maximum of 0.82 m by the late in the 21st century (Table 12.6).

Many of the impacts resulting from climate change, primarily those resulting from temperature increases and variability, change in precipitation patterns and amount, and sea level rises can be substantially reduced over the course of the 21st century through global-scale reductions in greenhouse emissions (USGCRP, 2018). The effect of mitigation in avoiding climate change impacts becomes evident by 2020, but early and substantial mitigation offers increased chances of avoiding adverse impacts (Table 12.7). The costs of avoidance of climate change risks from mitigation actions are dependent, in part, upon the adaptive changes by populations and their associated costs. Adaptation can require substantial upfront costs and long-term maintenance commitments (e.g., sea walls). They had better be prepared, for the sea is coming.

Coastal flooding is an immediate concern. The Union of Concerned Scientists estimates that 2.5 million present-day coastal properties in the United States alone, worth $1.1 trillion, could be as risk of flooding as frequently as every two weeks (The Economist, August 17, 2019). New York is paying $800 million for the Big U, a necklace of walls, parks and elevated roads to shield lower Manhattan. Mumbai is building 4 coastal sea walls, and Bamgladesh is doubliing its coastal embankment system. Indonesia intends a $40 billion wall to seal off Jakarta. Over the centuries, the Dutch have already invested in a system, of dykes, levees and seawalls to protect the 27% of their lands that are below sea level. This inlcudes the massive Maeslant barrier across a 360-meter-wide canal at Rotterdam. These are just a few examples. More will follow.

The complex interaction between mitigation and adaptation are inherent in the economic damage estimates summarized in Table 12.7. Annual damages in

**TABLE 12.6** Future sea level rise (in meters) projected from different radiative forcing function scenarios from assumptions of different GHG emissions.

| Forcing Function Scenario | 2046 − 2065 Mean and likely range (meters) | 2081- 2100 Mean and likely range (meters) |
|---|---|---|
| RCP2.6 | 0.24 (0.17 to 0.32) | 0.40 (0.26 to 0.55) |
| RCP4.5 | 0.26 (0.19 to 0.33) | 0.47 (0.32 to 0.63) |
| RCP6.0 | 0.25 (0.18 to 0.32) | 0.48 (0.33 to 0.63) |
| RCP8.5 | 0.30 (0.22 to 0.38) | 0.63 (0.45 to 0.82) |

Source: IPCC, 2013.

2015 dollars for year 2090 are derived from the high RCP scenario, RCP8.5, relative to a no-change scenario. RCP8.5 projects an atmospheric concentration of $\sim 1225$ $CO_2$-eq. ppm by year 2100. The benefits obtained from a decrease in damages from mitigation strategies yielding a lower RCP pathway and an atmospheric concentration of $CO_2$-eq. near 575 ppm in year 2100 (RCP4.5) are shown in the second column of Table 12.7. Reductions in damages range from 11% in agriculture to 107% for winter recreation, with a substantial avoidance in human temperature-induced mortality by 58%, electricity supply by 63%, inland flooding by 47%, coastal property damage by 22%, and water supplies of 33%. Damages costs exceed 480 billion dollars annually under RCP8.5 for just the United States. Damages that can be avoided put the costs of mitigation and decarbonization of our energy systems into perspective.

Global-scale mitigation does not come without effect upon GDP (Table 12.8), but the costs are offset by the cost avoidance of climate change impacts resulting from doing nothing. What do you do with this carbon dioxide once you have it? Carbon Engineering (2016) is working on an air to fuels pathway. Three direct air capture plants are operating, together capturing 1100 metric tons of $CO_2$ dioxide per year (Climeworks, 2019). In Iceland the direct air capture plant is turning its captured $CO_2$ into basalt rock, while in Switzerland the gas is used as a fertilizer in a greenhouse, and in Italy the $CO_2$ is used to make methane fuel for trucks. However, more than 800,000 times the current annual direct air capture capacity will be needed by 2100, to rely on this method alone to limit warming to 1.5°C. Other capture technologies have been developed and deployed. A plant in Canada captures $\sim 1$ metric ton of $CO_2$ day$^{-1}$, removing $CO_2$ from the atmosphere since 2015 and turning it into fuel since 2017 (Carbon Engineering, 2016). But getting governments and companies to invest in these technologies requires a price on carbon.

**TABLE 12.7** Projected annual economic damage estimates (in 2015 $) in the United States by 2090.

| Sector | Annual damages under RCP8.5 | Damages avoided under RCP4.5 |
|---|---|---|
| Labor | $155B | 48% |
| Extreme temp. Mortality | $141B | 58% |
| Coastal property | $118B | 22% |
| Roads | $20B | 59% |
| Electricity supply/demand | $9B | 63% |
| Inland flooding | $8B | 47% |
| Urban ddrainage | $6B | 26% |
| Rail | $6B | 36% |
| Water quality | $5B | 35% |
| Coral rreefs | $4B | 12% |
| West nile vvirus | $3B | 47% |
| Freshwater fish | $3B | 44% |
| Winter recreation | $2B | 107% |
| Bridges | $1B | 48% |
| Munic. and indus. Water supply | $316M | 33% |
| Harmful aalgal blooms | $199M | 45% |
| Alaskan infrastructure | $174M | 53% |
| Shellfish | $23M | 57% |
| Agriculture | $12M | 11% |
| Aeroallergens | $1M | 57% |
| Wildfire[a] | -$106M | −134% |

[a]Wildfire shows a small positive effect due to projected landscape-scale shifts to vegetation with longer fire-return intervals.
**Source:** USGCRP, Martinich et al., 2018.

Direct air capture, for example, would be especially useful for offsetting some of the hardest sectors to decarbonize, such as air travel. However, these companies estimate that it would cost about $100 per ton to withdraw $CO_2$ from

**TABLE 12.8** Estimated global macro-economic costs in 2030[a] relative to the baseline for least-cost trajectories toward different, long-term stabilization levels[b,c].

| Stabilization levels (ppm $CO_2$ eq) | Median GDP Reduction[d], % | Range of GDP Reduction[d,e], % | Reduction of average GDP growth rates[d,f] (percentage points) |
|---|---|---|---|
| 590–710 | 0.2 | 0.6 - 1.2 | <0.05 |
| 535–590 | 0.6 | 0.2 - 2.5 | <0.1 |
| 445–535[g] | Not available | <3 | <0.12 |

[a] For a given stabilization level, GDP reduction would increase over time in most models after 2030. long-term costs also become more uncertain. [Figure 3.25].
[b] Results based on studies using various baselines.
[c] Studies vary in terms of the point in time stabilization is achieved; generally this is in 2100 or later.
[d] This is global GDP based market exchange rates.
[e] The median and the 10th and 90th percentile range of the analyzed data are given.
[f] The calculation of the reduction of the annual growth rate is based on the average reduction during the period till 2030 that would result in the indicated GDP decrease in 2030.
[g] The number of studies that report GDP results is relatively small and they generally use low baselines.
**Source:** IPCC Climate Change 2007c.

the air, so a carbon price would have to be higher than that. Or, the technology has to become much, much cheaper (Irfan, 2018; *The Economist*, Dec. 1, 2018).

Governments face serious economic challenges in making decisions about how many financial resources need to be diverted from raising the standard of living and used to limit global warming (Table 12.8). Developing nations aspire to attain the economic prosperity that other nations have acquired using fossil fuels that have caused our problem today. International equity drives decisions as to how the pain of decarbonization should be shared among nations. But decarbonizing the global energy system need not be all pain. New research by the C40 group of big cities and the Global Covenant of Mayors representing more than 9000 municipalities, finds that climate policies such as increasing energy efficiency and decarbonizing public transportation and power generation could create 14 million new jobs and prevent 1.3 million pollution-related deaths by the year 2030 (*The Economist*, Sept. 2018). Climate change is already being shown to threaten American's health (Salas et al., 2018).

**The Discount Factor.** Because the benefits of reduced greenhouse gas emissions last for centuries, while many mitigation costs are borne today, climate change has presented a very difficult economic and political challenge. But issues with very long horizons do not just raise political questions; they pose thorny analytical questions as well. Deciding how quickly and aggressively to respond depends on how much we think a given amount spent on

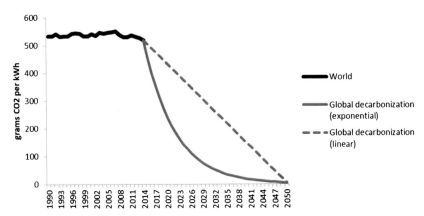

**FIGURE 12.5** Carbon intensity of electricity: history and forward trends necessary to reach a zero-carbon electricity grid by mid-century. *Source: Clean Air Task Force from International Energy Agency data, after Cohen, 2017, Courtesy Clean Air Task Force.*

mitigation today will benefit future generations. Converting future costs and benefits into today's currency requires the process of discounting, a calculation that accounts for the fact that people value present costs and benefits more than future costs and benefits. The U.S. Environmental Protection Agency and U.S. Office of Management and Budget have clearcut discounting guidelines when it comes to assessing the costs and benefits of environmental policies that have effects lasting a few decades, but those lines blur in the case of climate change, which will affect future generations. Yet in benefit–cost analyses, the rate at which future benefits and costs are discounted can have enormous implications for policy prescription. For example, the central issue separating the aggressive policy recommendations advocated in Nicholas Stern's 2007 report on climate change from the estimates of the damages associated with climate change by U.S. federal agencies is Stern's controversial decision to discount future damages at a much lower rate.

## 12.6 What has been the impedance?

In order to stabilize the planet at warming of no more than 2°C above pre-industrial levels, a zero-carbon-emitting electric grid will have to be reached by 2050 (Fig. 12.5). This will be necessary because electricity contributes approximately a third of global energy-related carbon emissions, and reductions in carbon emissions in the transport and industrial sectors also will require increased electrification, powered by an expanded zero-carbon grid. This will be a tremendous challenge to technological deployment and political leadership by the world economies.

In his 1994 book on the economics of climate change Nordhaus (1994) recommended a gradual reduction in carbon emissions, while a report (Stern et al., 2006) for Britain's government recommended dramatic efforts by the world's advance economies amounting to 1%−2% of GDP. The sharp difference is mainly due to different assumptions in discount rates. The discount rate is the rate of return on investment, or in other words how much we think a given amount spent on mitigation today will benefit future generations. Given a 5% discount rate, one life today is worth 132 a century later (*The Economist*, Dec. 2018). But the longer the delay, the more difficult it will be to correct the problem. The benefits of strong and early action far outweigh the economic costs of not acting.

What progress is being made in decarbonization of the electric grid (Fig. 12.6)? France and Sweden are very close to decarbonizing their electrical grids due to their large shares of nuclear energy (France, 75%; Sweden, 40%) and Sweden's hydroelectric resources. Canada's carbon intensity remains low. Development is driving Brazil, with a low intensity, to greater carbon emissions. Some regions and countries are making modest progress (Ritchie and Roser, 2017; Cohen, 2017), such as in the European Union (EU), although the rate of EU fossil displacement slowed in 2015. The most progress was made in China (which had the highest carbon intensity to begin with), due to an increase in hydroelectric, wind, solar, and nuclear energy deployment for new energy-production facilities, and in the United States, where the substitution of gas for coal is responsible for 75% of the reduction in coal for of electricity production in recent years. Russian utilization of gas is also resulting in improvements. But natural gas is an economic and resource convenience, and not based upon a long-term decarbonization strategy. The Middle East and Africa are making no progress. Japan, India, and Indonesia, all with high carbon intensities of electrical production, are heading in the wrong direction. Even

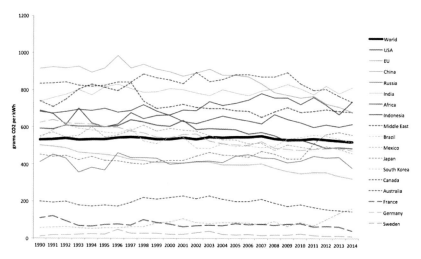

**FIGURE 12.6**   The world's economies vary considerably in how efficiently their GDPs utilize carbon-based fuels. *Image credit: Ritchie and Roser, 2017. Courtesy, Our World in Data.*

with solar and wind investments reaching nearly $300 billion per year globally, the overall carbon intensity of electricity production for the world has not changed substantially.

If we are to have a hope of meeting critical climate targets, the world will need to radically improve the pace of decarbonization, with efficiencies in electrical transmission and storage. Numerous studies have warned that we can have a greater chance of achieving these goals only if we expand a more affordable, practical, and broad range of technological options for zero-carbon electricity that can scale up by mid-century, including energy efficiency, renewables, and nuclear energy, as well as carbon capture and storage (Jenkins and Thernstrom, 2017; Clack et al., 2017; Loftus et al., 2015).

Nathaniel Rich in a work of history in the *New Yorker Magazine*, addressed the 10-year period from 1979 to 1989 as the decisive decade when humankind first came to a broad understanding of the causes and dangers of climate change.

*"We knew everything we needed to know, and nothing stood in our way;* nothing, that is, except ourselves."

(Rich, 2018)

The U.S. scientific community has long led the world in research on public health, environmental science, and other issues affecting the quality of life. Our scientists have produced landmark studies on the dangers of DDT, tobacco smoke, acid rain, and global warming. But at the same time, a small yet potent subset of this community leads the world in vehement denial of these dangers. In their new book, *Merchants of Doubt*, historians Naomi Oreskes and Erik Conway (Oreskes and Conway, 2010) explain:

*"...how a loose−knit group of high-level scientists, with extensive political connections, ran effective campaigns to mislead the public and deny well-established scientific knowledge over 4 decades. In seven compelling chapters addressing tobacco, acid rain, the ozone hole, global warming, andDDT, Oreskes and Conway roll back the rug on this dark corner of the American scientific community, showing how the ideology of free market fundamentalism, aided by a too-compliant media, has skewed public understanding of some of the most pressing issues of our era."*

(Oreskes and Conway, 2010)

Andrew Rowell, a freelance journalist and green campaigner put it succinctly:

*"Many will know the reason for our collective failure: In those 30 years we have witnessed one of the most sophisticated, expensive and devious corporate public relations campaigns ever undertaken by the oil industry to deny the science, delay regulatory action, and obfuscate the truth over climate change. As many repeatedly pointed out, including me in articles, blogs, and books, the oil*

*industry campaign learnt much of tactics from the masters of deception and manipulation, the tobacco industry. If the oil industry had spent even a tiny proportion of the money it had funding fake front groups and scientists, and poured it into renewable energy, then we would not be facing such a climate crisis as we are now."*

(Rowell, 2018)

The world knows the problem it faces, the stakes are huge, and solutions are within reach. Besides the special interest lobby that has bred a culture which rejects the evidence, another obstacle is that the world has no history or institutions to deal with such a formidable problem (The Economist, Dec. 1, 2018).

## 12.7 Is it too late to act?

Little of the basic information about the global carbon cycle and the consequences of climate change is new. We have known about the essentials for over 40 years; and science predicted what is now unfolding. Wally Broecker first sounded the alarm in 1975 in the journal *Science* with "Global Change: Are We on the Brink of a Pronounced Global Warming?" where he predicted the current rise in global temperature as a result of increases in atmospheric $CO_2$ and popularized the term "global warming." Thirty years later, in his book "An Inconvenient Truth: The Planetary Emergency of Global Warming and What We Can Do about It" (Gore, 2006), Vice President Al Gore wrote in a section called "The Politicization of Global Warming":

*"As for why so many people still resist what the facts clearly show, I think, in part, the reason is that the truth about the climate crisis is an inconvenient that means we are going to have to change the way we live our lives."*

(Gore, 2006)

The IPCC 2018 report has been issued (IPCC, 2018). In just one sentence, Gavin Schmidt, director of NASA's Goddard Institute for Space Studies, captured the essence of the report, stating during an interview with PBS News Hour:

*"The key thing to remember is that it's clear that the best time to have reduced emissions was 25 years ago, but the second best time to reduce emissions is right now."*

(Schmidt, 2018)

Meanwhile, the warning signs are appearing more frequently worldwide. In just the United States, consider sea level rise and the forthcoming disappearance of Tangier Island in Chesapeake Bay and the threatened Pacific Islands. Consider the coastal storm flooding at Sandy Hook, New Jersey, and elsewhere along the Atlantic seaboard. There has been increase in storm events (Lewis,

2018), and Atlantic hurricane frequencies and strength due to the higher energy in warmer ocean waters. Drought frequencies are increasing and aquifers are vanishing in the western U.S. grain belt. Has anyone not noticed the rapid rise in severity and frequency of forest fires? What about the increasing salt water intrusion of the city of Miami's ground water aquifers? And the growing municipal water shortages in California and elsewhere? The list goes on and on. Governments are beginning to respond, but the city of Boston's proposed 10−20 billion-dollar investment in sea walls, and New York City's recently announced $10 billion climate change damage averting plan, seem singular and costly consequences of adaptation at just one location for society's appetite for carbon-rich power generation— a high adaptive cost when a globally prorated decarbonization investment might be more economical.

Tired of waiting for national leadership, local governments are beginning to take action. California passed a law in September 2018 establishing the year 2045 as the deadline for the state's complete transition to renewable and zero-based carbon electricity. CDP, a climate action consortium (CDP, 2019) reports that 620 cities and 122 regions have reported climate actions; 800 firms worth $17 trillion have joined the We Mean Business coalition to reduce carbon footprints; and NAZCA (Non-State Actor Zone for Climate Action), a UN operated information repository, lists more than 12,500 pledges by 2500 cities, 209 regions, over 2100 firms, and nearly 500 investors (*The Economist*, Sept., 2018).

Pledges and plans are in progress, but what is needed are accomplishments. Meanwhile, what more can we learn from ecological energetics and our scientific understanding of the global carbon cycle that can contribute to our understanding of the causes of these problems and their possible solutions? And what confidence can we have that humanity will use this knowledge when it has shown such a tepid response to date?

## 12.8 Recommended reading

Gore, A., 2006. An Inconvenient Truth: The Planetary Emergency of Global Warming and What We Can Do About it. Rodale, Emmaus, PA. USA, 325 pp. https://www.worldcat.org/title/inconvenient-truth-the-planetary-emergency-of-global-warming-and-what-we-can-do-about-it/oclc/900477783.

Hansen, J., 2009. Storms of My Grandchildren: The Truth about the Coming Climate Catastrophe and Our Last Chance to Save Humanity. Bloomsbury Publishing. https://www.bloomsbury.com/us/storms-of-my-grandchildren-9781608195022.

IPCC, 2007a. Climate Change 2007: Synthesis Report. Contribution of Working Groups I, II and III to the Fourth Assessment Report of the Intergovernmental Panel on Climate Change Core Writing Team. Pachauri, R.K., Reisinger, A. (Eds.). IPCC, Geneva, Switzerland, 104 pp.

National Laboratory Directors, 1997. Technology Opportunities to Reduce Greenhouse Gas Emissions. Appendix B as a Separate Volume. http://www.ornl/climatechange/.

NASEM, National Academies of Sciences, Engineering, and Medicine, 2018. National Emissions Technologies and Reliable Sequestration: A Research Agenda. National Academies Press, Wash., D.C., 351+pp. https://org/10.17226/25259/

Nordhaus, W.D., 1992. An optimal transition path for controlling greenhouse gases. Science 258 (5086), 1315–1319. https://doi.org/10.1126/science.258.5086.1315.

Nordhaus, W.D., 1994. Managing the Commons: The Economics of Climate Change. The MIT Press, Cambridge, MA. USA, 213 pp.

Rich, N., Aug. 1, 2018. Losing Earth: the Decade We Almost Stopped Climate Change. We Knew Everything We Needed to Know, and Nothing Stood in Our Way. Nothing, That is, Except Ourselves. A Tragedy in Two Acts. New Yorker Magazine.

Oreskes, N., Conway, E.M., 2010. Merchants of Doubt: How a Handful of Scientists Obscured the Truth on Issues from Tobacco Smoke to Global Warming. Bloomsbury Press, New York, 368 pp.

Reichle, D., Houghton, J., Kane, B., Ekman, J., Benson, S., Clarke, J., Dahlman, R., Hendry, G., Herzog, H., Hunter-Cevera, J., Jacobs, G., Judkins, R., Ogden, J., Palmisano, A., Socolow, R., Stringer, J., Surles, T., Wolsky, A., Woodward, N., York, M., 1999. Carbon Sequestration Research and Development. U. S. Dept of Energy. DOE/SC/FE-1, Office of Sci. and Tech. Info. Oak Ridge, TN, USA. www.ornl.gov/carbon_sequestration/.

Stern, N., Peters, S., Bakhshi, V., Bowen, A., Cameron, C., Catovsky, S., Crane, D., Cruickshank, S., Dietz, S., Edmonson, N., Garbett, S.-L., Hamid, L., Hoffman, G., Ingram, D., Jones, B., Patmore, N., Radcliffe, H., Sathiyarajah, R., Stock, M., Taylor, C., Vernon, T., Wanjie, H., Zenghelis, D., 2006. Stern Review: The Economics of Climate Change. HM Treasury, London, 579 pp. https://onlinelibrary.wiley.com/doi/abs/10.1111/j.1728-4457.2006.00153.x/.

USGCRP, 2018. In: Reidmiller, D.R., Avery, C.W., Easterling, D.R., Kunkel, K.E., Lewis, K.L.M., Maycock, T.K., Stewart, B.C. (Eds.), Impacts, Risks, and Adaptation in the United States: Report-In-Brief: Fourth National Climate Assessment, Volume II. U.S. Global Change Research Program, Washington, DC, USA, 186 pp. https://www.globalchange.gov/nca4/; full report online:/nca2018.globalchange.gov/.

# Postscript

The global carbon cycle has been changing through the decades due to human influence and, as a consequence, so has the Earth's physical, chemical, and biological environment. As this book goes to press, The American Meteorological Society reports on the "State of the Climate in 2018" (American Meteorological Society, 2019), the most recent year for which there is a complete year's record. The report finds that:

In 2018 the globally averaged annual atmospheric $CO_2$ level was 407.4 ppm. This concentration was 2.4 ppm greater than that of 2017, about 124 ppm and 41% higher than at the beginning of the 19th century, and the higher than in ice core records dating back 800,000 years.

Global surface temperature approached a record high, averaging 0.30 °C to 0.40 °C above the 1981−2010 average, while the four warmest years on record have all occurred since 2015.

Sea level was the highest on record for the seventh consecutive year, 8.1 cm (3.2 inches) above the global average in 1993 when satellite altimeter records began. Global sea level is rising at the average rate of 3.1 cm (1.2 inches) per decade.

The Arctic continued to warm and sea ice coverage was at a near-record low. In the Antarctic it was warmer than average with near-record low sea ice.

Sea surface temperature was near-record high, cooling since the 2016 El Niño year, but far above the 1981−2010 mean by $0.33° \pm 0.05$ °C for 2018. The deeper ocean continues to warm year after year, as the upper ocean heat content reached a record high in 2018. Tropical cyclones across all ocean basins continue to increase in number.

The scientific evidence of the altered global carbon cycle is abundant, and the effects upon climate are now readily apparent. Are we paying attention? How will we respond?

The Global Carbon Cycle and Climate Change. https://doi.org/10.1016/B978-0-12-820244-9.00013-5

# Suggested classroom uses of this book

A seminar on the foundations of ecological research in the 1950–2000s.
A course on ecological energetics.
A seminar on energy flow in food chains and trophic levels.
A course on energy flow in ecological systems.
A course on the global carbon cycle.
A seminar on energy technology options for climate change.

**The Global Carbon Cycle and Climate Change.** https://doi.org/10.1016/B978-0-12-820244-9.00014-7

# Bibliography

Abatzoglou, J.T., Williams, A.P., 2016. Impact of anthropogenic climate change on wildfire across western U.S. forests. Proceedings of National Academy of Science of the United States 113 (4), 11770−11775. https://doi.org/10.1073/pnas.1607171113/.

Agroforestry Research Trust. https://www.agroforestry.co.uk.

Agroforestry Research Trust. /https://www.agroforestry.co.uk.

Ajtay, G.L., Ketner, P., Duvigneaud, P., 1979. Terrestrial primary production and phytomass. In: Bolin, B., Degens, E.T., Kempe, S., Ketner, P. (Eds.), The Global Carbon Cycle. SCOPE 13. John Wiley and Sons, New York, pp. 129−182, 491 pp. https://www.researchgate.net/publication/40170880_The_Global_Carbon_Cycle_SCOPE_Report_13.

Allee, W.C., Emerson, A.E., Park, O., Park, T., Schmidt, K.P., 1949. Principles of Animal Ecology. W. B. Saunders Co., Philadelphia and London, 837 pp.

American Meteorological Society, 2019. State of the Climate in 2018. In: Blunden, J., Arndt, D.S. (Eds.), Special Supplement to the Bulletin of the American Meteorological Society, Vol. 100. No. 9, September 2019. https://www.ncei.noaa.gov/news/reporting-state-climate-2018.

AMERIFLUX. 2018./https://tes.science.energy.gov/program/index.shtml.

AMIA, 2015. Food for Fruit: Nutrition Management in Mangoes. https://www.slideshare.net/AustralianMangoes/food-for-fruit-nutrition-management-in-mangoes.

Amthor, J.S., 1995. Terrestrial higher-plant response to increasing atmospheric $CO_2$ in relation to the global carbon cycle. Global Change Biology 1, 243−274. https://doi.org/10.1111/j.1365-2486.1995.tb00025.x.

Anderies, J.M., Carpenter, S.R., Steffen, W., Rockström, J., 2013. The topology of non-linear global carbon dynamics: from tipping points to planetary boundaries. Environmental Research Letters 8 (4). https://doi.org/10.1088/1748-9326/8/4/044048.

Andersson, E., Kumblad, L., 2006. A carbon budget for an oligotrophic clearwater lake in mid-Sweden. Aquatic Sciences 68 (1), 52−64. https://doi.org/10.1007/s00027-005-0807-0.

Andrews, R., Coleman, D.C., Ellis, J.E., Singh, J.S., 1974. Energy flow relationships on a short-grass prairie ecosystem. In: Proc. 1st International Congress of Ecology, Wageningen, Netherlands, pp. 222−228, 422 pp. https://scholar.google.com/scholar?q=Andrews%2C%20R.%2C%20D.%20C.%20Coleman%2C%20J.%20E.%20Ellis%20and%20J.%20S.%20Singh.%201974.%20Energy%20flow%20relationships%20in%20a%20shortgrass%20prairie%20ecosystem.%20Proc.%201st%20Int.%20Congr.%20Ecol.Proc.%201st%20Int.%20Congr.%20Ecol.%20Pudoc%2C%20Wageningen.

Anthony, W.K., von Deimling, T.S., Nitze, I., Frolking, S., Emond, A., Daanen, R., Anthony, P., Lindgren, P., Jones, B., Grosse, G., 2018. 21st-century modeled permafrost carbon emissions accelerated by abrupt thaw beneath lakes. Nature Communications 9 (1). https://doi.org/10.1038/s41467-018-05738-9/.

Archer, D., 2009. The Long Thaw: How Humans Are Changing the Next 100,000 Years of Earth's Climate. Princeton Univ. Press, Princeton, NJ, 192 pp. https://press.princeton.edu/books/paperback/9780691169064/the-long-thaw.

Archer, D., 2010. The Global Carbon Cycle. Princeton Univ. Press, Princeton, 216 pp. https://press. princeton.edu/books/ebook/9781400837076/the-global-carbon-cycle.

ARM, 2018. https://www.arm.gov/data.

Armentano, T.V., 1979. The Role of Organic Soils in the World Carbon Cycle.

Armentano, T.V., 1979. The role of organic soils in the world carbon cycle. In: Carbon Dioxide Effects Research and Assessment, CONF, 7905135. Washington, D.C. U.S. Department of Energy, Assistant Secretary for Environment, Office of Health and Environmental Research, Springfield, VA. Available from National Technical Information Service. https://www. worldcat.org/title/role-of-organic-soils-in-the-world-carbon-cycle-problem-analysis-and-research-needs/oclc/8097226.

Audubon, Mass, 2017. State of the Birds Report. https://www.massaudubon.org/content/download/ 21633/304821/file/mass-audubon_state-of-the-birds-2017-report.pdf.

AWEA, 2019. The Voice of Wind Energy. American Wind Energy Association. https://www.awea.org.

Baes, C.F., Killough, G.G., 1986. Chemical and biological processes in $CO_2$-ocean models. In: Trabalka, J.R., Reichle, D.E. (Eds.), The Changing Carbon Cycle: A Global Analysis. Springer Verlag, New York, pp. 329−347. https://link.springer.com/book/10.1007/978-1-4757-1915-4, 592 pp.

Baes, C.F., Goeller, H.E., Olson, J.S., R, M., Rotty, 1976. The Global Carbon Dioxide Problem (ORNL-5194). Oak Ridge National Laboratory, Oak Ridge, Tennessee, USA.

Baes, C.F., Goehler, H.E., Olson, J.S., Rotty, R.M., 1977. Carbon Dioxide and Climate: The Uncontrolled Experiment. American Scientist 65 (3), 310−320. http://adsabs.harvard.edu/abs/ 1977AmSci..65..310B.

Baitrago, M.F., 2016. Changes in thermal infrared spectra of plants caused by temperature and water stress. ISPRS Journal of Photogrammetry and Remote Sensing 111, 22−31. https://doi. org/10.1016/j.isprsjprs.2015.11.003.

Baker, T.R., Phillips, O.L., Malhi, Y., Almeida, S., Arroyo, L., Di Fiore, A., Erwin, T., Killeen, T.J., Laurance, S.G., Laurance, W.F., Lewis, S.L., Lloyd, J., Monteagudo, A., Neill, D.A., Patiño, S., Pitman, N.C.A., Silva, J.N.M., Martínez, R.V., 2004. Variation in wood density determines spatial patterns in Amazonian forest biomass. Global Change Biology 10 (5), 545−562. https://doi.org/10.1111/j.l529-8817.2003.00751.x.

Baker, T.R., Phillips, O.L., Malhi, Y., et al., 2004. Increasing biomass in Amazonian forest plots. Philosophical Transactions of the Royal Society of London B 359, 353−365. https://doi.org/ 10.1098/rstb.2003.1422.

Balch, W., Evans, R., Brown, J., Feldman, G., McClain, C., Esaias, W., 1992. The remote sensing of ocean primary productivity: use of new data compilation to test satellite algorithms. Journal of Geophysical Research 97, 2279−2294. https://agupubs.onlinelibrary.wiley.com/doi/pdf/10. 1029/93JC01314/.

Baldocchi, D.D., Hicks, B.B., Meyers, T.P., 1988. Measuring biosphere:atmosphere exchanges of biologically related gases with micrometeorological methods. Ecology 69, 1331−1340. https://doi.org/10.2307/1941631.

Baldochi, D.D., 2003. Assessing the eddy covariance technique for evaluating carbon dioxide exchange rates of ecosystems: past, present and future. Global Change Biology 9, 479−492. https://doi.org/10.1046/j.1365-2486.2003.00629.x.

Bartee, L., Shriner, W., 2016. MHC Principles of Biology: Biology 211, 212, 213. The Structure of the Chloroplast. Retrieved from OpenStax College. https://openoregon.pressbooks.pub/ mhccmajorsbio/chapter/8-2-main-structures-and-summary-of-photosynthesis.

Bastin, J-F., Finegold, Y., Garcia, C., Mollicone, D., Rezende, M., Routh, D., Zohner, C.M., Crowther, T.W., 2019. The global tree restoration potential. Science 365 (6448), 76−79. https://doi.org/10.1126/science.aax0848.

Benke, A.C., Huryn, A.D., 2017. Secondary production and quantitative food webs, pp. 235−254. In: Hauer, F.R., Lamberti, G.A. (Eds.), Methods in Stream Ecology, third ed. Academic Press. https://doi.org/10.1111/fwb.12131.

Björkstrom, A., 1979. A model of $CO_2$ interaction between atmosphere, oceans, and land biota. In: Bolin, B., Degens, E.T., Kempe, S., Ketner, P. (Eds.), The Global Carbon Cycle. SCOPE 13. John Wiley and Sons, New York, pp. 403−457, 491 pp. https://www.researchgate.net/publication/40170880_The_Global_Carbon_Cycle_SCOPE_Report_13.

Björkstrom, A., 1986. One-dimensional and two-dimensional ocean models for predicting the distribution of $CO_2$ between the ocean and the atmosphere. In: Trabalka, J.R., Reichle, D.E. (Eds.), The Changing Carbon Cycle: A Global Analysis. Springer Verlag, Berlin-Heidelberg-New York, pp. 258−278, 592 pp. https://www.springer.com/gp/book/9781475719178.

Blackman, F.F., Matthaei, G., 1905. A Quantitative Study of Carbon-Dioxide Assimilation and Leaf-Temperature in Natural Illumination. Proc. Royal Soc., London.

Blankenship, R.E., 2010. Early evolution of photosynthesis. Plant Physiology 154, 434−438. https://doi.org/10.1104/pp.110.161687.

Boden, T.A., Marland, G., Andres, R.J., 2017. Global, Regional, and National Fossil-Fuel $CO_2$ Emissions. Carbon Dioxide Information Analysis Center, Oak Ridge National Laboratory, U.S. Department of Energy, Oak Ridge, TN, USA. https://doi.org/10.3334/CDIAC/00001_V2017.

Bohn, H.L., 1976. Estimate of organic carbon in world soils. Soil Science Society of America Proceedings 40, 468−470.

Bohn, H.L., 1982. Estimate of organic carbon in world soils: II. Soil Science Society of Arizona Journal 4, 1118−1119.

Bokhari, U.G., 1978. Total nonstructural carbohydrates in the vegetation components of a short-grass prairie ecosystem under stress conditions. Journal of Range Management 31 (3), 224−229. https://journals.uair.arizona.edu/index.php/jrm/issue/view/387.

Bolin, B., 1970. The Carbon Cycle. Scientific American 23, 124−132. https://doi.org/10.1038/scientificamerican0970-124.

Bolin, B., Degens, E.T., Kempe, S., Ketner, P. (Eds.), 1979. The Global Carbon Cycle. SCOPE, vol. 13. John Wiley and Sons, New York, 491 pp. https://www.researchgate.net/publication/40170880_The_Global_Carbon_Cycle_SCOPE_Report_13.

Bolin, B., Degens, E.T., Duvigneaud, P., Kempe, S., 1979. The global biogeochemical carbon cycle. In: Bolin, B., Degens, E.T., Kempje, S., Ketner, P. (Eds.), The Global Carbon Cycle, SCOPE 13. John Wiley & Sons, New York, pp. 1−56, 491 pp. https://www.researchgate.net/publication/40170880_The_Global_Carbon_Cycle_SCOPE_Report_13.

Bolin, B. (Ed.), 1981. Carbon Cycle Modelling, SCOPE 16. John Wiley & Sons, New York, 404 pp. https://www-legacy.dge.carnegiescience.edu/SCOPE/SCOPE_16/SCOPE_16.html.

Bonham, C.D., 1989. Measurements of Terrestrial Vegetation. John Wiley Sons, New York, NY, 252 pp. https://doi.org/10.1002/9781118534540.

Bopp, L., Le Quéré, C., Heimann, M., Manning, A.C., Monfray, P., 2002. Climate-induced oceanic oxygen fluxes: implications for the contemporary carbon budget. Global Biogeochemical Cycles. 16 (2), 6-1 - 6-8. https://doi.org/10.1029/2001GB001445.

Borenstein, S., 2019. Earth's future being written in melting Greenland. Associated Press, 2019. https://www.witf.org/news/2019/08/earths-future-is-being-written-in-fast-melting-greenland.php.

Borges, A.V., Schiettecatte, L.-S., Abril, G., Dolille, B., Gazean, F., 2006. Carbon dioxide in European and coastal waters. Estuarine, Coastal and Shelf Science 70, 375−387.

Bosch, R. Dave, L. A. Meyer (Eds). Cambridge University Press, Cambridge, United Kingdom and New York, NY, USA. https://www.ipcc.ch/site/assets/uploads/2018/02/ar4-wg3-spm-1.pdf.

Botkin, D.B., Simpson, L.G., 1990. Biomass of the North American boreal forest. Biogeochemistry 9, 161−174. https://www.jstor.org/stable/1468542.

BrainyMedia Inc, 2019. Donald Rumsfeld Quotes. BrainyQuote.com,/https://www.brainyquote.com/quotes/donald_rumsfeld_148142.

Breymeyer, A., 1971. Productivity investigations of two types of meadows in the Vistula Valley. XII. Some regularities in structure and function of the ecosystem. Ekologia Polska 19, 249−261.

Brock, T.D., Brock, L.M., 1968. Relationship between environmental temperature and optimum temperature of bacteria along a hot spring thermal gradient. Applied Microbiology 31 (Issue 1), 54−58.

Brock, T.D., Freeze, H., 1969. Thermus aquaticus, a nonsporulating extreme thermophile. Journal of Bacteriology 98 (1), 289−297. https://doi.org/10.1111/j.1365-2672.1968.tb00340.x.

Brody, S., 1945. Bioenergetics and Growth. Reinhold, New York, 1023 pp. https://doi.org/10.1002/ajpa.1330040117.

Broecker, W.S., Peng, T.-H., 1982. Tracers in the Sea. ELDIGIO Press, Palisades, New York, 690 pp.

Broecker, W.S., Peng, T.-H., 1984. The climate- chemistry connection. In: Hansen, J.E., Takahashi, T. (Eds.), Climate Processes and Climate Sensitivity. American Geophysical Union, Washington, D.C, pp. 327−336. https://doi.org/10.1029/GM029p0314/.

Broecker, W.S., 1973. Factors controlling $CO_2$ content in the oceans and atmosphere. In: Woodwell, G.M., Pecan, E.V. (Eds.), Carbon and the Biosphere, CONF-720510. National Technical Information Office, Springfield, VA, pp. 32−50, 392 pp. https://www.biodiversitylibrary.org/bibliography/4036#/summary.

Broecker, W.S., 1974. Chemical Oceanography. Harcourt Brace Jovanovich, New York. https://doi.org/10.4319/lo.1975.20.2.0299.

Broecker, W.S., 1975. Climatic change: are we on the brink of a pronounced global warming? Science 189 (4201), 460−463. https://www.jstor.org/stable/1740491.

Broecker, W.S., 2000. Was a change in thermohaline circulation responsible for the Little Ice Age? Proceedings of the National Academy of Sciences of the United States of America 97 (4), 1339−1342. https://doi.org/10.1073/pnas.97.4.1339.

Brown, M.A., Levine, M.D., 1997. Scenarios of U.S. Carbon Reductions: The Potential Impact of Energy-Efficient and Low-Carbon Technologies. Oak Ridge National Laboratory. ORNL/CON-444, 249 pp. https://digital.library.unt.edu/ark:/67531/metadc694703/m2/1/high_res_d/563139.pdf.

Brown, S., Lugo, A.E., 1982. The storage and production of organic matter in tropical forests and their role in the global carbon cycle. Biotropica 14, 161−187. https://doi.org/10.2307/2388024.

Budyko, M.T., 1955. Atlas of the Heat Balance. Leningrad, 256 pp.

Budyko, M.T., 1956. The Heat Balance of the Earth's Surface. Gidrometeorolicheskoe izdatel'stvo, Leningrad. Russian translation: PB131692, U. S. Department of Commerce, Office of Technical Services, Washington, D. C.

Buringh, P., 1984. Organic carbon in soils of the world, pp 91-109. In: Woodwell, G.M. (Ed.), The Role of Terrestrial Vegetation in the Global Carbon Cycle: Measurement by Remote Sensing,

SCOPE 23. John Wiley & Sons, New York. https://www-legacy.dge.carnegiescience.edu/SCOPE/SCOPE_23/SCOPE_23_3.1_chapter3_91-109.pdf.

Bylinsky, M., Mann, K.H., 1973. An analysis of factors governing productivity in lakes and reservoirs. Limnology and Oceanography 18 (1), 1−14. https://doi.org/10.4319/lo.1973.18.1.0001.

Caldeira, K., Hoffert, M.I., Jain, A., 2005. Simple ocean carbon cycle models. In: Wigley, T.M.L., Schimel, D.S. (Eds.), The Carbon Cycle. Cambrdge Univ. Press, Cambridge, UK, pp. 199−211, 292 pp. https://doi.org/10.1017/CBO9780511573095

Callendar, G.S., 1938. The artificial production of carbon dioxide and its influence on temperature. Quarterly Journal of the Royal Meteorological Society 64, 233−240. https://doi.org/10.1002/qj.49706427503.

Canadell, J., Le Quéré, C., Raupach, M., Field, C., Buitenhuis, E., Ciais P, P., et al., 2007. Contributions to accelerating atmospheric $CO_2$ growth from economic activity, carbon intensity, and efficiency of natural sinks. Proceedings of the National Academy of Sciences 104 (47), 18866−18870. https://doi.org/10.1073/pnas.0702737104.

Cao, S., Sanchez-Azofeifa, G.A., Duran, S.M., Calvo-Rodriguez, S., 2016. Estimation of aboveground net primary productivity in secondary tropical dry forests using the Carnegie−Ames−Stanford approach (CASA) mode. Environmental Research Letters 11 (7). https://doi.org/10.1088/1748-9326/11/7/075004/meta.

Carbon Engineering, 2016. A Case Study. Harvard Business School. https://www.hbs.edu/faculty/Pages/item.aspx?num=45759.

Carpi, A., 2003. Carbon Chemistry, Vol. CHE-2(4). Visionlearning. www.emerson.com/BTU-Analysis-Using-a-Gas-Chromatograph-en-72722.pdf.

Carvalhais, N., Forkel1, M., Khomik, M., Bellarby, J., Jung, M., Migliavacca, M., Mu, M., Saatchi, S., Santoro, M., Thurner, M., Weber, U., Ahrens, B., Beer, C., Cescatti, A., Randerson, J.T., Reichstein, M., 2014. Global covariation of carbon turnover times with climate in terrestrial ecosystems. Nature 514, 213−217. https://doi.org/10.1038/nature13731.

Carvalhais, N., Thurner, M., Forkel, M., Beer, C., Reichstein, M., 2016. Diagnosing turnover times of carbon in terrestrial ecosystems to address global climate co-variability and for model evaluation. Geophysical Research Abstracts 18. https://publik.tuwien.ac.at/files/PubDat_249002.pdf.

CDIAC not GCP -CDIAC, 2013. Global Carbon Project, Global Carbon Budget. https://cdiac.ess-dive.lbl.gov/GCP/carbonbudget/2013.

CDIAC, 2013. Tier-1 Global Biomass Carbon Map for the Year 2000. https://cdiac.ess-dive.lbl.gov/epubs/ndp/global_carbon/FINAL_DATASETS.jpg.

CDIAC, 2013. Global Carbon Project, Global Carbon Budget. https://cdiac.ess-dive.lbl.gov/GCP/carbonbudget/2013.

CDP, 2019. https://www.cdp.net/en/climate.

Center for Global Development, 2007. Carbon Dioxide Emissions from Power Plants Rated Worldwide. https://www.sciencedaily.com/releases/2007/11/071114163448.htm.

Chameides, B., 2017. Environmental Defense Fund. http://blogs.edf.org/climate411/2007/06/21/human_cause-2.

Chapin, F.S., Woodwell, G.M., Randerson, J.T., Rastetter, E.B., Lovett, G.M., Baldocchi, D.D., Clark, D.A., Harmon, M.E., Schimel, D.S., Valentini, R., Wirth, C., Aber, J.D., Cole, J.J., Goulden, M.L., Harden, J.W., Heimann, M., Howarth, R.W., Matson, P.A., McGuire, A.D., Melillo, J.M., Mooney, H.A., Neff, J.C., Houghton, R.A., Pace, M.L., Ryan, M.G., Running, S.W., Sala, O.E., Schlesinger, W.H., Schulze, E.D., 2006. Reconciling carbon-cycle concepts, terminology, and methods. Ecosystems 9, 1041−1050. https://doi.org/10.1007/s10021-005-0105-7.

Cheng, L., Trenberth, K.E., Fasullo, J., Boyer, T., Abraham, J., Zhu, J., 2017. Improved estimates of ocean heat content from 1960 to 2015. Science Advances 3 (3), e1601545. https://doi.org/10.1126/sciadv.1601545.

Cheng, L., Abraham, J., Hausfather, Z., Trenberth, K.E., 2019. How fast are the oceans warming. Science 363 (6423), 128−129. https://science.sciencemag.org/content/363/6423/128.full. Erratum for the Perspective, 08 Mar., 2019, Vol. 363, (6431), eeax 1875.

Chmiel, H.E., Niggemann, J., Kokic, J., Ferland, M.-E., Dittmar, T., Sobek, S., 2015. Uncoupled organic matter burial and quality in boreal lake sediments over the Holocene. Journal of Geophysical Research: Biogeoscience 120, 1−13. https://esajournals.onlinelibrary.wiley.com/toc/19399170/71/2.

Choi, Y., Wang, Y., 2004. Dynamics of carbon sequestration in a coastal wetland using radiocarbon measurements. Global Biogeochemical Cycles 18 (4), GB4016. https://doi.org/10.1029/2004GB002261.

Ciais, P., Tans, P.P., Trolier, M., White, J.W.C., Francy, R.J., 1995. A large northern hemisphere terrestrial $CO_2$ sink indicated by the $^{13}C/^{12}C$ ratio of atmospheric $CO_2$. Science 269, 1098−1102. https://doi.org/10.1126/science.269.5227.1098.

Ciais, P., Reichstein, M., Viovy, N., Granier, A., Ogée, J., Allard, V., Aubinet, M., Buchmann, N., Bernhofer, C., Carrara, A., Chevallier, F., De Noblet, N., Friend, A.D., Friedlingstein, P., Grünwald, T., Heinesch, B., Keronen, P., Knohl, A., Krinner, G., Loustau, D., Manca, G., Matteucci, G., Miglietta, F., Ourcival, J.M., Papale, D., Pilegaard, K., Rambal, S., Seufert, G., Soussana, J.F., Sanz, M.J., Schulze, E.D., Vesala, T., Valentini, R., 2005. Europe-wide reduction in primary productivity caused by the heat and drought in 2003. Nature 437 (7058), 529−533. https://doi.org/10.1038/nature03972.

Clack, C.T.M., Qvist, S.A., Apt, J., Bazilian, M., Brandt, A.R., Caldeira, K., Davis, S.J., Diakov, V., Handschy, M.A., Hines, P.D.H., Jaramillo, P., Kammen, D.M., Long, J.C.S., Morgan, M.G., Reed, A., Sivaram, V., Sweeney, J., Tynan, G.R., Victor, D.G., Weyant, J.P., Whitacre, J.F., 2017. Evaluation of a proposal for reliable low-cost grid power with 100% wind, water, and solar. Proceedings of the National Academy of Sciences, USA 114 (26), 6722−6727. https://doi.org/10.1073/pnas.1610381114.

Clark, W.C. (Ed.), 1982. Carbon Dioxide Review: 1982. Oxford University Press, New York, p. 469. https://doi.org/10.2307/3323339.

Clarke, L., Edmonds, J., Jacoby, H., Pitcher, H., Reilly, J., Richels, R., 2007. Scenarios of Greenhouse Gas Emissions and Atmospheric Concentrations. Sub-report 2.1A of Synthesis and Assessment Product 2.1 by the U.S. Climate Change Science Program and the Subcommittee on Global Change Research. Department of Energy, Office of Biological & Environmental Research, Washington, DC, USA, 154 pp. https://globalchange.mit.edu/publication/14399.

Climate Reality Project. http://www.climaterealityproject.org.

Climeworks, 2019. https://www.facebook.com/climeworks.

CMIP5, 2018. Coupled Model Intercomparison Project. http://www.wcrp-climate.org/index.php/wgcm-cmip/about-cmip.

Coady, D., Parry, I.W.H., Sears, L., Shang, B., 2015. How Large Are Global Energy Subsidies? IMF Working Paper No. 15/105, 42 pp. https://www.imf.org/en/Publications/WP/Issues/2016/12/31/How-Large-Are-Global-Energy-Subsidies-42940.

Cohen, A., 2017. Clean Air Task Force. https://www.catf.us/2017/09/electricity-decarbonization.

Cohen-Rengifo, M., Garcia, E., Hernandez, C.A., Hernandez, J.C., Clemente, S., 2013. Global warming and ocean acidification affect fertilization and early development of the sea urchin *Paracentrotus lividus*. Cahiers de Biologie Marine 54, 667−675. https://www.nhbs.com/

cahiers-de-biologie-marine-volume-544-special-issue-14th-international-echinoderm-conference-book/.

Coleman, D.C., 2010. Big Ecology: The Emergence of Ecosystem Science. Univ. Calif. Press, Berkeley, Los Angeles-London, 236 pp. https://epdf.pub/big-ecology-the-emergence-of-ecosystem-science.html.

Cook, E., 1971. The flow of energy in an industrial society. Scientific American 224 (3), 134−147. https://www.jstor.org/stable/24923122.

Coupland, R.T., 1975. Productivity of Grassland Ecosystems, pp. 44−49. In: Reichle, D.E., Franklin, J.F., Goodall, D.W. (Eds.), Productivity of World Ecosystems. The National Academies Press, Washington, DC, 166 pp. http://nap.edu/20114.

Craig, H., 1957. The natural distribution of radiocarbon and the exchange time of carbon dioxide between atmosphere and sea. Tellus 9, 1−17. https://doi.org/10.1111/j.2153-3490.1957.tb01848.x.

Cramer, W.P., Solomon, A.M., 1993. Climate classification and future global redistribution of agricultural land. Climate Research 3, 97−110. https://doi.org/10.1088/1748-9326/6/1/014014/meta.

Crawford, M., 2019. Is Burning Wood Rewally a Long-term Energy Descent Strategy? Agroforestry Research Trust. Retrieved Feb. 2019 from. https://www.agroforestry.cp.uk. https://www.transitionculture.org/2008/05/19/is-burning-wood-really-a-long-term-energy-descent-strategy.

Crisp, D.J., 1968. Grazing in Terrestrial and Marine Ecosystems. Blackwell Science Pubs., Oxford.

Crisp, D.J., 1975. Secondary production in the sea, pp 71- 89. In: Reichle, D.E., Franklin, J.F., Goodall, D.W. (Eds.), Productivity of World Ecosystems. The National Academies Press, Washington, DC, 166 pp. http://nap.edu/20114.

Cristian, G., Sobek, S., Bastviken, D., Koehler, B., Tranvik, L., 2011. Mineralization of Organic Carbon in Lake Sediments: Temperature Sensitivity and a Comparison to Soils. Digitala Vetenskapliga Arkivet. http://urn.kb.se/resolve?urn=urn:nbn:se:uu:diva-150714.

Crossley, D.A., Reichle, D.E., 1969. Analysis of transient behavior of radioisotopes in insect food chains. BioScience 19 (4), 341−343. https://doi.org/10.2307/1294518.

Cullen, J., 2018. BP Technology Outlook: 2018. London, UK, 72 pp. https://www.bp.com/content/dam/bp/en/corporate/pdf/technology/bp-technology-outlook-2018.pdf.

Cummins, K.E., 1975. The importance of different energy sources in freshwater ecosystems. In: Reichle, D.E., Franklin, J.F., Goodall, D.G. (Eds.), Productivity of World Ecosystems. The National Academies Press, Washington, DC, pp. 50−54, 166 pp. http://nap.edu/20114.

Curtis, P.S., Hansson, P.J., Bolstad, P., Bartford, C., Randolph, J.C., Schmid, H.P., Wilson, K.B., 2002. Biometric and eddy-covariance based estimates of annual carbon storage in five eastern North American deciduous forests. Agricultural and Forest Meteorology 113, 3−19. https://doi.org/10.1016/S0168-1923(02)00099-0.

Cushing, D.H., 1959. The seasonal variation in oceanic production as a problem in population dynamics. Journal du Conseil/Conseil Permanent International pour l'Exploration de la Mer 24, 455−464. https://doi.org/10.1093/icesjms/24.3.455.

Dahlman, R.C., Kucera, C.L., 1968. Tagging native grassland vegetation with carbon-14. Ecology 49 (6), 1199−1203. https://doi.org/10.2307/1934516.

Dai, A., Trenberth, K.E., 2002. Estimates of freshwater discharge from continents: latitudinal and seasonal variations. Journal of Hydrometeorology 3, 660−697. https://doi.org/10.1175/1525-7541(2002)003<0660:EOFDFC2.0.CO;2.

Damm, H.C., Goldwyn, A.J., Thomas, C.A., 1966. The Handbook of Biochemistry and Biophysics. The World Pub. Co., 736 pp.

Daniels, F., Alberty, R.A., 1980. Physical Chemistry. Wiley and Sons, New York, 744 pp. https://doi.org/10.1002/aic.690070434.

de Vooys, C.G.N., 1979. Primary production in aquatic environments. In: Bolin, B., Degens, E.T., Kempe, S., Ketner, P. (Eds.), SCOPE, 13: The global carbon cycle. John Wiley and Sons, Chichester, UK, 491 pp.

de Vooys, C.G.N., 1979. Primary production in aquatic environments, pp. 259−292. In: Bolin, B., Degens, E.T., Kempe, S., Ketner, P. (Eds.), The Global Carbon Cycle. John Wiley and Sons, New York, 491 pp. https://www.researchgate.net/publication/40170880_The_Global_Carbon_Cycle_SCOPE_Report_13.

DeFries, R., Field, C.B., Fung, I., Justice, C.O., Matson, P.A., Matthews, M., Mooney, H.A., Potter, C.S., Prentice, K., Sellers, P.J., Townshend, J., Tucker, C.J., Ustin, S.L., Vitousek, P.M., 1995. Mapping the land surface for global atmosphere-biosphere models: toward continuous distributions of vegetation's functional properties. Journal of Geophysical Research 100, 20867−20882.

DeFries, R.S., Bounoua, L., Collatz, G.J., 2002. Human modification of the landscape and surface climate in the next fifty years. Global Change Biology 8 (5), 438−458. https://doi.org/10.1046/j.1365-2486.2002.00483.x.

DeFries, R.S., Houghton, R.A., Hansen, M.C., Field, C.B., Skole, D., Townshend, J., 2002. Carbon emissions from tropical deforestation and regrowth based on satellite observations for the 1980's and 1990's. Proceedings of the National Academy of Sciences 99, 14256−14261. https://doi.org/10.1073/pnas.182560099.

Delcourt, H.R., Harris, W.F., 1980. Carbon budget of the southeastern U. S. biota: analysis of historical change in trend from source to sink. Science 210, 321−323. https://doi.org/10.1126/science.210.4467.321.

Derner, J.D., Schuman, G.E., 2007. Carbon sequestration and rangelands: a synthesis of land management and precipitation effects. Journal of Soil and Water Conservation 62 (2), 77−85. http://www.jswconline.org/content/62/2/77.

DeWitt, C.B., 1967. Precision of thermoregulation and its relation to environmental factors in desert iguana, *Disposaurus dorsalis*. Physiological Zoology 40, 40−66. https://doi.org/10.1086/physzool.40.1.30152438.

DiFranco, J.L., 2014. Protocols for sampling aquatic macroinvertebrates in freshwater wetlands. In: Standard Operating Procedure Bureau of Land and Water Quality. Maine Dept. Environmental Protection. DEPLW0640A-2014, 10 pp. https://www.maine.gov/dep/water/monitoring/biomonitoring/materials/sop_wetland_invertebrates.pdf.

Dill, D.B. (Ed.), 1964. Adaptation to the Environment. American Physiological Society, 1056 pp. https://doi.org/10.1126/science.148.3671.832.

Dixon, K., Brown, W.S., Houghton, R.A., Solomon, M., Trexler, M.C., Wisnicwski, J., 1994. Carbon pools and flux in global forest ecosystems. Science 263, 185−190. https://doi.org/10.1126/science.263.5144.185.

Dold, N., Khan, A., Kohn, J., Kaufmann, C., 2014. Small leaves buffer ambient temperature. The Biomimicry Institute. https://asknature.org/strategy/small-leavesbuffer-ambient-temperature/#.W3eIK2a0W70.

Dolman, A.J., van der Werf, G.R., van der Molen, M.K., Ganssen, G., Erisman, [1]J.-W., Strengers, B., 2010. A carbon cycle science update since IPCC AR-4. Ambio 39 (5−6), 402−412. https://doi.org/10.1007/s13280-010-0083-7.

Domingues, C.M., White, N.J., Church, J.A., Gleckler, P.J., 2008. Improved estimates of upper-ocean warming and multi-decadal sea-level rise. Nature 453 (7198), 1090−1093. https://doi.org/10.1038/nature07080.

Downing, J.A., Anderson, M.R., 2011. Estimating the standing biomass of aquatic macrophytes. Canadian Journal of Fisheries and Aquatic Sciences 42 (12), 1860−1869. https://downing. public.iastate.edu/tier%202/jadpdfs/1985%20CJFAS%20Estimating%20the%20standing% 20biomass%201860-1869.pdf.

Dressler, A., 2012. Introduction to Modern Climate Change. Cambridge Univ. Press, Cambridge, UK, 257 pp. https://doi.org/10.1017/S0014479712000841.

Drozdz, A., 1967. Food preference, food digestibility and the natural food supply of small rodents, pp. 323−334. In: Petrusewicz, K. (Ed.), Secondary Production of Terrestrial ecosystems. Panstwowe Wydawnictwo Naukowe, Warsaw, 879 pp. http://agris.fao.org/agris-search/search. do?recordID=US201300166082.

Dunbar, M.J., 1975. Productivity of marine ecosystems. In: Reichle, D.E., Franklin, J.F., Goodall, D.W. (Eds.), Productivity of World Ecosystems. The National Acadamies Press, Washington, DC, pp. 27−32, 166 pp. http://nap.edu/20114.

Durban, E.G., Durban, A.G., 1983. Energy and Nitrogen Budgets for the Atlantic Menhaden, *Brevoorita tyrannus* (Pisces:Clupeidae), a Filter-Feeding Planktivore./https://www.st.nmfs. noaa.gov/spo/FishBull/81-2/durbin.pdf.

Duvigneaud, Denaeyer-De Smet, P.S., Ambroes, P., Trimperman, J., 1971. Recherches sur l'éco- système fort. Biomasse, productivité et cycle des polyéléments dans l'écosystème 'chenaie caducifoliée.' Essai de phytogéochemie forestière. Institut Roy. Sci. naturelles de Belgique. Mémoire No. 164.

Duvigneaud, P., 1971. Productivity of Forest Ecosystems. UNESCP, Paris, 707 pp. https://www. worldcat.org/title/productivity-of-forest-ecosystems-productivite-des-ecosystemes-forestiers- proceedings-of-the-brussels-symposium-organized-by-unesco-and-the-international- biological-programme-27-31-october-1969/oclc/651164.

EcologicalPyramids.jpg, 2016. Wikimedia Commons, the free media repository. Retrieved 04:17, August 18, 2019 from. https://commons.wikimedia.org/w/index.php?title=File:Ecological Pyramids.jpg&oldid=217291673.

Edenhofer, O., Pichs- Madruga, R., Sokona, Y., Farahani, E., Kadner, S., Seyboth, K., Adler, A., et al. (Eds.), 2014. Climate Change 2014: Mitigation of Climate Change. Contribution of Working Group III to the Fifth Assessment Report (AR5) of the Intergovernmental Panel on Climate Change. Cambridge University Press, UK. https://doi.org/10.1017/ CBO9781107415416.

Climate Change 2014: Mitigation of Climate Change. In: Edenhofer, O., Pichs- Madruga, R., Sokona, Y., Farahani, E., Kadner, S., Seyboth, K., Adler, A., et al. (Eds.), 2014. Contribution of Working Group III to the Fifth Assessment Report (AR5) of the Intergovernmental Panel on Climate Change. Cambridge University Press, UK. https://doi.org/10.1017/ CBO9781107415416.

Edmonds, J., Richels, R., Wise, M., 2005. Future fossil fuel carbon emissions without policy intervention: a review, pp. 171−189. In: Wigley, T.M.L., Schimel, D.S. (Eds.), The Carbon Cycle. Cambridge Univ. Press, Cambridge, UK, 292 pp. https://doi.org/10.1017/ CBO9780511573095.

Edreva, A., 2005. Agriculture, Ecosystems & Environment 106, 135−146. https://www. sciencedirect.com/journal/agriculture-ecosystems-and-environment/vol/106/issue/2.

Edwards, N.T., Harris, W.F., 1977. Carbon cycling in a mixed deciduous forest. Ecology 58 (2), 431−437. https://doi.org/10.2307/1935618.

Egerton, F.N., 2017. History of Ecological Sciences, Part 59: Niches, Biomes, Ecosystems, and Systems. https://www.researchgate.net/publication/320227603_History_of_Ecological_ Sciences_Part_59_Niches_Biomes_Ecosystems_and_Systems.

Eggers, D.M., Bartoon, N.W., Rickard, A., Nelson, R.E., Wissmarr, R.C., Burgner, L., Devoll, A.H., 1978. The Lake Washington ecosystem: the perspective from the fish community production and forage base. Journal of the Fisheries Research Board of Canada 35, 1553−1571. https://swfsc.noaa.gov/publications/cr/1978/7810.pdf.

Ehhalt, D., Dentener, M., Derwent, F., Dlugokencky, R., Edward, J., Holland, E., Isaksen, I., Katima, J., Kirchhoff, V., Matson, P., Midgley, P., Wang, M., Berntsen, T., Bey, I., Brasseur, G., Buja, L., Collins, W.J., Daniel, J.S., DeMore, W.B., Derek, N., Dickerson, R., Etheridge, D., Feichter, J., Fraser, P., Friedl, R., Fuglestvedt, J., Gauss, M., Grenfell, L., Grubler, A., Harris, N., Hauglustaine, D., Horowitz, L., Jackman, C., Jacob, D., Jaegle, L., Jain, A., Atul, K., Kanakidou, M., Karlsdottir, S., Ko, M., Kurylo, M., Lawrence, M., Logan, J.A., Manning, M., Mauzerall, D., McConnell, J., Mickley, L.J., Montzka, S., Muller, J.F., Olivier, J., Pickering, K., Pitari, G., Roelofs, G.-J., Rogers, H., Rognerud, B., Smith, S.J., Solomon, S., Staehelin, J., Steele, P., Stevenson, D.S., Sundet, J., Thompson, A., van Weele, M., von Kuhlmann, R., Wang, Y., Weisenstein, D.K., Wigley, T.M., Wild, O., Wuebbles, D.J., Yantosca, R., Joos, F., McFarland, M., 2001. IPCC Third Assessment Report Climate Change 2001, Chapter 4: Atmospheric Chemistry and Greenhouse Gases, PNNL-SA-39647. Pacific Northwest National Lab. (PNNL), Richland, WA, United States. https://www.osti.gov/biblio/901482.

Ekholm, P., 2008. N:P Ratios in Estimating Nutrient Limitation in Aquatic Systems. Finnish Environment Institute, 4 pp. https://www.cost869.alterra.nl/FS/FS_NPratio.pdf.

Elton, C., 1927. Animal Ecology. Macmillan Co, New York, 200 pp. https://openlibrary.org/books/OL13993167M/Animal_ecology.

Emanuel, W.R., Shugart, H.H., Stevenson, M.P., 1985. Climate change and the broad-scale distribution of terrestrial ecosystem complexes. Climate Change 7, 29−43. https://link.springer.com/article/10.1007%2FBF00139439.

Emerson. https://www.emerson.com/...BTU-Analysis-Using-a-Gas-Chromatograph-en-72722.pdf.

Encyclopedia Britannica, 2018. Paleocene-Eocene Thermal Maximum. Retrieved March 16, 2019. https://www.britannica.com/science/Paleocene-Eocene-Thermal-Maximum.

Encylopedia Britanica, 2019. https://www.britannica.com/science/body-temperature.

Energy Information Administration, 1998. Annual Energy Review 1997. DOE/EIA-0384(97). U.S. Department of Energy, Washington, D.C. https://doi.org/10.2172/1212301.

Energy Information Administration, 1998. Emissions of Greenhouse Gases in the United States 1997. DOE/EIA-0573(97). U.S. Department of Energy, Washington, D.C. https://doi.org/10.2172/348897.

Enting, I.G., Wigley, T.M.L., Heinman, M., 1994. Future Emissions and Concentratins of Carbon Dioxide: Key Ocean/atmosphere/land Analyses. CSIRO Division of Atmospheric Research Technical Paper No. 31. https://www.researchgate.net/publication/228608466_Future_Emissions_and_Concentrations_of_Carbon_Dioxide_Key_OceanAtmosphereLand_Analyses.

Eswaran, H., Bliss, N., Lytle, D., Lammers, D., 1993. Major Soil Regions of the World. USDA-SCS. U.S. Gov. Print. Office, Washington, DC.

Eugester, W., Siegrist, F., 2000. The influence of nocturnal $CO_2$ advection on $CO_2$ flux measurement. Basic and Applied Ecology 1, 177−188. https://doi.org/10.1078/1439-1791-00028.

Evans, F.C., 1951. Notes on a population of the striped ground squirrel (*Citellus tridecemlineatus*) in an abandoned field in southeastern Michigan. Journal of Mammology 32, 437−449. https://doi.org/10.2307/1375792.

Evans, F.C., 1964. The food of vesper, field, and chipping sparrows nesting in an abandoned field in southeastern Michigan. American Midland Naturalist 72, 57−75.

Evans, F.C., 1967. The significance of investigations in secondary terrestrial productivity. In: Petrusewicz, K. (Ed.), Secondary Productivity of Terrestrial Ecosystems. Institute of Ecology, Polish Academy of Sciences, Warsaw, pp. 1−15, 879 pp. http://agris.fao.org/agris-search/search.do?recordID=US201300166082.

FAO (Food and Agricultural Organization)./http://cdiac.ess-dive.lbl.gov/epubs/ndp/global_carbon/ecofloristic_zones_sm.gif.

FAO, 1997. Renewable biological systems for unsustainable energy production. Agricultural Services Bulletin. http://agris.fao.org/agris-search/search.do?recordID=XF1998080893.

FAO, 2003. Food and nutrition paper 77, Food energy - methods of analysis and conversion factors. In: Report of a Technical Workshop, Rome, 3-6 December 2002. Food and Agriculture Organization of the United Nations, Rome. http://www.fao.org/3/Y5022E/y5022e00.htm.

FAO, 2005. Global Forest Resources Assessment Report. Food and Agriculture Organization, United Nations, Geneva. http://www.fao.org/forest-resources-assessment/past-assessments/fra-2005/en.

FAO, 2015. The Global Forest Resource Assessment, 2015. http://www.fao.org/forest-resources-assessment/en.

Farmer, G., Cook, J., 2013. Climate Change Science: A Modern Synthesis. In: Physical Climate, vol. 1. Springer Verlag, Berlin-Heidelberg-New York, 564 pp. https://www.springer.com/gp/book/9789400757561.

Feely, R.A., Sabine, C.L., Takahashi, T., Wanninkhof, R., 2001. Uptake and Storage of Carbon Dioxide in the Ocean: The Global CO2 Survey. Oceanography 14 (4), 18−32. https://www.pmel.noaa.gov/pubs/outstand/feel2331/feel2331.shtml.

Field, C.B., Behrnfeld, M.J., Randerson, J.T., Falkowski, P., 1998. Primary production of the biosphere: Integrating terrestrial and oceanic components. Science 281, 237−240. https://doi.org/10.1126/science.281.5374.237.

Findlay, S., Pace, M.L., Lints, D., Howe, K., 1992. Bacterial metabolism of organic carbon in the tidal freshwater Hudson Estuary. Marine Ecology Progress Series 89, 117−153. https://doi.org/10.3354/meps089147.

Fisher, S.G., Likens, G.E., 1973. Energy flow in Bear Brook, New Hampshire: an integrative approach to stream ecosystem metabolism. Ecological Monographs 43 (4), 421−439. https://doi.org/10.2307/1942301.

Fisher, S.G., 1977. Organic matter processing by a stream-segment ecosystem: fort river, Massachusetts, USA. Internationale Revue der Gesamten Hydrobiologie 62 (6), 701−727. https://doi.org/10.1002/iroh.1977.3510620601.

Fletcher, C., 2013. Climate Change: What the Science Tells Us. John Wiley & Sons, Hoboken, New Jersey, 265 pp. https://www.wiley.com/en-us/Climate+Change%3A+What+the+Science+Tells+Us-p-9781118057537.

Fraser, P.J., Elliott, W.P., Watterman, L.S., 1986. Atmospheric $CO_2$ record from direct chemical measurements during the 19th century, pp. 66−88. In: Trabalka, J.R., Reichle, D.E. (Eds.), The Changing Global Carbon Cycle: A Global Analysis. Springer Verlag, New York, 592 pp. https://link.springer.com/book/10.1007/978-1-4757-1915-4.

Freyer, H.D., 1986. Interpretation of the northern hemispheric record of $^{13}C/^{12}C$ trends of atmospheric $CO_2$ in tree rings. In: Trabalka, J.R., Reichle, D.E. (Eds.), The Changing Carbon Cycle: A Global Analysis. Springer Verlag, New York, pp. 125−150, 592 pp. https://link.springer.com/book/10.1007/978-1-4757-1915-4.

Friggens, M.M., Williams, M.I., Bagne, K.E., Wixom, T.T., Cushman, S.A., 2018. Chapter 9: effects of climate change on terrestrial animals. In: Halofsky, J.E., et al. (Eds.), Climate Change Vulnerability and Adaptation in the Intermountain Region, Parts I and II, USDA

Forest Service RMRS-GTR-375, pp. 264−315, 513 pp. www.fs.usda.gov/treesearch/pubs/56101.

Fruton, J.S., Simmonds, S., 1953. General Biochemistry. John Wiley & Sons, 1077 pp. https://doi.org/10.1126/science.119.3081.101-a.

Fujino, J., Nair, R., Kainuma, M., Masui, T., Matsuoka, Y., 2006. Multi-gas mitigation analysis on stabilization scenarios using AIM global model. Multigas Mitigation and Climate Policy. Energy Journal 3 (Special Issue), 343−353. https://econpapers.repec.org/article/aenjournl/2006se_5fweyant-a17.htm.

Fung, I.Y., 1986. Analysis of the seasonal and geographical patterns of atmospheric $CO_2$ distributions with a three-dimensional tracer model. In: Trabalka, J.R., Reichle, D.E. (Eds.), The Changing Carbon Cycle: A Global Analysis. Springer Verlag, Berlin-Heidelberg-New York, pp. 459−473, 592 pp. https://link.springer.com/book/10.1007/978-1-4757-1915-4.

Gaille, B., 2018. The Pros of a Cap Trade. https://brandongaille.com/12-cap-and-trade-pros-and-cons.

Galoux, A., Benecke, P., Gietl, G., Hager, H., Kayser, C., Kiese, O., Knoerr, K.R., Murphy, C.E., Schnock, G., Sinclair, T.R., 1981. Radiation, heat, water, and carbon dioxide balances. In: Reichle, D.E. (Ed.), Dynamic Properties of Forest Ecosystems. Cambridge Univ. Press, London, pp. 87−201, 683 pp. https://www.worldcat.org/title/dynamic-properties-of-forest-ecosystems/oclc/1025081165.

Gammon, R.H., Sundquist, E.T., Fraser, P.J., 1985. History of carbon dioxide in the atmosphere. In: Trabalka, J.R. (Ed.), Atmospheric Carbon Dioxide and the Global Carbon Cycle. U. S. Dept. of Energy, DOE/ER-0239, Washington, D.C., pp. 25−62, 316 pp. https://www.osti.gov/servlets/purl/6048470

Gammon, R.H., Komhyr, W.D., Peterson, J.T., 1986. The global atmospheric $CO_2$ distribution 1968-1983: interpretation of the results of the NOAA/GMCC measurement program, pp. 1-15. In: Trabalka, J.R., Reichle, D.E. (Eds.), The Changing Global Carbon Cycle: A Global Analysis. Springer Verlag, New York, 592 pp. https://link.springer.com/book/10.1007/978-1-4757-1915-4.

Gardner, R.H., Trabalka, J.R., 1985. Methods of uncertainty analysis for a global carbon dioxide model. U.S. Dept. of Energy, Carbon Dioxide Research Division. Technical Report TR024, 42 pp.

Gates, D.M., Tautraporn, W., 1952. The reflectivity of deciduous trees and herbaceous plants in the infrared to 25 microns. Science 115, 613−616. https://doi.org/10.1007/BF02315993.

Gates, D., Johnson, B., Yocum, C.S., Lommen, P.W., 1969. Geophysical factors affecting plant productivity. In: Proc. International Symposium "Productivity of Photosynthetic Systems," Part II: Theoretical Foundations of Optimization of the Photosynthetic Productivity. U. S. S. R. http://www.dtic.mil/dtic/tr/fulltext/u2/705517.pdf.

Gates, D.M., 1962. Energy Exchange in the Biosphere. Harper and Row, New York, 151 pp. https://www.worldcat.org/title/energy-exchange-in-the-biosphere/oclc/1180343.

Gates, D.M., 1965. Energy, plants and ecology. Ecology 46, 1−13. https://doi.org/10.2307/1935252.

Gates, D.M., 1965. Radiant energy, its receipt and disposal. Meteorological Monographs 6, 1−26. https://doi.org/10.1007/978-1-940033-58-7_1.

Gates, D.M., 1968. Energy exchange in the biosphere. In: Eckhardt, F.F. (Ed.), Proceedings of Copenhagen Symposium. UNESCO, Paris. https://instaar.colorado.edu/research/publications/instaar/P6060.

Gates, D.M., 1968. Energy exchange between organism and environment. In: Biometeorology. Oregon State University Press, pp. 1−22. https://archive.org/details/biometeorologyth 0000biol.

Gates, D.M., 1971. The flow of energy in the biosphere. Scientific American 224 (3), 88−103. https://www.jstor.org/stable/24923119.

Gigilio, L., Randerson, J.T., van der Werf, G.R., 2013. Analysis of daily, monthly, and annual burned area using fourth generation global fire emissions database (GFED4). Journal of Geophysical Research: Biosciences 118 (1), 317−328. https://doi.org/10.1002/jgrg.20042.

Gilbert, J.A., Henry, C., 2015. Predicting ecosystem emergent properties at multiple scales. Environmental Microbiology Reports 7 (1), 20−22. https://doi.org/10.1111/1758-2229.12258.

Glazier, D.S., 2008. Effects of metabolic level on the body size scaling of metabolic rate in birds and mammals. Proceedings of Royal Society B 275 (1641), 1405−1419. https://doi.org/10.1098/rspb.2008.0118.

Global CCS Institute. hub.globalccsinstitute.com/ … /23-algae-biomass-productivity/.

Global CCS Institute. hub.globalccsinstitute.com/…/23-algae-biomass-productivity.

Golley, F.B., 1960. Energy dynamics of a food chain of an old-field community. Ecology Monographs 30 (2), 187−206. https://doi.org/10.2307/1948551.

Golley, F.B., 1961. Energy values of ecological materials. Ecology 42 (3), 581−584. https://doi.org/10.2307/1932247.

Golley, F.B., 1968. Secondary productivity in terrestrial communities. American Zoologist 8 (1), 53−59. https://doi.org/10.1093/icb/8.1.53.

Goodman, A., 2018. Ecosystems and the Physical Environment. Strayer University. http://slideplayer.com/slide/2473763.

Gordon, D.C., Sutcliffe, W.H., 1973. A new dry combustion method for the simultaneous determination of total organic carbon and nitrogen in sea water. Marine Chemistry 1, 231−244. https://doi.org/10.1016/0304-4203(73)90006-6.

Gore, A.J.P., Olson, J.S., 1967. Preliminary models for accumulation of organic matter in an *Eriophorum Calluna* ecosystem. Aquilo Ser Botanica 6, 297−313.

Gore, A., 2006. An Inconvenient Truth: The Planetary Emergency of Global Warming and what We Can Do about it. Rodale, Emmaus, PA, 325 pp. https://www.worldcat.org/title/inconvenient-truth-the-planetary-emergency-of-global-warming-and-what-we-can-do-about-it/oclc/900477783.

Gorham, E., 1995. The biogeochemistry of northern peatlands and its possible responses to global warming, pp. 169−187. In: Woodwell, G.M., Mackenzie, F.T. (Eds.), Biotic feedbacks in the global climatic system. Oxford University Press, Oxford, UK, 436 pp. https://global.oup.com/academic/product/biotic-feedbacks-in-the-global-climatic-system-9780195086409?cc=us&lang=en&/.

Gornitz, V., Lebedeff, S., Hansen, J., 1982. Global sea level rise in the past century. Science 215, 1611−1614. https://doi.org/10.1126/science.215.4540.1611.

Grace, J., Zhang, R., 2006. Predicting the effect of climate change on global plant productivity and the carbon cycle. In: Morison, J.I.L., Morecroft, M.D. (Eds.), Plant Growth and Climate Change. Blackwell, Oxford, pp. 187−208, 213 pp. https://doi.org/10.1002/9780470988695.fmatter

Green, G.W., 1967. Weather and Insects, Biometeorology, pp. 81−111. http://www.wmo.int/pages/prog/wcp/agm/publications/technical_notes.php.

Griffith, H., Jarvis, P., 2005. The Carbon Balance of Forest Biomes. Taylor Francis eBooks, Florence, Kentucky, 720 pp. https://www.bookdepository.com/Carbon-Balance-Forest-Biomes-57-Howard-Griffith/9781859962145.

Griggs, M.B., 2016. Popular Science. June issue. www.popsci.com/snow-algae-is-melting-glaciers-in-arctic.

Hall, R.O., Tank, J.L., Baker, M.A., Rosi-Marshall, E.J., Hotchkiss, E.R., 2016. Metabolism, Gas Exchange, and Carbon Spiraling in Rivers. Ecosystems 19 (1), 73−86. https://doi.org/10.1007/s10021-015-9918-1.

Halofsky, J.E., Peterson, D.L., Prendeville, H.R. (Eds.), 2018. Climate Change Vulnerability and Adaptation in the Intermountain Region, Parts I and II. USDA Forest Service RMRS-GTR-375, 513 pp. www.fs.usda.gov/treesearch/pubs/56101.

Hambrook-Berkman, J.A., Canova, M.G., 2007. Algal biomass indicators. Chap. A7, sec. 7.4., 36 pp. In: U.S. Geological Survey Techniques of Water-Resources Investigations. Book 9. Accessed from. http://pubs.water.usgs.gov/twri9A7.

Hamilton, E. (Ed.), 2017. Environmental Biochemistry. ML Books International, 277 pp.

Handbook of Chemistry and Physics, 53rd ed., 1971. Chemical Rubber Publishing Co.

Hansen, J., Sato, M., 2016. Regional climate change and national responsibilities. Environmental Research Letters 11 (3). https://doi.org/10.1088/1748-9326/11/3/034009.

Hansen, J., Sato, M., Ruedy, R., Kharecha, P., Lacis, A., Miller, R., Nazarenko, L., Lo, K., Schmidt, G.A., Russell, G., Aleinov, I., Bauer, S., Baum, E., Cairns, B., Canuto, V., Chandler, M., Cheng, Y., Cohen, A., Del Genio, A., Faluvegi, G., Fleming, E., Frien, A., Hall, T., Jackman, C., Jonas, J., Kelley, M., Kiang, N.Y., Koch, D., Labow, G., Lerner, J., Menon, S., Novakov, T., Oinas, V., Perlwitz, J., Perlwitz, J., Rind, D., Romanou, A., Schmunk, R., Shindell, D., Stone, P., Sun, S., Streets, D., Tausnev, N., Thresher, D., Unger, N., Yao, M., Zhang, S., 2007. Dangerous human-made interference with climate: a GISS model study. Atmospheric Chemistry and Physics 7, 2287−2312. https://doi.org/10.5194/acp-7-2287-2007.

Hansen, J., Kharecha, P., Sato, M., 2013. Doubling Down on Our Faustian Bargain. https://www.countercurrents.org/hansen010413.htm.

Hansen, J., 2009. Storms of My Grandchildren: The Truth about the Coming Climate Catastrophe and Our Last Chance to Save Humanity. Bloomsbury Publishing. https://www.bloomsbury.com/us/storms-of-my-grandchildren-9781608195022.

Harris, W.F., Sollins, P., Edwards, N.T., Dinger, B.E., Shugart, H.H., 1975. Analysis of carbon flow and productivity in a temperate deciduous forest ecosystem. In: Reichle, D.E., Franklin, J.F., Goodall, D.W. (Eds.), Productivity of World Ecosystems. The National Academies Press, Washington, DC, pp. 116−122, 166 pp. http://nap.edu/20114.

Harvey, L.D.D., 2005. Box models of the terrestrial biosphere. In: Wigley, T.M.L., Schimel, D.S. (Eds.), The Carbon Cycle. Cambridge Univ. Press, Cambridge, UK, pp. 238−2447, 292 pp. http://catdir.loc.gov/catdir/samples/cam031/00023735.pdf.

Harvey, C., 2019. $CO_2$ Levels Just Hit Another Record—Here's Why It Matters. Scientific American E&E News, 2019. https://www.scientificamerican.com/article/co2-levels-just-hit-another-record-heres-why-it-matters.

Haskin, C., 2013. Determination of the concentration of atmospheric gases by gas chromatography. McNair Scholars Research Journal 6 (1), 37−52. http://commons.emich.edu/mcnair/vol6/iss1/631/00023735.pdf.

Hatcher, B.G., Mann, K.H., 1975. Above-ground production of marsh cordgrass (*Spartina alterniflora*) near the northern end of its range. Journal of the Fisheries Research Board of Canada 32, 83−87. https://doi.org/10.1139/f75-013.

Hayden, B.J., 2008. Comparison of assimilation efficiency on diets of nine phytoplankton species of the greenshell mussel, *Perna canaliculus*. Journal of Shellfish Research. 887−892. https://doi.org/10.2983/0730-8000(2006)25[887:COAEOD]2.0.CO;2.

Heal, O.W., MacLean, S.F., 1975. Comparative productivity in ecosystems − secondary productivity. In: Van Dobben, W.H., Lowe-McConnell, R.H. (Eds.), Unifying Concepts in Ecology. W. Junk Publ., The Hague, pp. 89−108, 302 pp. https://www.worldcat.org/title/unifying-concepts-in-ecology-report-of-the-plenary-sessions-of-the-first-international-congress-of-ecology-the-hague-the-netherlands-september-8-14-1974/oclc/1941730.

Hedges, J.I., 1992. Global biogeochemical cycles: progress and problems. Marine Chemistry 39, 67−93. https://doi.org/10.1016/0304-4203(92)90096-S.

Heimann, M., Keeling, C.D., Fung, I.Y., 1986. Simulating the atmospheric carbon dioxide distribution with a three-dimensional tracer model. In: Trabalka, J.R., Reichle, D.E. (Eds.), The Changing Carbon Cycle: A Global Analysis. Springer Verlag, Berlin-Heidelberg-New York, pp. 16−49, 592 pp. https://link.springer.com/book/10.1007/978-1-4757-1915-4.

Heldt, H.-W., Piechulla, B., 2011. Plant Biochemistry, fourth ed. Elsevier, 622 pp. https://epdf.pub/plant-biochemistry-4th-edition.html.

Hestir, E.L., Brando, V.E., Bresciani, M., Giardino, C., Matta, E., Villa, P., Dekker, A.G., 2015. Measuring freshwater aquatic ecosystems: the need for a hyperspectral global mapping satellite mission. Remote Sensing of Environment 167, 181−195. https://doi.org/10.1016/j.rse.2015.05.023.

Hijioka, Y., Matsuoka, Y., Nishimoto, H., Masui, M., Kainuma, M., 2008. Global GHG emissions scenarios under GHG concentration stabilization targets. Journal of Global Environmental Engineering 13, 97−108. http://www.iiasa.ac.at/web-apps/tnt/RcpDb/dsd?Action=htmlpage&page=welcome.

Hill, R., 1937. Oxygen evolved by isolated chloroplasts. Nature 139 (3525), 881−882. https://www.nature.com/articles/139881a0.

Hoellein, T.J., Bruesewitz, D.A., Richardson, D.C., 2013. Reevisiting Osum (1956): a synthesis of aquatic ecosystem metabolism. Limnology & Oceanography 58 (6), 2089−2100. https://doi.org/10.4319/lo.2013.58.6.2089.

Holdridge, H.L., 1967. Life Zone Ecology. Tropical Science Center, San Jose, California, USA. https://openlibrary.org/books/OL3895575M/Life_zone_ecology.

Holland, H.D., 1978. The Chemistry of the Atmosphere and Oceans. Wiley, New York, 351 pp. http://agris.fao.org/agris-search/search.do?recordID=XF2015047918.

Hollinger, D.Y., Ollimger, S.V., Richardson, A.D., 2010. Albedo estimates for land surface models and support for a new paradigm based on foliage nitrogen concentration. Global Change Biology 16, 696−710. https://doi.org/10.1111/j.1365-2486.2009.02028.x.

Houghton, R.A., Skole, D.L., 1990. Carbon. In: Turner, B.L., Clark, W.C., Kates, R.W., Richards, J.F., Mathews, T.T., Meter, W.B. (Eds.), The Earth as Transformed by Human Action. Cambridge Univ. Press, Cambridge, UK and New York, pp. 393−408. https://www.amazon.com/Earth-Transformed-Human-Action-Biosphere/dp/0521446309.

Houghton, R.A., Hobbie, J.E., Melillo, J.M., Moore, B., Peterson, B.J., Shaver, G.R., Woodwell, G.M., 1983. Changes in the carbon content of terrestrial biota and soils between 1860 and 1980: a net release of $CO_2$ to the atmosphere. Ecological Monographs 53, 235−262. https://doi.org/10.2307/1942531.

Houghton, R.A., Schlesinger, W.H., Brown, S., Richards, J.F., 1985. Carbon dioxide exchange between the atmosphere and terrestrial ecosystems. In: Trabalka, J.R. (Ed.), Atmospheric Carbon Dioxide and the Global Carbon Cycle. U. S. Dept. of Energy, DOEE/ER-0239, Office of Technical and Scientific Information, Oak Ridge, TN, pp. 141−174, 316 pp. https://www.osti.gov/servlets/purl/6048470.

Houghton, R.A., 1986. Estimating changes in the carbon content of terrestrial ecosystems from historical data. In: Trabalka, J.R., Reichle, D.E. (Eds.), The Changing Global Carbon Cycle: A

Global Analysis. Springer Verlag, New York, pp. 175−193, 592 pp. https://link.springer.com/book/10.1007/978-1-4757-1915-4.

Houghton, R.A., 2003. Revised estimates of the annual net flux of carbon to the atmosphere from changes in land use and land management 1850-2000. Tellus 55B (2), 378−390. https://doi.org/10.1034/j.1600-0889.2003.01450.x.

Hutchinson, G.E., 1954. The biochemistry of the terrestrial atmosphere, pp. 371−433. In: Kuiper, G.P. (Ed.), The Solar System: Vol. 2 The Earth as a Planet. Univ. Chicago Press, Chicago, IL, USA, 751 pp. http://adsabs.harvard.edu/abs/1954eap..book..371H.

IEA (International Energy Agency), 2010. https://www.iea.org/publications/freepublications/publication/hydropower_essentials.pdf.

IEA (international Energy Agency), 2019. Global Energy and $CO_2$ Status Report. https://webstore.iea.org/global-energy-co2-status-report-2018.

Ingram, M., 2013. Mark Ingram Blog. http://apbiomarkip7.blogspot.com/201555/11/investigation-5-photosynthesis.html.

IPCC, 2014. Climate change 2014. In: Pachauri, R.K., Meyer, L.A. (Eds.), Synthesis Report. Contribution of Working Groups I, II and III to the Fifth Assessment Report of the Intergovernmental Panel on Climate Change [Core Writing Team. IPCC, Geneva, Switzerland, 151 pp. https://www.ipcc.ch/report/ar5/syr.

IPCC, 1994. In: Houghton, L.T., Meira Filho, L.C., Bruce, J., Lee, H., Callander, B.A., Haites, E.F., Harris, N., Marshall, K. (Eds.), Climate Change 1994: Radiative Forcing of Climate Change and an Evaluation of the IPCC IS92 Emission Scenarios. Cambridge Univ. Press, Cambridge UK and New York. https://archive.ipcc.ch/publications_and_data/publications_and_data_reports.shtml#1.

IPCC, 2000. In: Nakićenović, N., Swart, R. (Eds.), Special Report on Emissions Scenarios: A special report of Working Group III of the Intergovernmental Panel on Climate Change. Cambridge University Press, Cambridge, UK and New York, NY, USA. https://en.wikipedia.org/wiki/Special_Report_on_Emissions_Scenarios.

IPCC, 2007. In: Pachauri, R.K., Reisinger, A. (Eds.), AR4 Climate Change 2007: Synthesis Report: Contribution of Working Groups I, II and III to the Fourth Assessment Report of the Intergovernmental Panel on Climate Change, Core Writing Team. IPCC, Geneva, Switzerland. https://www.ipcc.ch/report/ar4/syr, 104 pp.

IPCC, 2012. In: Edenhofer, O., Pichs-Madruga, R., Sokona, Y., Seyboth, K., Matschoss, P., Kadner, S., Zwickel, T., Eickemeier, P., Hansen, G., Schlömer, S., von Stechow, C. (Eds.), Intergovernmental Panel on Climate Change Special Report on Renewable Energy Sources and Climate Change Mitigation. Cambridge University Press, Cambridge, UK and New York, NY, 1075 pp. http://srren.ipcc-wg3.de/report.

IPCC, 1995. Climate change 1995: the science of climate change. In: Houghton, L.T., Meira Filho, L.C., Bruce, J., Lee, H., Callander, B.A., Haites, E.F., Harris, N., Kattenberg, A., Marshall, K. (Eds.), Contribution of Working Group 1 to the Second Assessment Report of the IPCC. Cambridge Univ. Press, Cambridge UK and New York, NY, USA. https://archive.ipcc.ch/publications_and_data/publications_and_data_reports.shtml#1.

IPCC, 2001. Climate change 2001. In: Houghton, J.T., Ding, Y., Griggs, D.J., Ngusese, M., and der Linden, R.J., Dai, X., Maskell, K., Johnson, C.A. (Eds.), The Scientific Basis: Contribution of Working Group 1 to the Third Assessment Report of the Intergovernmental Panel on Climate Change. Cambridge University, Cambridge, UK and New York, NY, USA. https://www.ipcc.ch/report/ar3/wg1/.

IPCC, 2007. AR4 Climate change 2007: new developments in knowledge of the carbon cycle since the third assessment report climate change 2007: the physical science basis. In: Solomon, S.,

Qin, D., Maning, M., Chen, Z., Marquis, M., Avery, K.B., Tingor, M., Miller, H.L. (Eds.), Contribution of Working Group I to the Fourth Emissions Report of the Intergovernmental Panel on Climate Change. Cambridge Press, Cambridge, UK and New York, NY, USA. https://www.ipcc.ch/report/ar4/syr/.

IPCC, 2007. Climate change 2007. Table SPM.4, pg. 27. In: Mitigation of Climate Change. Working Group III contribution to the Fourth Assessment Report of the Intergovernmental Panel on Climate Change. https://www.ipcc.ch/site/assets/uploads/2018/02/ar4-wg3-spm-1.pdf. B. Metz, O.R. Davidson, P.R.

IPCC, 2013. Climate change 2013: Table SPM.2. In: Summary for Policymakers: the physical science basis. In: Stocker, T.F., Qin, D., Plattner, G.-K., Tignor, M., Allen, S.K., Boschung, J., Nauels, A., Xia, Y., Bex, V., Midgley, P.M. (Eds.), Contribution of Working Group I to the Fifth Assessment Report of the Intergovernmental Panel on Climate Change. Cambridge University Press, Cambridge, United Kingdom and New York, NY, USA, 1538 pp. https://www.ipcc.ch/report/ar5/wg1/.

IPCC, 2018. Summary for policymakers. In: Masson-Delmotte, V., Zhai, P., Pörtner, H.-O., Roberts, D., Skea, J., Shukla, P.R., Pirani, A., Moufouma-Okia, W., Péan, C., Pidcock, R., Connors, S., Matthews, J.B.R., Chen, Y., Zhou, X., Gomis, M.I., Lonnoy, E., Maycock, T., Tignor, M., Waterfield, T. (Eds.), Global Warming of 1.5°C. An IPCC Special Report on the Impacts of Global Warming of 1.5°C above Pre-industrial Levels and Related Global Greenhouse Gas Emission Pathways, in the Context of Strengthening the Global Response to the Threat of Climate Change, Sustainable Development, and Efforts to Eradicate Poverty. World Meteorological Organization, Geneva, Switzerland, 32 pp. https://www.ipcc.ch/2018/10/08/summary-for-policymakers-of-ipcc-special-report-on-global-warming-of-1-5c-approved-by-governments.

IPSO, 2013. Introduction to the special issue: the global state of the ocean; interactions. Marine Pollution Bulletin 74 (2). http://www.sciencedirect.com/science/journal/0025326X/74/and/ http://www.stateoftheocean.org/.

Irfan, U., 2018. Updated Oct 25, 2018, Vox Media. https://www.vox.com/.

Ishii, M., Fukuda, Y., Hirahara, S., Yasui, S., Suzuki, T., Sato, K., 2017. Accuracy of global upper ocean heat content estimation expected from present observational data sets. Scientific Online Letters on the Atmosphere 13, 163–167. https://doi.org/10.2151/sola.2017-030.

Islam, K.R., Wright, S.W., 2015. Microbial Biomass Measurement Methods. www.researchgate.net/publication/277131720_Microbial/.

Jackson, C.O., Johnson, C.W., 1981. City behind A Fence: Oak Ridge, Tennessee, 1942–1946. Univ. of Tennessee Press, Knoxville, 248 pp. https://www.abebooks.com/book-search/title/city-behind-a-fence/author/charles-johnson.

Jenkins, J., Thernstrom, S., 2017. Deep Decarbonization of the Electric Power Sector: Insights from Recent Literature. Energy Innovation Reform Project. https://www.innovationreform.org/wp-content/uploads/2018/02/EIRP-Deep-Decarb-Lit-Review-Jenkins-Thernstrom-March-2017.pdf/.

Jenkinson, D.S., Powlson, D.S., 1976. The effects of biocidal treatments on metabolism in soil. V. A method for measuring soil biomass. Soil Biology and Biochemistry 8, 209–213. https://doi.org/10.1016/0038-0717(76)90005-5.

Jennings, S., Kaiser, M.J., Reynolds, J.D., 2001. Marine Fisheries Ecology. Blackwell Science, Oxford, 221 pp. https://www.scirp.org/(S(i43dyn45teexjx455qlt3d2q))/reference/References Papers.aspx?ReferenceID=903971.

Johansson, T.B., Patwardhan, A., Nakicenovic, N., Gomez-Echeverri, L. (Eds.), 2012. Global Energy Assessment: Toward a Sustainable Future. Cambridge University Press, UK and the International Institute for Applied Systems Analysis, Laxenberg, Austria, 1865 pp. https://doi.org/10.1017/CBO9780511793677.

Johnson, K.E., Del-Giudice-Tuttle, Anderson, S., 2012. Chapter 10, terrestrial invertebrates. In: Baseline Assessment Program Report, Ballona Wetlands Ecological Reserve, Santa Monica Bay Restoration Committee. California State Coastal Conservancy, pp. 1−9. https://www.wildlife.CA.gov/regions/5/BallonaEIR.

Jorgensen, S.E., 2008. Ecosystem theory: fundamental laws in ecology, pp. 1697−1701. In: Encyclopedia of Ecology. Elsevier, 2780 pp. https://www.elsevier.com/books/encyclopedia-of-ecology/jorgensen/978-0-444-52033-3/,https://www.elsevier.com/books/ecosystem-ecology/jorgensen/978-0-444-53466-8.

Kaiser, D., 2001. Dr. Kaiser 's Microbiology Home Page, 2001. Fig. 3. Electron Transport and Chemiosmosis during Photosynthesis. Creative Commons. http://faculty.ccbcmd.edu/~gkaiser/welcome.html/.

Kanemasu, E.T., Thurtell, G.W., Tanner, C.B., 1969. Design calibration and field use of a stomatal diffusion porometer. Plant Physiology 44 (6), 881−885. https://doi.org/10.1104/pp.44.6.881.

Karlberg, L., Gustafsson, D., Jansson, P., 2007. Modeling Carbon Turnover in Five Terrestrial Ecosystems in the Boreal Zone Using Multiple Criteria of Acceptance. AMBIO, pp. 448−458. https://doi.org/10.1579/0044-7447(2006)35[448:MCTIFT]2.0.CO;2.

Keeling, R.F., Severinghaus, J.P., 2000. Atmospheric oxygen measurements and the carbon cycle, pp. 134−140. In: Wigley, T.M.L., Schimel, D.S. (Eds.), The Carbon Cycle. Cambridge University Press, UK. http://catdir.loc.gov/catdir/samples/cam031/00023735.pdf.

Keeling, C.D., Whorf, T.P., 2005. Atmospheric $CO_2$ records from sites in the SIO air sampling network. In: Trends of Data on Global Change. Carbon Dioxide Information and Analysis Center. Oak Ridge National Laboratory, U. S. Dept. of Energy, Oak Ridge, TN. http://cdiac.esd.ornl.gov/trends/co2/sio-keel-flask/sio-keel-flask.html/.

Keeling, R.F., Piper, S.C., Heimann, M., 1989. A three-dimensional model of atmospheric $CO_2$ transport based on observed winds. 4. Mean annual gradients and interannual variations. In: Peterson, D.H. (Ed.), Aspects of Climate Variability in the Pacific and Western Americas. Amer. Geophysical Union, Wash., D. C, pp. 305−363. https://doi.org/10.1029/GM055p0305.

Keil, R.S., 2005. Effects of ocean circulation change on atmospheric $CO_2$. In: Wigley, T.M.L., Schimel, D.S. (Eds.), The Carbon Cycle. Cambridge Univ. Press, Cambridge, UK, pp. 229−237, 292 pp. http://catdir.loc.gov/catdir/samples/cam031/00023735.pdf.

Kerrick, D.M., 2001. Present and past nonanthropogenic $CO_2$ degassing from the solid earth. Reviews of Geophysics 39 (4). https://doi.org/10.1029/2001RG000105.

Kicklighter, D.W., Bondeau, A., Schloss, A.L., Kaduk, J., McGuire, A.D., et al., 1999. Comparing global models of terrestrial net primary productivity (NPP): global pattern and differentiation by major biomes. Global Change Biology 5 (1), 16−24. https://doi.org/10.1046/j.1365-2486.1999.00003.x.

Killough, G.S., Emanuel, W.R., 1981. A comparison of several models of carbon turnover in the ocean with respect to their distribution of transit time and age, and responses to atmospheric $CO_2$ and $^{14}C$. Tellus 33, 274−290. https://doi.org/10.3402/tellusa.v33i3.10715.

Kincaid, A., March 22, 2018. 2017 Was a Bad Year for Hilton Head's Sea Turtles. 2018 Could Be Worse. The Island Packet. https://www.islandpacket.com/news/local/news-columns-blogs/untamed-lowcountry/article206353759.html#storylink=cpy/.

Kinerson, R.S., Ralston, C.H., Wells, C.G., 1977. Carbon cycling in a loblolly pine plantation. Oecologia 29, 1−10. https://doi.org/10.1007/BF00345358.

Kira, T., 1975. Primary production of forests. In: Cooper, J.P. (Ed.), Photosynthesis and Productivity in Different Environments. Cambridge University Press, Cambridge, London, New York, pp. 5−41. https://scholar.google.com/scholar_lookup?title=Photosynthesis+and+Productivity+in+Different+Environments&publication+year=1975&author=Kira+T.&author=Cooper+J.+P.&pages=1-40.

Kirby, A., 2001. Science Academies Back Kyoto. BBC News.

Kitazawa, Y., 1977. Ecosystem metabolism of the subalpine coniferous forest of the Shigayama IBP area. In: Kitazawa, Y. (Ed.), Ecosystem Analysis of the Subalpine Coniferous Forest of the Shigayama IBP Area, Central Japan, JIBP Synthesis 15. University of Tokyo Press, pp. 181−199. https://r.search.aol.com/_ylt=A2KLfSFCm5td7r4AXUJpCWVH;_ylu=X3o DMTBzdWd2cWI5BGNvbG8DYmYxBHBvcwMxMAR2dGlkAwRzZWMDc3I-/RV=2/RE= 1570507715/RO=10/RU=https%3a%2f%2fir.library.oregonstate.edu%2fdownloads%2f1c18 dh048/RK=0/RS=FJPqLYusgl20hnIN66FOr8tLEHQ-/.

Kleiber, M., 1961. The Fire of Life: An Introduction to Animal Energetics. John Wiley & Sons, Inc., New York, 454 pp. https://www.cabdirect.org/cabdirect/abstract/19621404881 and https://doi.org/10.1126/science.134.3495.2033.

Knauer, J.A., 1993. Productivity and new production in the ocean systems, pp. 211−231. In: Wollast, R., MacKenzie, F.T., Chou, L. (Eds.), Interactions of C, N, P, and S: Biogeochemical Cycles and Global Change. Springer Verlag, New York, 521 pp. https://link.springer.com/chapter/10.1007/978-3-642-76064-8_8.

Knorr, W., 2009. Is the airborne fraction of anthropogenic $CO_2$ emissions increasing? Geophysical Research Letters 36 (21), L21710. https://doi.org/10.1029/2009GL040613.

Koch, A., Brierley, C., Maslin, M.M., Lewis, S.L., 2018. Earth system impacts of the European arrival and great dying in the Americas after 1492. Quaternary Science Reviews 207, 13−36. https://doi.org/10.1016/j.quascirev.2018.12.004.

Kondratyev, K., 1969. Radiation in the Atmosphere. International Geophysics Series, vol. 12. Academic Press, New York, 916 pp. https://www.elsevier.com/books/radiation-in-the-atmosphere/kondratyev/978-0-12-419050-4.

Kozlovsky, D.G., 1976. A critical evaluation of the trophic level concept. I. Ecological efficiencies. In: Wiegert, R.G. (Ed.), Ecological Energetics. Dowden. Hutchinson & Ross, Stroudsburg, PA, pp. 137−149, 457 pp. https://link.springer.com/chapter/10.1007/978-1-4612-3842-3_3.

Krey, V., Luderer, G., Clarke, L., Kriegler, E., 2014. Getting from here to there − energy technology transformation pathways in the EMF27 scenarios. Climatic Change 123 (3−4), 369−382. https://doi.org/10.1007/s10584-013-0947-5.

Kumar, B.M., Nair, P.K.R. (Eds.), 2011. Carbon Sequestration in Agroforestry Systems. Springer Verlag, Berlin-Heidelberg-New York, 307 pp.

Lawson, J.W., Miller, E.H., Noseworthy, E., 1997. Variation in assimilation efficiency and digestive efficiency of captive harp seals (*Phoca groenlandica*) on different diets. Canadian Journal of Zoology 75 (8), 1285−1291. https://doi.org/10.1139/z97-152.

Le Quéré, C., Aumont, O., Bopp, L., Bousquet, P., Ciais, P., Francey, R., Heimann, M., Keeling, C.D., Keeling, R.F., Kheshgi, H., Peylin, P., Piper, S.C., Prentice, I.C., Rayner, P.J., 2003. Two decades of ocean $CO_2$ sink and variability. Tellus 55 (2), 649−656. https://doi.org/10.1034/j.1600-0889.2003.00043.x.

Le Quéré, C., Harrison, S.P., Prentice, I.C., Buitenhuis, E.T., Aumont, O., Bopp, L., et al., 2005. Ecosystem dynamics based on plankton functional types for global ocean biogeochemistry models. Global Change Biology 11 (11), 2016−2040. https://doi.org/10.1111/j.1365-2486. 2005.1004.x.

Le Quéré, C., Raupach, Michael R., Canadell, Josep G., Gregg, Marland, Bopp, Laurent, Ciais, Philippe, Conway, Thomas J., Doney, Scott C., Feely, Richard A., Foster, Pru, Friedlingstein, Pierre, Gurney, Kevin, Houghton, Richard A., House, Joanna I., Huntingford, Chris, Levy, Peter E., Lomas, Mark R., Joseph, Majkut, Metzl, Nicolas, Ometto, J.P., Peters, G.P., Prentice, I.C., Randerson, J.T., Running, S.W., Sarmiento, J.L., Schuster, U., Sitch, S., Takahashi, T., Viovy, N., van der Werf, G.R., Woodward, F.I., 2009. Trends in the sources and sinks of carbon dioxide. Nature Geosciences 2, 831−836. https://doi.org/10.1038/ngeo689.

Le Quéré, C., Raupach, M.R., Canadell, J.G., Marland, G., Bopp, L., Ciais, P., Conway, T.J., Doney, S.C., Feely, R., Foster, P., Friedlingstein, P., Gurney, K., Houghton, R.A., House, J.I., Huntingford, C., Levy, P.E., Lomas, M.R., Majkut, J., Metzl, N., Levine, J.S., Cofe, W.R., Pinto, J.P., 1993. Biomass burning. In: Khalil, M.A.K. (Ed.), Atmospheric Methane Sources, Sinks, and Role in Global Change. Springer Verlag, New York, pp. 299−313. https://www.worldcat.org/title/atmospheric-methane-sources-sinks-and-role-in-global-change/oclc/607896749.

Lehmann, J., Mempel, F., Coumou, D., 2018. Increased occurrence of record-wet and record-dry months reflect changes in mean rainfall. Geophysical Research Letters 45. https://doi.org/10.1029/2018GL079439.

Levine, J.S., Cofe, W.R., Pinto, J.P., 1993. Biomass burning, pp. 299−313. In: Khalil, M.A.K. (Ed.), Atmospheric Methane Sources, Sinks, and Role in Global Change. Springer Verlag, New York, 563 pp.

Levitus, S., Antonov, J.I., Boyer, T.P., Baranova, O.K., Garcia, H.E., Locarnini, R.A., Mishonov, A.V., Reagan, J.R., Seidov, D., Yarosh, E.S., Zweng, M.M., 2012. World ocean heat content and thermosteric sea level change (0−2000 m), 1955−2010. Geophysical Research Letters 39 (10), L10603. https://doi.org/10.1029/2012GL051106.

Lewis, M., 2018. Coming Storm. Audiobook by Audio Studios.

Li, W., Dickinson, R.E., Fu, R., Niu, G.-Y., Tang, Z.-L., Canadell, J.G., 2007. Future precipitation changes and their implications for tropical peatlands. Geophysical Research Letters 34 (1). https://doi.org/10.1029/2006GL028364.

Liang, S., Li, X., Wang, J. (Eds.), 2012. Advanced Remote Sensing: Chapter 16, Vegetation Production in Terrestrial Ecosystems, 16.2.2 Eddy Covariance. Academic Press, pp. 501−531, 800 pp. https://books.google.com/books?hl=en&lr=&id=NLRiZiXe8CEC&oi=fnd&pg=PP1 &ots=kISs-BBPKZ&sig=2KShNTi-0_uqolOxZJdjbopU9Cg#v=onepage&q&f=false.

Liang, S. Li, X., Wang, J. (Eds.), 2006. Advanced sensing: Eddy covariance, pp. 501-531. In: Vegetation Production in Terrestrial Ecosystems. Academic Press, Oxford, UK. 800 pp

LibreTexts, 2019. Ecological Pyramids. https://bio.libretexts.org/Bookshelves/Introductory_and_General_Biology/Book%3A_General_Biology_(Boundless)/46%3A_Ecosystems/46.2%3A_Energy_Flow_through_Ecosystems/46.2D%3A_Ecological_Pyramids/.

Lieth, H., 1974. Phenology and Seasonality Modeling. Springer Verlag, Berlin-Heidelberg-New York, 444 pp. https://www.springer.com/gp/book/9783642518652.

Lieth, H., 1975. Modeling the primary production of the world. In: Lieth, H., Whittaker, R.H. (Eds.), Primary Productivity of the Biosphere. Springer Verlag, Berlin-Heidelberg-New York, pp. 237−263, 339 pp. https://www.springer.com/gp/book/9783642809156.

Lieth, H., 1975. Primary production of the major vegetation units of the world. In: Lieth, H., Whittaker, R.H. (Eds.), Primary Productivity of the Biosphere. Springer Verlag, Berlin-Heidelberg-New York, pp. 203−215, 339 pp.

Lieth, H., 1975. Modeling the primary production of the world. In: Lieth, H., Whittaker, R.H. (Eds.), Primary Productivity of the Biosphere. Springer Verlag, Berlin-Heidelberg-New York, pp. 237−263, 339 pp. https://www.springer.com/gp/book/9783642809156.

Lieth, H., 1975. Primary production in ecosystems: comparative analysis of global patterns. In: Lowe-McConnell, R.H. (Ed.), Unifying Concepts in Ecology, W. H. Van Dobben and. W. Junk, The Hague, pp. 67−88, 302 pp. https://link.springer.com/book/10.1007/978-94-010-1954-5.

Likens, G.E., 1975. Primary production of inland aquatic ecosystems, p. 185−202. In: Lieth, H., Whittaker, R.H. (Eds.), Primary productivity of the biosphere. Springer-Verlog, New York, 339 pp. https://www.springer.com/gp/book/9783642809156.

Linacre, E.T., 1964. Determinaions of the heat transfer coefficient of a leaf. Plant Physiology 39 (4), 687−690. https://doi.org/10.1104/pp.39.4.687.

Linacre, E.T., 1969. Net radiation to various surfaces. Journal of Ecology 6, 61−73. https://doi.org/10.2307/2401301. https://www.jstor.org/stable/2401301/.

Lindeman, R., 1942. The trophic dynamic aspect of ecology. Ecology 23 (4), 399−418. https://doi.org/10.2307/1930126.

Link, J.S., Griffis, R., Busch, S. (Eds.), 2015. NOAA Fisheries Climate Science Strategy. U.S. Dept. of Commerce. NOAA Technical Memorandum NMFS-F/SPO-155, 70 pp. https://www.st.nmfs.noaa.gov/Assets/ecosystems/climate/documents/NCSS_Final.pdf.

Lisboa, L.K., Thomas, S., Moulton, T.P., 2016. Reviewing carbon spiraling approach to understand organic matter movement and transformation in lotic ecosystems. Acta Limnologica Brasiliensia 28, e14. https://doi.org/10.1590/S2179-975X2116.

Lockau, W., Nitschke, W., 1993. Photosystem I and its bacterial counterparts. Physiologia Plantarum 88 (2), 372−381. https://doi.org/10.1111/j.1399-3054.1993.tb05512.x.

Loftus, P.J., Cohen, A.M., Long, J.C.S., Jenkins, J.D., 2015. A critical review of global decarbonization scenarios: what do they tell us about feasibility? WIREs Climate Change 6 (1), 93−112. https://doi.org/10.1002/wcc.324.

Lovejoy, T.E., October 4,1984. Aid debtor nation's ecology. N.Y. Times A31.

Lovelock, J.E., Margulis, L., 1974. Atmospheric homeostasis by and for the biosphere: the Gaia hypothesis. Tellus. Series A. Stockholm: International Meteorological Institute 26 (1−2), 2−10. https://doi.org/10.3402/tellusa.v26i1-2.9731.

Lovelock, J.E., 1972. Gaia as seen through the atmosphere. Atmospheric Environment 6 (8), 579−580. https://doi.org/10.1016/0004-6981(72)90076-5.

Lucarini, V., Ragone, F., 2011. Energetics of climate models: net energy balance and meridional enthalpy transport. Reviews of Geophysics 49. http://www.bgc-jena.mpg.de/bgc-theory/index.php/Pubs/2008-EncycEcol-AK/.

Lugo, A.E., 2004. H. T. Odum and the liquillo experimental forest. Ecological Modeling 178, 65−74. https://www.fs.fed.us/global/iitf/pubs/ja_iitf_2004_lugo003.pdf.

Luyssaert, S., Schulze, E.D., Börner, A., Knohl, A., Hessenmöller, D., Law, B.E., Clais, P., Grace, J., 2008. Old-growth forests as global carbon sinks. Nature 455 (7210), 213−215. https://doi.org/10.1038/nature07276.

Macfadyen, A., 1964. Energy flow in ecosystems and its exploitation by grazing, pp. 3-20. In: Crisp, D.J. (Ed.), Grazing in Terrestrial and Marine Environments, British Ecological Soc. Symp. Blackwell, Oxford, 322 pp.

Macfadyen, A., 1970. Soil metabolism in relation to ecosystem energy flow and primary and secondary production. In: Methods of Study in Soil Ecology. UNESCO, Paris, pp. 167−172. https://www.cabdirect.org/cabdirect/abstract/19711903864.

Maier-Reeimer, E., Hasselmann, K., 1987. Transport and storage in the ocean − an inorganic ocean-circulation carbon cycle model. Climate Dynamics 2, 63−90. https://doi.org/10.1007/BF01054491.

Manning, A.C., Keeling, R.F., 2006. Global oceanic and land biotic carbon sinks from Scripps atmospheric oxygen flask sampling network. Tellus 58B (2), 95−116. https://doi.org/10.1111/j.1600-0889.2006.00175.x.

Marland, G., Boden, T.A., Griffin, R.C., Huang, S.F., Kanciruk, P., Nelson, T.R., 1989. Estimates of $CO_2$ Emissions from Fossil Fuel Burning and Cement Manufacturing, Based on the United Nationals Energy Statistics and the U.S. Bureau of Mines Cement Manufacturing Data. Report No. #ORNL/CDIAC-25. Carbon Dioxide Information Analysis Centre, Oak Ridge National Laboratory, Oak Ridge, Tennessee, USA. https://cdiac.ess-dive.lbl.gov.

Marland, G., Boden, T.A., Andres, R.J., 2006. Global, regional, and national $CO_2$ emissions. In: Trends: A Compendium of Data on Global Change. Carbon Dioxide Information Analysis Center, Oak Ridge National Laboratory, U. S. Dept. of Energy, Oak Ridge, TN. http://cdiac.esd.ornl.govtrends/emis/tre_glob.htm/.

Marland, G., 2005. The future role of reforestation in reducing buildup of atmospheric $CO_2$. In: Wigley, T.M.L., Schimel, D.S. (Eds.), The Carbon Cycle. Cambridge Univ. Press, Cambridge, UK, pp. 190−198. http://catdir.loc.gov/catdir/samples/cam031/00023735.pdf, 292 pp.

Martin, J.H., Gordon, R.M., 1988. Northeast Pacific iron distribution in relation to phytoplankton productivity. Deep Sea Research 35, 177−196. https://doi.org/10.1016/0198-0149(88)90035-0.

Martin, J.H., Coale, K.H., Johnson, K.S., Fitzwater, S.E., Gordon, R.M., Tanner, S.J., Tindale, N.W., 1994. Testing the iron hypothesis in ecosystems of the equatorial Pacific Ocean. Nature 371 (6493), 123−129. https://doi.org/10.1038/371123a0/.

Mascali, N.M., 2017. Boston ponders sea wall as rising water levels pose future threat. https://www.metro.us/boston/boston-ponders-sea-wall-as-rising-water-levels-pose-future-threat/zsJqbs—7oBwYr4MOgNvw/.

Martinich, J., Crimmins, C., 2019. Climate damages and adaptation potential across diverse sectors of the United States. Nature Climate Change 9, 397−404. https://www.nature.com/articles/s41558-019-0444-6.

Martinich, J., DeAngelo, B., Diaz, D., Ekwurzel, B., Franco, G., Frisch, C., McFarland, J., O'Neill, B., 2018. Impacts, Risks, and Adaptation in the United States: Report-In-Brief: Fourth National Climate Assessment, vol. II. USGCRP (Chapter 29). https://www.globalchange.gov/nca4.

Mason, B., 1958. Principles of Geochemistry. John Wiley & Sons, New York, NY, USA, 310 pp. https://openlibrary.org/books/OL21999238M/Principles_of_geochemistry_by_Brian_Mason_and_Carleton_B._Moore.

Mass Audubon, 2017. State of the Birds Report. https://www.massaudubon.org/content/download/21633/304821/file/mass-audubon_state-of-the-birds-2017-report.pdf.

Mathworks, 2019. Estimated Constrained Values Lookup Table. https://www.mathworks.com/help/sldo/ug/estimate-constrained-values-of-a-lookup-table.html/.

Matthews, E., Fung, I., 1987. Methane emission from natural wetlands: global distribution, area and environmental characteristics of sources. Global Biogeochemical Cycles 1, 61−86. https://doi.org/10.1029/GB001i001p00061.

Mayr, E., 1982. The Growth of Biological Thought. Belkamp Press, 992 pp.

McBrayer, J.F., Reichle, D.E., 1971. Trophic structure and feeding rates of forest soil invertebrate populations. Oikos 22, 381−388. https://doi.org/10.2307/3543862.

McCollum, D., Krey, V., Kolp, P., Nagai, Y., Riahi, K., 2014. Transport electrification: a key element for energy system transformation and climate stabilization. Climatic Change 123 (3−4), 651−664. https://doi.org/10.1007/s10584-013-0969-z.

McConathy, R.C., McLaughlin, S.B., 1978. Transpirational Relationships of Tulip-Poplar. Oak Ridge National Laboratory Environmental Sciences Division, Oak Ridge, TN, 22 pp.

McCree, K.J., 1971-1972. The action spectrum, absorbance and quantum yield of photosynthesis in crop plants. Agricultural Meteorology 9, 191−216. https://doi.org/10.1016/0002-1571(71)90022-7.

McCullough, D.R., 1970. Secondary production of birds and mammals. In: Reichle, D.E. (Ed.), Analysis of Temperate Forest Ecosystems. Springer Verlag, Berlin-Heidelberg-New York, pp. 107−130, 304 pp. https://www.worldcat.org/title/analysis-of-temperate-forest-ecosystems/oclc/102449.

McInnes, L., 2018. Radiative-forcing Components. Wikipedia Commons. https://en.wikipedia.org/wiki/File:Radiative-forcings.svg/.

Medwecka-Kornas, A., Lomnicki, A., Bandola-Ciolczyk, E., 1973. Energy flow in the deciduous woodland ecosystem, Ispina Project, Poland. In: Kern, L. (Ed.), Modeling Forest Ecosystems. Report of International Woodlands Workshop, International Biological Program/PT Section, August 14-26, 1972. Oak Ridge National Laboratory, Oak Ridge, TN, pp. 144−150, 339 pp.

Medwecka-Kornas, A., Lomnicki, A., Bandola-Ciolczyk, E., 1974. Energy flow in the oak-Hornbeam forest (IBP project Ispina). Bulletin de L'academie Polonaise Des Sciences, Serie des sciences biologiques cl. II 22, 563−567.

Meehl, G.A., Washington, W.M., Collins, W.D., Arblaster, J.M., Hu, A., Buja, L.E., Strand, W.G., Teng, H., 2005. How much more global warming and sea level rise. Science 307 (5716), 1769−1772. https://doi.org/10.1126/science.1106663.

Meentemeyer, V., Box, E.O., Thompson, R., 1982. World patterns and amounts of terrestrial plant litter production. BioScience 32, 125−128. https://doi.org/10.2307/1308565.

Meinshausen, M., Meinshausen, N., Hare, W., Raper, S.C.B., Frieler, K., Knutti, R., Frame, D.J., Allen, M.R., 2009. Greenhouse-gas emission targets for limiting global warming to 2°C. Nature 458, 1158−1162. https://doi.org/10.1038/nature08017.

Melillo, J., McGuire, A.D., Kicklighter, D.W., Moore, B., Vorosmarty, C.J., Schloss, A.I., 1993. Global climatic change and terrestrial net production. Nature 363, 234−240. https://www.nature.com/articles/363234a0.

Meshek, M., Sandor, D. (Eds). 4 vols. NREL/TP-6A20-52409. Golden, CO. /http://www.nrel.gov/analysis/re_futures.

Meshek, M., Sandor, D. (Eds.), vol. 4. NREL/TP-6A20-52409. Golden, CO. http://www.nrel.gov/analysis/re_futures/.

Meybeck, M., Ragu, A., 2012. GEMS-GLORI World River Discharge Database. Laboratoire de Géologie Appliquée, Université Pierre et Marie Curie, Paris, France, PANGAEA. https://doi.org/10.1594/PANGAEA.804574/.

Meyer, R.T., Edwards, J.L., 1990. Ecosystem metabolism and turnover of organic carbon along a blackwater river continuum. Ecology 71 (2). https://doi.org/10.2307/1940321.

Miller, S.L., Urey, H.C., 1959. Organic compound synthesis on the primitive earth. Science 130 (3370), 245−251. https://doi.org/10.1126/science.130.3370.245. https://en.wikipedia.org/wiki/Digital_object_identifier.

Monteith, J.L., Szeicz, G., Yabuki, K., 1964. Crop photosynthesis and the flux of carbon dioxide below the canopy. Journal of Applied Ecology 1, 321−337. https://doi.org/10.2307/2401316.

Moomaw, W., Burgherr, P., Heath, G., Lenzen, M., Nyboer, J., Verbruggen, A., 2011. Annex II: methodology. In: Edenhofer, O., Pichs-Madruga, R., Sokona, Y., Seyboth, K., Matschoss, P., Kadner, S., Zwickel, T., Eickemeier, P., Hansen, G., Schlömer, S., von Stechow, C. (Eds.), IPCC Special Report on Renewable Energy Sources and Climate Change Mitigation. Cambridge University Press, Cambridge, UK, pp. 973−1000. http://www.ipcc.ch/pdf/special1011reports/srren/srren_full_report.pdf/.

Mooney, C., October 8, 2015. Scientists say a dramatic worldwide coral bleaching event is now underway. The Washington Post. https://www.washingtonpost.com/news/energy-environment/wp/2015/10/08/scientists-say-a-dramatic-worldwide-coral-bleaching-event-is-now-underway/?utm_term=.66e4e318e2ef/.

Mörner, N., Etiope, G., 2002. Carbon degassing from the lithosphere. Global and Planetary Change 33 (1−2), 185−203. https://doi.org/10.1016/S0921-8181(02)00070-X/.

Mororwitz, H.J., 1968. Energy Flow in Biology. Academic Press, London & New York, 179 pp. https://doi.org/10.1002/jobm.19700100414.

Morowitz, H.J., 1970. Entrophy for Biologists. Academic Press, 195 pp. https://books.google.com/books?hl=en&lr=&id=JSrLBAAAQBAJ&oi=fnd&pg=PP1&ots=QaAM4BSjNz&sig=jFAYu5hagX6HTST-xSCq1094Mbk#v=onepage&q&f=false.

Moss, R., Babiker, M., Brinkman, S., Calvo, E., Carter, T., Edmonds, J., Elgizouli, I., Emori, S., Erda, L., Hibbard, K., Jones, R., Kainuma, M., Kelleher, J., Lamarque, J.F., Manning, M., Matthews, B., Meehl, J., Meyer, L., Mitchell, J., Nakicenovic, N., O'Neill, B., Pichs, R., Riahi, K., Rose, S., Runci, P., Stouffer, R., van Vuuren, D., Weyant, J., Wilbanks, T., van Ypersele, J.P., Zurek, M., 2008. Towards New Scenarios for Analysis of Emissions, Climate Change, Impacts, and Response Strategies, Intergovernmental Panel on Climate Change. Geneva, p. 132. http://www.igbp.net/publications/publishedarticlesandbooks/synthesisandoverviewpapers/synthandoverviewpapers/towardsnewscenariosforanalysisofemissionsclimatechangeimpactsandresponsestrategies.5.316f1832132347017758000121l.html.

Muessig, 2016. ADP ATP Cycle − By Muessig (Own work) Wikimedia Commonshttps://commons.wikimedia.org/wiki/File%3AADP_ATP_cycle.png/.

Munro, J.K., Olson, J.S., 1982. A box-diffusion model of the global carbon dioxide cycle including marine organic compartments and a polar ocean, pp. 37−44. In: Mitsch, W.M., Bosserman, R.M., Klopatek, J.M. (Eds.), Energy and Ecological Modelling. Eldsevier Press, Amsterdam, The Netherlands.

Narby, K.A., 2005. Field metabolic rate and body size. Journal of Experimental Biology 208, 1621−1625. https://doi.org/10.1242/jeb.01553.

NAS, 2019. U.S. National Academy of Sciences Organized Collections: The International Biological Program (IBP), 1964−1974. http://www.nasonline.org/about-nas/history/archives/collections/ibp-1964-1974-1.html/.

NAS, 2019. U.S. National Academy of Sciences Organized Collections: The International Biological Program (IBP), 1964−1974. http://www.nasonline.org/about-nas/history/archives/collections/ibp-1964-1974-1.html.

NASA, 2006. Carbon Cycle. Retrieved from. https://commons.wikimedia.org/wiki/File:Carbon_cycle-cute_diagram.jpeg.

NASA, 2008. Atmospheric $CO_2$ Measurements, by NASA/JPL −http://photojournal.jpl.nasa.gov/catalog/PIA11194/, Public Domain, https://commons.wikimedia.org/w/index.php?curid=5066057/.

NASA, 2018. HyspIRI 2018 Science and Applications Workshop. https://hyspiri.jpl.nasa.gov.

NASA, 2018. Scientific Consensus: Earth's Climate is Warming. Climate Change: Vital Signs of the Planet. Retrieved. https://climate.nasa.gov/scientific-consensus/.

NASA, 2018. NASA GIS & NOAA NCEI Global Temperature Abnormalities (°F) Averaged and Adjusted to Early Industrial Baseline (1881-1910). https://medialibrary.climatecentral.org/resources/2017-global-temperature-review/.

NASA, 2019. Global Climate Change: Scientific Consensus: Earth's Climate Is Warming. https://climate.nasa.gov/scientific-consensus/.

NASEM, National Academies of Sciences, 2019. Engineering, and medicine. National Emissions Technologies and Reliable Sequestration: A Research Agenda. National Academies Press, Wash., D.C., 351+ pp. https://org/10.17226/25259/

National Laboratory Directors, 1977. Technology Opportunities to Reduce Greenhouse Gas Emissions. Appendix B as a separate volume. http://www.ornl/climatechange/.

National Park Service, 2016. Series: Montane Forests of the Southwest Ecosystem Drivers in Southwest Montane Forests. https://www.nps.gov/articles/montane-ecosystem-drivers.htm/.

National Renewable Energy Laboratory, 2012. In: Hand, M.,M., Baldwin, S., DeMeo, E., Reilly, J.M., Mai, T., Arent, D., Porro, G., Meshek, M., Sandor, D. (Eds.), Renewable Electricity Futures Study, 4 vols. NREL/TP-6A20-52409. Golden, CO. http://www.nrel.gov/analysis/re_futures/.

NEON, 2018. http://www.neonscience.org/_http://www.neonscience.org/.

Nepstad, D.C., de Carvalho, C.R., Davidson, E.A., Jipp, P.H., Lefabvre, P.A., Negreiros, G.H., da Silva, E.D., Stone, T.A., Trumbore, S.E., Vieira, S., 1994. The role of deep roots in the hydrological and carbon cycles of Amazonian forests and pastures. Nature 372, 666−669.

Nepstad, D.C., de Carvalho, C.R., Davidson, E.A., Jipp, P.H., Lefabvre, P.A., Negreiros, G.H., da Silva, E.D., Stone, T.A., Trumbore, S.E., Vieira, S., 1994. The role of deep roots in the hydrological and carbon cycles of Amazonian forests and pastures. Nature 372, 666−669. https://doi.org/10.1038/372666a0.

Neubold, P.J., 1967. Primary Production of Forests. IBP Handbook No. 2. Blackwell Scientific Publications, Oxford, UK, 62 pp. http://citeseerx.ist.psu.edu/viewdoc/download?doi=10.1.1.475.3375&rep=rep1&type=pdf.

Newbold, J.D., Mulholl, P.J., Elwood, J.W., O'Neill, R.V., 1982. Organic carbon spiraling in stream ecosystems. Oikos 38 (3), 266−272. https://doi.org/10.2307/3544663.

Nixon, S.W., 1980. Between coastal marshes and coastal waters − a review of twenty years of speculation and research on the role of salt marshes in estuarine productivity and water chemistry, pp. 437−525. In: Hamilton, P., MacDonald, K.B. (Eds.), Estuarine and Wetland Processes. Plenum, New York, 653 pp. https://doi.org/10.1007/978-1-4757-5177-2_20.

NOAA ESRL Global Monitoring Division, 2014. Updated Annually. Atmospheric Carbon Dioxide Dry Air Mole Fractions from Quasi-Continuous Measurements at Mauna Loa, Hawaii. Compiled by K.W. Thoning, D.R. Kitzis, and A. Crotwell. National Oceanic and Atmospheric Administration (NOAA). Earth System Research Laboratory (ESRL), Global Monitoring Division (GMD), Boulder, Colorado, USA. Version 2014-08. https://doi.org/10.7289/V54X55RG/.

NOAA, 2014. National Marine Fisheries Service. Fisheries of the United States, 2014. NOAA Current Fishery Statistics No. 2014. https://www.st.nmfs.noaa.gov/commercial-fisheries/fus/fus14/index/.

NOAA, 2017. NOAA Tides and Currents/. https://tidesandcurrents.noaa.gov/.

NOAA, 2017. Global and Regional Sea Level Rise Scenarios for the United States. NOAA Tech. Rept. NOS CO-OPS 083. https://tidesandcurrents.noaa.gov/publications/techrpt83_Global_and_Regional_SLR_Scenarios_for_the_US_final.pdf.

NOAA, 2018. Coral Reefs and Climate Change. https://oceanservice.noaa.gov/facts/coralreef-climate.html/.

NOAA, 2018. Earth System Research Laboratory, Global Monitoring Division. https://www.esrl. noaa.gov/gmd/ccgg/trends/global.html/.

NOAA, 2019. U.S. National Oceanic and Atmospheric Administration. National Weather Service. https://www.weather.gov/jetstream/energy/.

Norby, R.J., Wullschleger, S.D., Gunderson, C.A., Johnson, D.W., Ceulemans, R., 1999. Tree responses to rising $CO_2$ in field experiments: implication for the future forest. Plant, Cell and Environment 22, 683−714. https://doi.org/10.1007/3-540-31237-4_13.

Norby, R.J., Warren, J.M., Iversen, C.M., Medlyn, B.E., McMurtrie, R.E., 2010. $CO_2$ enhancement of forest productivity constrained by limited nitrogen availability. Proceedings of the National Academy of Sciences of the United States of America 107 (45). https://doi.org/10.1073/pnas. 1006463107.

Nordhaus, W.D., 1992. An optimal transition path for controlling greenhouse gases. Science 258 (5086), 1315−1319. https://doi.org/10.1126/science.258.5086.1315.

Nordhaus, W.D., 1994. Managing the Commons: The Economics of Climate Change. The MIT Press, Cambridge, MA, 213 pp. https://doi.org/10.2458/v2i1.20166.

Oak Ridge National Laboratory, 2019. https://public.ornl.gov/site/gallery/detail.cfm?id=326#credits/.

Oak Ridge National Laboratory, 2019. https://public.ornl.gov/site/gallery/detail.cfm?id=326#credits.

Oberbauer, S.F., Olivas, P.C., Tweedie, C., Oechel, W.C., 2019. https://soa.arcus.org/files/sessions/ 3-3-arctic-system-change/pdf/3-3-10-oberbauer-steven.pdf/.

Odum, H.T., 1956. Primary production in flowing waters. Limnology & Oceanography 1 (2), 102−117. https://doi.org/10.4319/lo.1956.1.2.0102.

Odum, H.T., 1957. Trophic structure and productivity of silver springs, Florida. Ecological Monographs 25, 55−112. https://doi.org/10.2307/1948571.

Odum, E.P., 1959. Fundamentals of Ecology, 2nd Ed. W. B. Sanders Co., Philadelphia and London. 546 pp. https://s3.amazonaws.com/academia.edu.documents/30234225/odum_barrett-ch1-sections1-4.pdf?response-content-disposition=inline%3B%20filename%3DFundamental_of_ ecology.pdf&X-Amz-Algorithm=AWS4-HMAC-SHA256&X-Amz-Credential=AKIAIWOW YYGZ2Y53UL3A%2F20191008%2Fus-east-1%2Fs3%2Faws4_request&X-Amz-Date= 20191008T054328Z&X-Amz-Expires=3600&X-Amz-SignedHeaders=host&X-Amz-Signature=7ee238cb1bb89407f73576d1dd290ab0d19c00bd89f96d9556ebee334cc5cf1f.

Odum, E.P., 1960. Organic production and turnover in old field succession. Ecology 41 (1), 34−49. https://doi.org/10.2307/1931937.

Odum, E.P., 1963. Ecology. Modern Biol. Series. Holt, Rinehart and Winston Inc., New York. https://biblio.co.uk/book/ecology-modern-biology-series-odum-eugene/d/159490673, 152 pp.

Odum, H.T., 1970. Summary, an emerging view of the ecological system at El Verde. In: Odum, H.T., Pigeon, R.F. (Eds.), A Tropical Rain Forest. U.S. Atomic Energy Commission. Division of Technical Information, Oak Ridge, TN, pp. I191−I289. https://dev.openlibrary. org/books/OL5738630M/A_Tropical_rain_forest.

Odum, H.T., 1970. Summary, an emerging view of the ecological system at El Verde. In: Odum, H.T., Pigeon, R.F. (Eds.), A Tropical Rain Forest. U.S. Atomic Energy Commission. Division of Technical Information, Oak Ridge, TN, pp. I191−I289. https://dev.openlibrary. org/books/OL5738630M/A_Tropical_rain_forest.

Odum, E.P., 1974. Energetics and ecosystems, pp. 599−602. In: Reimold, R.J., Queen, W.H. (Eds.), Ecology of Halophytes. Academic Press, London-New York, 605 pp.

Oechel, W.C., Cowles, S., Grulke, N., Hastings, S.J., Lawrence, B., Prudhomme, T., Riechers, G., Strain, B., Tissue, D., Vourlites, G., 1994. Transient nature of $CO_2$ fertilization in arctic tundra. Nature 371, 500−503. https://www.nature.com/articles/371500a0.

Oeschger, H., Stauffer, B., 1986. Review of the history of atmospheric $CO_2$ recorded in ice cores. In: Trabalka, J.R., Reichle, D.E. (Eds.), The Changing Global Carbon Cycle: A Global Analysis. Springer Verlag, New York, pp. 89−108, 592 pp. https://link.springer.com/book/10.1007/978-1-4757-1915-4.

Office of Naval Research, 2019. Oceanography. Retrieved from. http://www.onr.navy.mil/focus/ocean/.

Office of Naval Research, Oceanography. http://www.onr.navy.mil/focus/ocean/.

O'Hara, F. (Ed.), 1990. Glossary: Carbon Dioxide and Climate, ORNL/CDIAC-39, Carbon Dioxide Information Analysis Center. Oak Ridge National Laboratory, Oak Ridge, Tennessee. https://cdiac.ess-dive.lbl.gov/pns/glossary.html.

Oke, T.R., 1987. Boundary Layer Climates. Meutheun Co., London, UK, New York, NY, USA, 335 pp. https://doi.org/10.1002/qj.49711448412.

Olson, J.S., Watts, J.A., Allison, L.J., 1983. Carbon in Live Vegetation in Major World Ecosystems. DOE/NBB-0037. National Technical Information Services, Springfield, VA. https://www.osti.gov/biblio/5963568-carbon-live-vegetation-major-world-ecosystems.

Olson, J.S., Garrels, R.M., Berner, R.A., Armentano, T.V., Dyer, M.I., Yaalon, D.H., 1985. The natural carbon cycle. In: Trabalka, J.R. (Ed.), Atmospheric Carbon Dioxide and the Global Carbon Cycle. U. S. Dept. of Energy, DOE/ER-0239, Washington, D.C., pp. 175−213, 316 pp. https://www.osti.gov/servlets/purl/6048470/

Ometto, J.P., Peters, G.P., Prentice, I.C., Randerson, J.T., Running, S.W., Sarmiento, J.L., Schuster, U., Sitch, S., Takahashi, T., Viovy, N., van der Werf, G.R., Woodward, F.I., 2009. Trends in the sources and sinks of carbon dioxide. Nature Geosciences 2, 831−836.

O'Neill, R.V., Reichle, D.E., 1980. Dimensions of ecosystem/theory. In: Waring, R.H. (Ed.), Forests: Fresh Perspectives from Ecosystem Analysis. Oregon State University Press, Corvallis, OR, pp. 11−26, 199 pp. https://search.aol.com/aol/search?s_it=webmail-searchbox&q=O'Neill%2C%20R.V.%20and%20D.%20E.%20Reichle.%201980.%20Dimensions%20of%20ecosystem%2Ftheory%2C%20pp.11-26.%20%20In%2C%20Forests%3A%20fresh%20perspectives%20from%20ecosystem%20analysis.%20R.H.%20Waring%2C%20Ed.%20Oregon%20State%20University%20Press%2C%20Corvallis%2C%20OR%20199%20pp.

O'Neill, R.V., 1976. Ecosystem persistence and heterotrophic regulation. Ecology 57, 1244−1253. https://doi.org/10.2307/1935048.

OpenStax CNX. http://cnx.org/contents/185cbf87-c72e-48f5-b51e-f14f21b5eabd@10.8/.

Oreskes, N., Conway, E.M., 2010. Merchants of Doubt: How a Handful of Scientists Obscured the Truth on Issues from Tobacco Smoke. Global Warming Bloomsbury Press, New York, 368 pp. https://www.bloomsbury.com/us/merchants-of-doubt-9781608193943.

Orr, J.C., 1993. Accord between ocean models predicting uptake of atmospheric CO2. Water, Air, and Soil Pollution 70, 465−481. https://doi.org/10.1007/978-94-011-1982-5_32.

Osborne, E., Richter-Menge, J., Jefferies, M., 2018. Arctic Report Card 2018. http://www.arctic.noaa.gov/Report-Card/.

Osmond, C.B., 1987. Photosynthesis and carbon economy of plants. New Phytologist 106 (1), 161−175. https://doi.org/10.1111/j.1469-8137.1987.tb04688.x.

Osmond, C.B., 1989. Photosynthesis from the molecule to the biosphere: a challenge for integration. In: Briggs, W.R. (Ed.), Photosynthesis. Alan R. Liss Inc., New York, pp. 5−17, 524 pp. http://agris.fao.org/agris-search/search.do?recordID=US9013328.

OurWorldinData.org/fossil-fuels.

OurWorldinData.org/fossil-fuels.

Parry, I., Mylonas, V., Vernon, N., 2018. IMF Working Paper: Mitigation Policies for the Paris Agreement: An Assessment for G20 Countries. International Monetary Fund. https://www.

imf.org/en/Publications/WP/Issues/2018/08/30/Mitigation-Policies-for-the-Paris-Agreement-An-Assessment-for-G20-Countries-46179/.

Penman, H.L., 1963. Vegetation and Hydrology. Tech. Comm. No. 53. Commonwealth Bureau of Soils. Harpenden, UK, 125 pp.

Pepper, D.A., Del Grosso, S.J., McMurtrie, R.E., Parton, W.J., 2005. Simulated carbon sink response of shortgrass steppe, tallgrass prairie and forest ecosystems to rising [$CO_2$], temperature and nitrogen input. Global Biogeochemical Cycles 19, 20 pp. https://doi.org/10.1029/2004GB002226.

Perruchoud, P.F., Joos, A.F., Hajdas, I., Bonani, G., 1999. Evaluating timescales of carbon turnover in temperate forest soils with radiocarbon data. Global Biogeochemical Cycles 13 (2), 555−573. https://doi.org/10.1029/1999GB90003.

Peters, W., Jacobson, A.R., Sweeney, C., Andrews, A.E., Conway, T.,J., Masarie, K., Miller, J.B., Bruhwiler, L.M., Pétron, G., Hirsch, A.I., Worthy, D.E., van der Werf, G.R., Randerson, J.T., Wennberg, P.O., Krol, M.C., Tans, P.P., 2007. An atmospheric perspective on North American carbon dioxide exchange: CarbonTracker. Proceedings of the National Academy of Sciences of the United States of America 104 (48), 18925−18930. https://doi.org/10.1073/pnas.0708986104.

Petrides, G.A., Swank, W.G., 1966. Estimating the productivity and energy relations of an African elephant population. In: Proc. Internat. Grassland Congr., San Paulo, Brazil, vol. 9, pp. 831−942. In: https://scholar.google.com/scholar_lookup?title=Estimating%20the%20productivity%20and%20energy%20relations%20of%20an%20African%20elephant%20population&author=GA.%20Petrides&author=WG.%20Swank&journal=Proc.%20Internat.%20Grassland%20Cong.%2C%20Sao%20Paulo%2C%20Brazil%20%281965%29&volume=9&pages=831-942&publication_year=1966.

Petrusewicz, K., Macfayden, A., 1970. Productivity of Terrestrial Animals. IBP Handbook No. 13. Blackwell Scientific Pub., Oxford, 192 pp.

Petrusewicz, K. (Ed.), 1967. Secondary Productivity of Terrestrial Ecosystems. Institute of Ecology, Polish Academy of Sciences, Warsaw, 879 pp. https://scholar.google.com/scholar_lookup?title=Secondary%20productivity%20of%20terrestrial%20ecosystems&publication_year=1967.

Phillips, O.L., Malhi, Y., Higuchi, N., Laurance, W.F., Vasquez, P.V., Laurance, S.G., Ferreira, L.V., Stern, M., Brown, S., Grace, J., 1998. Changes in the carbon balance of tropical forests: evidence from long-term plots. Science 16 (282), 439−442. http://eprints.whiterose.ac.uk/340.

Phillipson, J., 1966. Ecological Energetics. St. Martin's Press, New York, 57 pp. https://www.worldcat.org/title/ecological-energetics/oclc/220312888/.

PIOMAS, 2018. http://psc.apl.uw.edu/research/projects/arctic-sea-ice-volume-anomaly/.

Pimentel, D., 2009. Energy Inputs in Food Crop Producction in Developing and Developed Nations. Energies 2 (1), 1−24. https://doi.org/10.3390/en20100001.

Ponge, J.-F., 2005. Emergent properties from organisms to ecosystems: towards a realistic approach. Biological Reviews of the Cambridge Philosophical Society 80 (3), 403−411. https://doi.org/10.1017/S146479310500672X.

Porter, W.B., Gates, D.M., 1969. Thermodynamic equilibria of animals with environment. Ecological Monographs 39 (3), 227−244. https://doi.org/10.2307/1948545.

Porter, W.P., Mitchell, J.W., Beckman, W.A., DeWitt, C.B., 1973. Behavioral implications of mechanistic ecology. Thermal and behavioral modeling of desert ectotherms and their microenvironment. Oecologia 13, 1−54. https://doi.org/10.1007/BF00379617.

Portland State University, November 7, 2018. Climate change causing more severe wildfires, larger insect outbreaks in temperate forests. ScienceDaily. www.sciencedaily.com/releases/2018/11/181107172914.htm/.

Post, W.M., Emanuel, W.R., Zinke, P.J., Strangeburger, A.G., 1982. Soil carbon pools and world life zones. Nature 298, 156–159. https://www.nature.com/articles/298156a0.

Potter, C.S., Richardson, J.T., Feld, C.B., Matson, P.A., Vitousek, P.M., Mooney, H.A., Klooster, S.A., 1993. Terrestrial ecosystem production: a process model based on global satellite and surface data. Global Biogeochemical Cycles 7, 811–841. https://doi.org/10.1029/93GB02725.

PoultryHub./http://www.poultryhub.org/family-poultry-training-course/trainers-manual/broiler-production.

Prather, M.J., Derwent, R., Ehhalt, D., Fraser, P., Sanhueza, E., Zhou, X., 1995. Other trace gases and atmospheric chemistry. In: Houghton, J.H., Meira, L.G., Filho, T., Bruce, H., Lee, B.A., Callender, E., Haites, N., Harris, K., Maskell, K. (Eds.), Climate Change 1995. Cambridge Univ. Press, Cambridge, pp. 73–126. https://archive.ipcc.ch/publications_and_data/publications_and_data_reports.shtml.

Precht, H., Christophersen, J., Hensel, H., Larcher, W., 1973. Temperature and Life. Springer-Verlag, Berlin, Heidelberg, 782 pp. https://www.springer.com/gp/book/9783642657108.

Prigogine, I., Stengers, I., 1997. The End of Certainty: Time, Chaos, and the New Laws of Nature. The Free Press, New York, 228 pp. https://philpapers.org/rec/PRITEO-3.

Privitera, D., Noli, M., Falugi, C., Chiantore, M., 2011. Benthis assemblages and temperature effects of *Paracentrotus lividus* and *Arbacia lixula* larvae and settlement. Journal Experimental. Marine Biology and Ecology 407, 6–11. https://doi.org/10.1016/j.jembe.2011.06.030.

Prosser, C.L., Brown, F.A., 1961. Comparative Animal Physiology. W. B. Sanders. https://www.abebooks.co.uk/book-search/title/comparative-animal-physiology/author/prosser-c-l/, 688 pp.

Rahman, S.M., Tsuchiya, Uehara, T., 2009. Effects of Temperature on Hatching Rate, Embryonic Development and Early Larval Survival of the Edible Sea Urchin, *Tripneustes gratilla*. Published online 07/17/2009. https://doi.org/10.2478/s11756-009-0135-2/.

Raich, J.W., Schlesinger, W.H., 1992. The global carbon dioxide flux in soil respiration and its relationship to vegetation and climate. Tellus 44 B, 81–99. https://doi.org/10.1034/j.1600-0889.1992.t01-1-00001.x.

Ramanathan, V., 1988. The greenhouse theory of climate change: a test by an inadvertent global experiment. Science 240, 293–299. https://doi.org/10.1126/science.240.4850.293.

Randal, R.G., Minns, C.K., 2000. Use of fish production per unit biomass ratios for measuring the productive capacity of fish habitats. Canadian Journal of Fisheries and Aquatic Sciences 57 (8), 1657–1667. https://doi.org/10.1139/f00-103.

Randerson, J.T., Chapin, F.S., Harden, J.W., Neff, J.C., Harmon, M.E., 2002. Net ecosystem production: a comprehensive measure of net carbon accumulation by ecosystems. Ecological Applications 12 (4). https://pubs.er.usgs.gov/publication/70024418.

Rao, S., Riahi, K., 2006. The role of non-$CO_2$ greenhouse gases in climate change mitigation: long-term scenarios for the 21st century. Multigas mitigation and climate policy. Energy Journal 3 (Special Issue), 177–200. http://pure.iiasa.ac.at/7872.

Raschke, K., 1960. Heat transfer between the plant and the environment. Annual Review of Plant Biology 11, 111–126. https://doi.org/10.1146/annurev.pp.11.0601160.000551.

Reichle, D.E., Auerbach, S.I., 1972. Analysis of ecosystems, pp. 260–280. In: Behnke, J.A. (Ed.), Challenging Biological Problems: Directions Toward Their Solution. Oxford University Press, New York, 502 pp. https://doi.org/10.1126/sc. ience.179.4068.58.

Reichle, D.E., Auerbach, S.I., 2003. U.S. RadioecologicL Research Programs of the Atomic Energy Commission in the 1950s. ORNL/TM-2003/280. Oak Ridge National Laboratory, Oak Ridge, TN. https://digital.library.unt.edu/ark:/67531/metadc892742/m2/1/high_res_d/885597. pdf/.

Reichle, D.E., Crossley, D.A., 1966. Radiocesium dispersion in a cryptozoan food web. In: Hungate, F.P. (Ed.), Radiation and Terrestrial Ecosystems. Pergamon Press, Belfast, Ireland, pp. 1375−1384, Also https://journals.lww.com/health-physics/toc/1965/12000#-2009087835. publ. 1965 in Health Physics, 11:1375−1384.

Reichle, D.E., Crossley, D.A., 1967. Investigations on heterotrophic productivity in forest insect communities. In: Petrusewicz, K. (Ed.), Secondary Productivity of Terrestrial Ecosystems. Institute of Ecology, Polish Academy of Sciences, Warsaw, pp. 563−584, 879 pp. https://scholar.google.com/scholar_lookup?title=Secondary%20productivity%20of%20terrestrial%20ecosystems&publication_year=1967.

Reichle, D.E., Van Hook Jr., R.I., 1970. Radiocesium dynamics in insect food chains. Manitoba Entomologist 4, 22−32. https://www.osti.gov/biblio/4298968.

Reichle, D.E., Dinger, B.E., Edwards, N.T., Harris, W.F., Sollins, P., 1973. Carbon flow and storage in a forest ecosystem. In: Woodwell, G.M., Pecan, E.V. (Eds.), Carbon and the Biosphere, CONF-720510. National Technical Information Service, Springfield, Virginia, pp. 345−365, 392 pp. https://www.biodiversitylibrary.org/bibliography/4036#/summary.

Reichle, D.E., O'Neill, R.V., Olson, J.S. (Eds.), 1973. Modeling Forest Ecosystems. EDFB/IBP 73/7, Oak Ridge National Lab., Oak Ridge, Tenn.

Reichle, D.E., Franklin, J.F., Goodall, D.W. (Eds.), 1975a. Productivity of World Ecosystems. National Academy of Sciences, 166 pp. ISBN 0-309-02317-3. PDF available at:/http://nap. edu/20114/.

Reichle, D.E., O'Neill, R.V., Harris, W.F., 1975. Principles of energy and material exchange in ecosystems. In: Van Dobben, W.H., Lowe-McConnell, R.H. (Eds.), Unifying Concepts in Ecology. W. Junk Pub, The Hague, pp. 27−43, 302 pp. https://link.springer.com/chapter/10. 1007/978-94-010-1954-5_3/.

Reichle, D.E., Trabalka, J.R., Solomon, A.M., 1985. Approaches to studying the global carbon cycle, pp. 15−24. In: Trabalka, J.R. (Ed.), Atmospheric Carbon Dioxide and the Global Carbon Cycle. U.S. Dept. of Energy DOE/ER-0239, Office of Technical and Scientific Information, Oak Ridge, TN, 316 pp. https://www.osti.gov/servlets/purl/6048470.

Reichle, D.E., Houghton, J., Kane, B., Ekman, J., Benson, S., Clarke, J., Dahlman, R., Hendry, G., Herzog, H., Hunter-Cevera, J., Jacobs, G., Judkins, R., Ogden, J., Palmisano, A., Socolow, R., Stringer, J., Surles, T., Wolsky, A., Woodward, N., York, M., 1999. Carbon Sequestration Research and Development. U. S. Dept of Energy, DOE/SC/FE-1, Office of Sci. and Tech. Info, Oak Ridge, TN, USA. www.ornl.gov/carbon_sequestration/.

Reichle, D.E., McBrayer, J.F., Ausmus, B.S., 1975. Ecological energetics of decomposers in a deciduous forest, 283-292. In: Vanek, J. (Ed.), Proc. 5th Internatl. Colloquia of Soil Zoology, Prague, Czechoslovakia, October 1973. Academia Publ. House, Prague, 630 pp. https://link.springer.com/chapter/10.1007/978-94-010-1933-0_31.

Reichle, D.E., 1967. Radioisotope turnover and energy flow in terrestrial Isopod populations. Ecology 48 (3), 351−366. https://doi.org/10.2307/1932671.

Reichle, D.E., 1968. Relation of body size to food intake, oxygen consumption, and trace element metabolism in forest floor arthropods. Ecology 49 (3), 538−542. https://doi.org/10.2307/1934119.

Reichle, D.E., 1969. Measurement of elemental assimilation by animals from radioisotope retention patterns. Ecology 50 (6), 1102−1104. https://doi.org/10.2307/1936907.

Reichle, D.E. (Ed.), 1970. Analysis of Temperate Forest Ecosystems. Springer-Verlag, New York. https://www.worldcat.org/title/analysis-of-temperate-forest-ecosystems/oclc/102449, 304 pp.

Reichle, D.E., 1971. Energy and nutrient metabolism of soil and litter invertebrates. In: Duvigneaud, P. (Ed.), Productivity of Forest Ecosystems. UNESCO, pp. 465−477, 707 pp. https://www.worldcat.org/title/productivity-of-forest-ecosystems-productivite-des-ecosystemes-forestiers-proceedings-of-the-brussels-symposium-organized-by-unesco-and-the-international-biological-programme-27-31-october-1969/oclc/651164.

Reichle, D.E., 1975. Advances in ecosystem science. BioScience 25, 257−264. https://doi.org/10.2307/1296988.

Reichle, D.E. (Ed.), 1981. Dynamic Properties of Forest Ecosystems. Springer Verlag, New York, 683 pp. https://www.worldcat.org/title/dynamic-properties-of-forest-ecosystems/oclc/1025081165.

Reifsnyder, W.E., Lull, H.W., 1965. Radiant energy in relation to forests. USDA Forest Service Technical Bulletin 1334. https://www.coursehero.com/file/pig42g/Reifsnyder-W-E-and-H-W-Lull-1965-Radiant-energy-in-rela-tion-to-forests-United.

Reiners, W.A., 1975. Terrestrial detritus and the carbon cycle. In: Woodwell, G.M., Pecan, E.V. (Eds.), Carbon and the Biosphere. National Technical Information Service, Springfield, Virginia, pp. 303−327, 392 pp. https://www.biodiversitylibrary.org/bibliography/4036#/summary/.

Resplandy, L., Keeling, R.F., Eddebbar, Y., Brooks, M.K., Wang, R., Bopp, L., Long, M.C., Dunne, J.P., Koeve, W., Oschlies, A., 2018. Quantification of ocean heat uptake from changes in atmospheric $O_2$ and $CO_2$ composition. Nature 563, 105−108. https://www.nature.com/articles/s41586-018-0651-8.

Revelle, R., Suess, H., 1957. Carbon dioxide exchange between atmosphere and ocean and the question of an increase of atmospheric $CO_2$ during the past decades. Tellus 9, 18−27. https://doi.org/10.1111/j.2153-3490.1957.tb01849.x.

Rhein, M., et al., 2013. In: Stocker, T.F., et al. (Eds.), Climate Change 2013: AR4, Chapter 3 the Physical Science Basis. Contribution of Working Group I to the Fifth Assessment Report of the Intergovernmental Panel on Climate Change. Cambridge Univ. Press, pp. 215−315. https://www.ipcc.ch/report/ar5/wg1/.

Riahi, K., Gruebler, A., Nakicenovic, N., 2007. Scenarios of long-term socio-economic and environmental development under climate stabilization. Technological Forecasting and Social Change 74 (7), 887−935. https://doi.org/10.1016/j.techfore.2006.05.026.

Rich, N., August 1, 2018. Losing Earth: the Decade We Almost Stopped Climate Change, we knew everything we needed to know, and nothing stood in our way. Nothing, that is, except ourselves. A tragedy in two acts. New Yorker Magazine. https://www.nytimes.com/by/nathaniel-rich/.

Richardson, C.J., Dinger, B.E., Harris, W.F., 1973. The Use of Stomatal Resistance, Photopigments, Nitrogen, Water Potential and Radiation to Estimate Net Photosynthesis in Liriodendron Tulipifera L. − a Physiological Index. In: Eastern Deciduous Forest Biome IBP-72-13. Oak Ridge National Laboratory.

Rigby, M., Prinn, R.G., Fraser, P.J., Simmonds, P.G., Langenfelds, R.L., Huang, J., Cunnold, D.M., Steele, L.P., Krummel, P.B., Weiss, R.F., O'Doherty, S., Salameh, P.K., Wang, H.J., Harth, C.M., Muehle, J., Porter, L.W., 2008. Renewed growth of atmospheric methane. Geophysical Research Letters 35, L22805. https://doi.org/10.1029/2008GL036037.

Riley, G.A., 1972. Patterns of production in marine ecosystems. In: Wiens, J.A. (Ed.), Ecosystem Structure and Function. Oregon State University Press, Corvallis, OR, pp. 91−112, 176 pp. https://science.sciencemag.org/content/176/4033/398.

Ripley, B.S., Pammenter, N.W., Smith, V.R., 1999. Function of leaf hairs revisited: the hair layer on leaves *Arctotheca populifolia* reduces photoinhibition, but leads to higher leaf temperatures caused by lower transpiration rates. Journal of Plant Physiology 155 (1), 78−85. https://doi.org/10.1016/S0176-1617(99)80143-6.

Ritchie, H., Roser, M., 2017. $CO_2$ and Other Greenhouse Gas Emissions. https://ourworldindata.org/co2-and-other-greenhouse-gas-emissions/.

Rockström, J., Gaffney, O., Rogelj, J., Meinshausen, M., Nakicenovic, N., Schellnhuber, H.J., 2017. A roadmap for rapid decarbonization. Science 355 (6331), 1269−1271. https://doi.org/10.1126/science.aah3443.

Rodin, L.E., Bazilevich, N.I., Rozov, N.N., 1975. Productivity of the world's main ecosystems, pp. 13−26. In: Reichle, D.E., Franklin, J.F., Goodall, D.W. (Eds.), Productivity of World Ecosystems. The National Acadamies Press, Washington, DC, 166 pp. http://nap.edu/20114.

Romm, J., 2008. https://thinkprogress.org/the-decarbonization-story-and-why-a-carbon-price-beats-technology-breakthroughs-44d9c4c28bb4/.

Rosenbloom, E., 2007. A Problem with Wind Power: Opposing Viewpoints to what Energy Sources Should Be Pursued. Greenhaven Press. http://www.aweo.org/windIPCC.html/.

Rotmans, J., den Elzen, M.G.J., 1993. Modelingfeedback mechanisms in the carbon cycle: balancing the carbon budget. Tellus 45B, 301−320. https://doi.org/10.1034/j.1600-0889.1993.t01-3-00001.x.

Rowell, A., June 18, 2018. http://priceofoil.org/2018/06/18/thirty-years-on-how-jim-hansen-was-proved-right-on-climate-change/.

Rubey, W.W., 1951. Geological history of sea water. Bull. Geol. Soc. Amer. 62 (9), 1111−1148. https://www.academia.edu/4895570/RUBEY_Willian_W._Geologic_History_of_Sea_Water._Geological_Society_of_America_v._62_no._9_p._1111-1148_1951.

Ruesch, A., Gibbs, H.K., 2008. New IPCC Tier-1 Global Biomass Carbon Map for the Year 2000. The Carbon Dioxide Information Analysis Center, Oak Ridge National Laboratory, Oak Ridge, Tennessee. Available online from: http://cdiac.ess-dive.lbl.gov/.

Ryszkowski, L., 1975. Energy and matter economy of ecosystems. In: Van Dobben, W.H., Lowe-McConnell, R.H. (Eds.), Unifying Concepts in Ecology. W Junk, Publ., The Hague, pp. 109−126, 302 pp. https://link.springer.com/chapter/10.1007/978-94-010-1954-5_3.

Sabine, C.L., Feely, R.A., Gruber, N., Key, R.M., Lee, K., Bullister, J.L., Wanninkhof, R., Wong, C.S., Wallace, D.W.R., Tilbrook, B., Millero, F.J., Peng, T.-H., Kozyr, A., Ono, T., Rios, A.F., 2004. The Ocean Sink for Anthropogenic $CO_2$. U.S. Dept. of Commerce/NOAA/OAR/PMEL/Publications.

Salas, R.N., Knappenberger, P., Hess, J., 2018. 2018 Lancet Countdown on Health and Climate Change Brief for the United States of America. https://apha.org/-/media/files/pdf/topics/climate/2018_us_lancet_countdown_brief.ashx?la=en&hash=99279F373B9F005C9EC364AB02EB7F636F1380CF/.

Salisbury, F.B., Ross, C., 1969. Plant Physiology. Wadsworth Publ. Co., Inc., Belmont, CA, 504 pp. https://books.google.com/books/about/Plant_Physiology.html?id=sdtqsT0OLf4C.

Sanderman, J., Creamer, C., Baisden, W.T., Farrell, M., Fallon, S., 2017. Greater soil carbon stocks and faster turnover rates with increasing agricultural productivity. Soil 3, 1−16. https://doi.org/10.5194/soil-3-1-2017.

Sanders, D., Moser, A., Newton, J., van Veen, F.J.F., 2016. Trophic assimilation efficiency markedly increases at higher trophic levels in four-level host−parasitoid food chain. Proceedings of the Royal Society B 283 (1626). https://doi.org/10.1098/rspb.2015.3043.

Sarmiento, J.L., Gruber, N., 2006. Ocean Biogeochemical Dynamics. Princeton University Press, 528 pp. https://press.princeton.edu/books/hardcover/9780691017075/ocean-biogeochemical-dynamics.

Sarmiento, J.L., Orr, J.C., Siegenthaler, U., 1992. A perturbation simulation of $CO_2$ uptake in an ocean general circulation model. Journal of Geophysics Research 97, 3621−3645. https://doi.org/10.1029/91JC02849.

Satchell, J.E., 1971. Feasibility study of an energy budget for Meathop. In: Duvigneaud, P. (Ed.), Productivity of Forest Ecosystems. UNESCO, Paris, pp. 619−630, 707 pp. https://www.worldcat.org/title/productivity-of-forest-ecosystems-productivite-des-ecosystemes-forestiers-proceedings-of-the-brussels-symposium-organized-by-unesco-and-the-international-biological-programme-27-31-october-1969/oclc/651164.

Satchell, J.E., 1973. Biomass model of a mixed oak forest, United Kingdom. In: Reichle, D.E., O'Neill, R.V., Olson, J.S. (Eds.), Modeling Forest Ecosystems. Oak Ridge National Laboratory, Oak Ridge, TN, USA. EDFB-IBP-72-7.

S.C. etv, April 4, 2019. Sea Change.

Schimel, D.S., Braswell Jr., J.B.H., Holland, E.A., McKeown, R., Ojima, D.S., Painter, T.H., Parton, W.J., Townsend, A.R., 1994. Climatic, edaphic and biotic controls over storage and turnover of carbon in soils. Global Biogeochemical Cycles 8, 279−293. https://doi.org/10.1029/94GB00993.

Schimel, D., Enting, I.G., Heinman, M., Wigley, T.M.L., Raynaud, D., Alves, D., Siegenthaler, U., 2005. $CO_2$ and the carbon cycle. In: Wigley, T.M.L., Schimel, D.S. (Eds.), The Carbon Cycle. Cambridge Univ. Press, Cambridge, UK, pp. 7−36, 292 pp. http://catdir.loc.gov/catdir/samples/cam031/00023735.pdf.

Schimel, J., Basler, T.C., Allenstein, M.W., 2007. Microbial stress-response physiology and its implications for ecosystem function. Ecology 88 (6), 1386−1394. https://doi.org/10.1890/06-0219.

Schlesinger, W.H., Bernhardt, E.S., 2013. Remote sensing of primary production and biomass. In: Biogeochemistry, third ed. Academic Press. 688 pp. https://sites.duke.edu/biogeochemistry2015/files/2015/08/3rd-edition-BGC-through-Ch-2.pdf.

Schlesinger, W.H., 1977. Carbon balances in terrestrial detritus. Annual Review of Ecology, Evolution, and Systematics 8, 51−81. https://doi.org/10.1146/annurev.es.08.110177.000411.

Schlesinger, W.H., 1984. Soil organic matter: a source of atmospheric $CO_2$, pp. 111−127. In: The role of terrestrial vegetation in the global carbon cycle. John Wiley & Sons, New York, NY, USA. https://www-legacy.dge.carnegiescience.edu/SCOPE/SCOPE_23/SCOPE_23_0.1_title pages.pdf.

Schlesinger, W.H., 1986. Changes in soil carbon storage and associated properties with disturbance and recovery, pp. 194−220. In: Trabalka, J.R., Reichle, D.E. (Eds.), The Changing Carbon Cycle: A Global Analysis. Springer Verlag, New York, USA, 592 pp. https://link.springer.com/book/10.1007/978-1-4757-1915-4.

Schlesinger, W.H., 1997. Biogeochemistry: An Analysis of Global Change. Academic Press, 588 pp. https://sites.duke.edu/biogeochemistry2015/files/2015/08/3rd-edition-BGC-through-Ch-2.pdf.

Schmidt, G., October 8, 2018. PBS Newshour. https://www.facebook.com/newshour/posts/10156845070218675/.

Schuur, E.A., Bockheim, G.J., Canadell, J.G., Euskirchen, E., Field, C.B., Goryachkin, S.V., Hagemann, S., Kuhry, P., Lafleur, P.M., Lee, H, Mazhitova, G., Nelson, F.E., Rinke, A., Romanovsky, V.E., Shiklomanov, N., Tarocai, C., Venevsky, S., Vogel, J.G., Zimov, S.A.,

2008. Vulnerability of permafrost carbon to climate change: implications for the global carbon cycle. BioScience 58 (8), 701−714. https://doi.org/10.1641/B580807/.

Schuur, E., Vogel, J., Crummer, K., Lee, H., Sickman, J., Osterkamp, T., 2009. The effect of permafrost thaw on old carbon release and net carbon exchange from tundra. Nature 459 (7246), 556−559. https://www.nature.com/articles/nature08031.

Scientists and Experts from Universities and Institutions in the Great Lakes Region 2019. An Assessment of the Impacts of Climate Change on the Great Lakes. Environmental Law and Policy Center. http://elpc.org/wp-content/uploads/2019/03/Great-Lakes-Climate-Change-Report.pdf.

Seitzinger, S.P., Harrison, J.A., Dumont, E., Beusen, A.H.W., Bouwman, A.F., 2005. Sources and delivery of carbon, nitrogen, and phosphorus to the coastal zone: An overview of Global Nutrient Export from Watersheds (NEWS) models and their application. Global Biogeochemical Cycles 9 (4). https://doi.org/10.1029/2005GB002606.

Semtner, A., 2005. Very high resolution estimates of global ocean circulation suitable for carbon cycle modeling. In: Wigley, T.M.L., Schimel, D.S. (Eds.), The Carbon Cycle. Cambridge Univ. Press, Cambridge, UK, pp. 212−228, 292 pp. http://catdir.loc.gov/catdir/samples/cam031/00023735.pdf.

Sévellec, F., Drijfhout, S.S., 2018. A novel probabilistic forecast system predicting anomalously warm 2018-2022 reinforcing the long-term global warming trend. Nature Communications 9 (1), 1−12. https://www.nature.com/articles/s41467-018-05442-8.

Shahidan, M.F., Salleh, E., Mustafa, K.M.S., 2007. Effects of tree canopy on solar radiation filtration in a tropical microclimatic environment. In: Proc. 24th Conference on Passive and Low Energy Architecture, Singapore, 22-24, Nov. 2007. PLEA 2007, pp. 400−406. https://ukm.pure.elsevier.com/en/publications/effects-of-tree-canopies-on-solar-radiation-filtration-in-a-tropi.

Shindell, D.T., et al., 2012. Simultaneously mitigating near-term climate change and improving human health and food security. Science 335, 183−189. https://doi.org/10.1126/science.1210026.

Shugart, H.H., Reichle, D.E., Edwards, N.T., Kercher, J.R., 1976. A model of calcium cycling in an East Tennessee Liriodendron forest: model structure, parameters, and frequency response analysis. Ecology 57, 99−109. https://doi.org/10.2307/1936401.

Siegenthaler, U., Joos, F., 1992. Use of a simple model for studying oceanic tracer distributions and the global carbon cycle. Tellus 44B, 186−207. https://doi.org/10.1034/j.1600-0889.1992.t01-2-00003.x.

Siegenthaler, U., Sarmiento, J.L., 1993. Atmospheric carbon dioxide and the ocean. Nature 365 (6442), 119−125. https://doi.org/10.1038/365119a0.

Sinsabaugh, R.L., Findley, S., 1995. Microbial production, enzyme activity, and carbon turnover in surface sediments of the Hudson River estuary. Microbial Ecology 3 (2), 127−141. https://doi.org/10.1007/BF00172569.

Slobdkin, L.B., Richman, S., 1961. Calories/gm in species of animals. Nature 191, 299. https://www.nature.com/articles/191299a0.

Smalley, A.E., 1959. The growth cycle of *Spartina* and its relation to the insect population in the marsh, pp/96-100. In: Proc. Salt Marsh Conf. At Sapelo Island, Georgia, 1958. Publ. Marine Inst. of Univ. Ga, Athens. http://www.archive.org/stream/proceedingssaltm00salt/proceedingssaltm00salt_djvu.txt.

Smalley, A.E., 1960. Energy flow of a salt marsh grasshopper. Ecology 41 (4), 672−677. https://doi.org/10.2307/1931800.

Smil, V., 2017. Energy Transitions: Global and National Perspective & BP Statistical Review of World Energy. OurWorlfibData.org/fossil-fuels/CC BY-SA.

Smith, T.M., Shugart, H.H., 1993. The transient response of terrestrial carbon storage in a perturbed climate. Nature 361, 523−526. https://www.nature.com/articles/361523a0.

Smith, S.J., Wigley, T.M.L., 2006. Multi-Gas forcing stabilization with the MiniCAM. Energy Journal 373−391. https://doi.org/10.2307/23297091.

Smith, F.E., 1968. The international biological program and the science of ecology. Proceedings of the National Academy of Sciences of the United States of America 60 (1), 5−11. https://www.ncbi.nlm.nih.gov/pmc/articles/PMC539127/.

Solhaug, K.A., Gauslaa, Y., Nybakken, L., Bilger, W., 2003. New Phytologist 158 (1), 91−100. https://doi.org/10.1046/j.1469-8137.2003.00708.x.

Solomon, A.M., Trabalka, J.R., Reichle, D.E., Voorhees, L.D., 1985. The global cycle of carbon. In: Trabalka, J.R. (Ed.), Atmospheric Carbon Dioxide and the Global Carbon Cycle. U.S. Dept. of Energy DOE/ER-0239, Office of Technical and Scientific Information, Oak Ridge, TN, pp. 1−14, 316 pp. https://www.osti.gov/servlets/purl/6048470/.

Speakman, J.R., Keijer, J., 2013. Measuring energy metabolism in the mouse − theoretical, practical, and analytical considerations. Frontiers in Physiology. https://doi.org/10.3389/fphys.2013.00034.

Spencer, R.-J., Thompson, M.B., Hume, D., 1998. Comparative Biochemistry and Physiology Part A: Molecular & Integrative Physiology 121 (4), 341−349. https://www.sciencedirect.com/journal/comparative-biochemistry-and-physiology-part-a-molecular-and-integrative-physiology/vol/121/issue/4.

Spinage, C., 1994. Elephants. T & AD Poyser Ltd., London, UK. https://www.worldcat.org/title/elephants/oclc/32968164.

Stanhill, G., 1970. The water flux in temperate forests: precipitation and evapotranspiration. In: Reichle, D.E. (Ed.), Analysis of Temperate Forest Ecosystems. Springer Verlag, Berlin, Heidelberg, New York, pp. 247−256, 303 pp. https://www.worldcat.org/title/analysis-of-temperate-forest-ecosystems/oclc/102449.

Statistica, 2019. Wind and Hydropower Production Statistics. https://www.statista.com/topics/2577/hydropower/.

Statistica, 2019b. https://www.statista.com/statistics/268363/installed-wind-power-capacity-worldwide/.

Stephens, G.L., Li, J., Wild, M., Clayson, C.A., Loeb, N., Kato, S., L'Ecuyer, T., Stackhouse, P.W., Lebsock, M., 2012. An update on Earth's energy balance in light of the latest global observations. Nature Geoscience 5 (10), 691−696. https://www.nature.com/articles/ngeo1580.

Stern, N., Peters, S., Bakhshi, V., Bowen, A., Cameron, C., Catovsky, S., Crane, D., Cruickshank, S., Dietz, S., Edmonson, N., Garbett, S.-L., Hamid, L., Hoffman, G., Ingram, D., Jones, B., Patmore, N., Radcliffe, H., Sathiyarajah, R., Stock, M., Taylor, C., Vernon, T., Wanjie, H., Zenghelis, D., 2006. Stern Review: The Economics of Climate Change. HM Treasury, London, 579 pp. https://onlinelibrary.wiley.com/doi/abs/10.1111/j.1728-4457.2006.00153.x/.

Stokes, G.M., Banard, J.C., 1986. Presentation of the $20^{th}$ century atmospheric $CO_2$ record in Smithsonian spectrographic plates. In: Trabalka, J.R., Reichle, D.E. (Eds.), The Changing Global Carbon Cycle: A Global Analysis. Springer Verlag, New York, pp. 5−65, 592 pp. https://link.springer.com/book/10.1007/978-1-4757-1915-4.

Stuiver, M., Burk, R.L., Quay, P.D., 1984. 13C/12C ratios and the transfer of biosphereic carbon to the atmosphere. Journal of Geophysical Research 89 (11), 731−748. https://doi.org/10.1029/JD089iD07p11731.

Stuiver, M., 1986. Ancient carbon cycle changes derived from tree-ring $^{13}$C and $^{14}$C. In: Trabalka, J.R., Reichle, D.E. (Eds.), The Changing Carbon Cycle: A Global Analysis. Springer Verlag, New York, pp. 109−124, 592 pp. https://link.springer.com/book/10.1007/978-1-4757-1915-4.

Sturman, A.P., Taper, N.J., 1996. The Weather and Climate of Australia and New Zealand. Oxford Univ. Press, 476 pp. https://catalogue.nla.gov.au/Record/3550680.

Suh, R., 2019. Natural Resources Defense Council. www.nrdc.org/.

Sverdup, H.V., Johnson, M.W., Flemning, R.H., 1942. The Oceans: Their Physics, Chemistry and General Biology. Prentice-Hall, New York, NY, USA, 1087 pp. http://ark.cdlib.org/ark:/13030/kt167nb66r.

Szent-Györgi, A., 1961. Introductory Comments, pp. 7−10. In: McElroy, W.D., Glass, B. (Eds.), Light and Life. Johns Hopkins Press, Baltimore, MD, USA. https://www.questia.com/library/1441301/a-symposium-on-light-and-life, 944 pp.

Taiz, L., Zeiger, E., 2010. Plant Physiology, 5th Edition. Sinauer Associates Inc., Sunderland, 782 pp. https://www.sinauer.com/media/wysiwyg/tocs/PlantPhysiology5.pdf.

Takahara, T., Minamoto, T., Yamanka, H., Doi, H., Kawabata, Z., 2012. Estimation of fish biomass using environmental DNA. PLoS One 7 (4), e35868. https://doi.org/10.1371/journal.pone.0035868/.

Tan, Z., Zhang, Y., Yu, G., Sha, L., Tang, J., Deng, X., Song, Q., 2010. Carbon balance of a primary tropical seasonal rain forest. Journal of Geophysical Research 115, D00H26. https://doi.org/10.1029/2009JD012913/.

Tang, J., Miller, P.A., Persson, A., Olefeldt, D., Pilesjö, P., Heliasz, M., Jackowicz-Korczynski, M., Yang, Z., Smith, B., Callaghan, T.V., Christensen, T.R., 2015. Carbon budget estimation of a subarctic catchment using a dynamic ecosystem model at high spatial resolution. Biogeosciences Discuss 12, 933−980. www.biogeosciences-discuss.net/12/933/2015/.

Tarnocai, C., Canadell, J.G., Schuur, E.A.G., Kuhry, P., Mazhitova, G., Zimov, S., 2009. Soil organic carbon pools in the northern circumpolar permafrost region. Global Biogeochemical Cycles 23, GB2023. https://doi.org/10.1029/2008GB003327.

Taylor, K.E., Stouffer, R.J., Meehl, G.A., 2012. An overview of CMIP5 and the experiment design. Bulletin of the American Meteorological Society 93, 485−498. https://doi.org/10.1175/BAMS-D-11-00094.1.

Taylor, F.G., 1969. Phenological Records of Vascular Plants at Oak Ridge, Tennessee. Oak Ridge National Laboratory, 46 pp. ORNL-IBP-69-1, U.S. DOE, (ref. # 4752457). www.OSTI.gov/.

Taylor, F.G., 1974. Phenodynamics of production in a mesic decisuous forest. In: Leith, H. (Ed.), Phenology and Seasonality Modeling, Ecological Stidies, vol. 8. Springer Verlag, Berlin-Heidelberg-New York, pp. 237−254, 444 pp. https://www.springer.com/gp/book/9783642518652.

Teal, J.M., 1962. Energy flow in the salt marsh ecosystem of Georgia. Ecology 43, 614−624. https://doi.org/10.2307/1933451.

Talapia Territory. http://houseoftilapia.blogspot.com/2008/10/feeding-behaviour-and-nutrition.html/.

Than, K., 2019. Cyclone, hurricane, typhoon: What's the Difference? Retrieved from. https://www.nationalgeographic.com/news/2019/6/130923-typhoon-hurricane-cyclone-primer-natural-disaster.

The Biomimicry Institute, 2018. The boundary layer between a leaf and its environment. https://asknature.org/strategy/small-leaves-buffer-ambient-temperature/.

The Economist, August 18, 2018. p. 66. https://www.economist.com

The Economist, August 25, 2018. p. 22.

The Economist, September 15, 2018. p. 68.

The Economist, September 22, 2018. p. 54.

The Economist, October 13, 2018. pp. 14, 25, 76.

The Economist, November 10, 2018. pp. 12, 29.

The Economist, December 1, 2018. pp 4, 5,11, 22.

The Economist, December 8, 2018. p. 75.

The Economist, January 5, 2019, pp. 10-13.

The Economist, March 30, 2019. pp.10, 81.

The Economist 11, August 17, 2019, 15−18.

The Economist, August 2, 2019 pg. 45.

The Economist, July 6, 2019, p. 24.

The Nature Conservancy, 2015. https://nature4climate.org/.

The Shift Project, 2010. Breakdown of GHG Emissions. Creative Commons. http://www.tsp-data-portal.org/Breakdown-of-GHG-Emissions-by-Sector-and-Gas#tspQvChart/.

The Stern Review and the Economics of Climate Change: An Editorial, 2019. https://www.researchgate.net/publication/225718755_The_Stern_Review_and_the_Economics_of_Climate_Change_An_Editorial_Essay.

Toi, R.S.J., 2018. The economic impacts of climate change. Rev. Environ. Economics and Policy 12 (1), 4−25. https://doi.org/10.1093/recp/rex027.

Trabalka, J.R., Reichle, D.E. (Eds.), 1986. The Changing Carbon Cycle: A Global Analysis. Springer Verlag, New York, 592 pp. https://link.springer.com/book/10.1007/978-1-4757-1915-4.

Trabalka, J.R. (Ed.), 1985. Atmospheric carbon dioxide and the global carbon cycle. DOE/ER-0239, Office of Technical and Scientific Information, Oak Ridge, TN, U.S. Dept. of Energy DOE/ER-0239, 315 pp. https://www.osti.gov/servlets/purl/6048470.

Trojan, P., 1967. Investigation on production of cultivated fields. In: Petrusewicz, K. (Ed.), Secondary Productivity of Terrestrial Ecosystems (Principles and Methods), vol. II. Polish Academy of Sciences, Warsaw, pp. 545−563, 879 pp. https://scholar.google.com/scholar_lookup?title=Secondary%20productivity%20of%20terrestrial%20ecosystems&publication_year=1967.

Trumbore, S.E., Harden, J.W., 1997. Accumulation and turnover of carbon in organic and mineral soils of the BOREAS northern study area. Journal of Geophysical Research 102 (D24). https://pubs.er.usgs.gov/publication/70019222.

Tucker, C.J., Townsend, J.R.G., Goff, T.E., Holben, B.N., 1986. Continental and global scale remote sensing of land cover. In: Trabalka, J.R., Reichle, D.E. (Eds.), The Changing Global Carbon Cycle: A Global Analysis. Springer Verlag, New York, pp. 221−241, 592 pp. https://link.springer.com/book/10.1007/978-1-4757-1915-4.

U. S. Environmental Protection Agency, 2019. Understanding Lake Ecology. https://cfpub.epa.gov/watertrain/pdf/limnology.pdf/.

U.S. Department of Energy. U.S. DOE, 2008. Carbon Cycling and Biosequestration: Report from the March 2008 Workshop. DOE/SC-108, U.S. Department of Energy Office of Science. Prepared by the Biological and Environmental Research Information System, Oak Ridge National Laboratory. genomicscience.energy.gov.

U.S. Environmental Protection Agency, 2012. The Greenhouse Effect. https://commons.wikimedia.org/wiki/File:Earth%27s_greenhouse_effect_(US_EPA,_2012).png.

U.S. Environmental Protection Agency, 2019. Understanding Lake Ecology. Retrieved 23 July 2019. https://cfpub.epa.gov/watertrain/pdf/limnology.pdf.

U.S. Global Change Research Program, Washington, DC, USA.2018 186 pp./https://www.globalchange.gov/nca4/; full report online:/nca2018.globalchange.gov/.

U.S. National Oceanic and Atmospheric Administration, National Weather Service. https://www.weather.gov/jetstream/energy.

UCAR, 2018. International Panel on Climate Change. IPCC AR4 WG1, cited by. https://scied.ucar.edu/imagecontent/carbon-cycle-diagram-ipcc/.

Union of Concerned Scientists (UCS), 2009. Clean Power Green Jobs. https://www.ucsusa.org/sites/default/files/legacy/assets/documents/clean_energy/Clean-Power-Green-Jobs-25-RES.pdf/.

Union of Concerned Scientists, 2011. Climate Hot Map. https://www.climatehotmap.org/global-warming-effects/drought.html/.

Union of Concerned Scientists, 2017. Benefits of Renewable Energy Use. https://www.ucsusa.org/clean-energy/renewable-energy/public-benefits-of-renewable-power/.

United Nations Climate Change, 2019. History of the United Nations Framework Convention on Climate Change. In: https://unfccc.int/process/the-convention/history-of-the-convention/.

United Nations Climate Change, 2019. History of the United Nations Framework Convention on Climate Change. In: https://unfccc.int/process/the-convention/history-of-the-convention.

Urbina, M.A., Glover, C.N., 2013. Relationship between fish size and metabolic rate in the oxyconforming inanga Galaxias maculatus reveals size-dependent strategies to withstand hypoxia. Physiological and Biochemical Zoology 86 (96), 740−749. https://doi.org/10.1086/673727.

USGCRP, 2009. Global Climate Change Impacts in the United States: 2009 Report. U.S. Global Change Research Program, Washington, DC, USA. https://nca2009.globalchange.gov.

USGCRP. 2014. Ch. 6: agriculture. Climate change impacts in the United States, by Hatfield, J., G. Takle, R. Grotjahn, P. Holden, R. C. Izaurralde, T. Mader, E. Marshall, and D. Liverman, pp.150-174. In, The Third National Climate Assessment, J. M. Melillo, T. C. Richmond, and G. W. Yohe, Eds., U.S. Global Change Research Program, Washington, DC, USA,https://nca2009.globalchange.gov

USGCRP, 2017. Climate science special report: fourth national climate assessment, vol. 1. In: Wuebbles, D.J., Fahey, D.W., Hibbard, K.A., Dokken, D.J., Stewart, B.C., Maycock, T.K. (Eds.), U.S. Global Change Research Program, Washington, DC, USA, 470 pp. science2017. globalchange.gov/;ful_report online:/nca2018.globalchange.gov.

USGCRP, 2018. Impacts, Risks, and Adaptation in the United States. In: Reidmiller, D.R., Avery, C.W., Easterling, D.R., Kunkel, K.E., Lewis, K.L.M., Maycock, T.K., Stewart, B.C. (Eds.), Report-In-Brief: Fourth National Climate Assessment, Volume II. U.S. Global Change Research Program, Washington, DC, USA, 186 pp. https://www.globalchange.gov/nca4/; full report online:/nca2018.globalchange.gov/.

van der Werf, G.R., Randerson, J.T., Giglio, L., van Leeuwen, T.T., Chen, Y., Rogers, B.M., Mu, M., van Marle, M.J.E., Morton, D.C., Collatz, G.J., Yokelson, R.J., Kasibhatla, P.S., 2017. Global fire emissions estimates during 1997−2016. Earth System Science Data 9, 697−720. https://doi.org/10.5194/essd-9-697-2017/.

Van Kooten, G.C., Grainger, A., Ley, E., Marland, G., Solberg, B., 1997. Conceptual issues related to carbon sequestration: uncertainty and time. Critical Reviews in Environmental Science and Technology 27 (special), S65−S82. https://doi.org/10.1080/10643389709388510.

van Vuuren, D.P., Eickhout, B., Lucas, P.L., den Elzen, M.G.J., 2006. Long-term multi-gas scenarios to stabilise radiative forcing — exploring costs and benefits within an integrated assessment framework. Multigas mitigation and climate policy. Energy Journal 3 (Special Issue), 201−234. https://ideas.repec.org/a/aen/journl/2006se_weyant-a10.html.

van Vuuren, D., den Elzen, M., Lucas, P., Eickhout, B., Strengers, B., van Ruijven, B., Wonink, S., van Houdt, R., 2007. Stabilizing greenhouse gas concentrations at low levels: an assessment of reduction strategies and costs. Climatic Change 81, 119−159. https://doi.org/10.1007/s10584-006-9172-9/.

VEMAP members, 1995. Vegetation/ecosystem modeling and analysis project: comparing biogeography and biochemistry models in a continental-scale study of terrestrial ecosystem responses to climate change and $CO_2$ doubling. Global Biogeochemical Cycles 9, 407−437. https://harvardforest.fas.harvard.edu/sites/harvardforest.fas.harvard.edu/files/publications/pdfs/VEMAP_GlobalBiogeochemicalCycles_1995.pdf.

Vernberg, F.J., Vernberg, W.B., 1970. The Animal and the Environment. Holt, Reinheart and Winston, 398 pp. https://www.amazon.com/Animal-Its-Environment-John-Vernberg/dp/0030796458.

Vitousek, P.M., Ehrlich, P.R., Ehrlich, A.H., Matson, P.A., 1986. Human appropriation of the products of photosynthesis. BioScience 36, 368−373.

Vollenweider, R., 1969. Primary Production in Aquatic Environments. IBP Handbook No. 12. Blackwell Scientific Publications, Oxford, UK, 213 pp. https://doi.org/10.1002/mmnz.19720480213.

Wang, J., Sun, J., Xia, J., He, N., Li, M., Niu, S., 2017. Soil and vegetation carbon turnover from tropical to boreal forests. British Ecological Society 32 (1), 71−82. https://doi.org/10.1111/1365-2435.12914.

Weart, S., 2008. The Discovery of Global Warming. Harvard University Press, Cambridge, MA, 240 pp. https://www.hup.harvard.edu/catalog.php?isbn=9780674031890.

Weart, S., 2012. The discovery of global warming. Scientific American 307 (2). August 17. https://www.scientificamerican.com/article/discovery-of-global-warming.

Wei, M., Jones, C., Greenblatt, J., McMahon, J., 2011. Carbon reduction potential from behavior change in future energy systems. In: Behavior, Energy and Climate Change Conference. Lawrence Berkeley National Laboratory. http://web.stanford.edu/group/peec/cgi-bin/docs/events/2011/becc/presentations/1%20The%20Importance%20and%20Potential%20-%20Max%20Wei.pdf/.

Weigert, R.G., Mitchell, R., 1973. Ecology of Yellowstone thermal effluent systems: intersects of blue-green algae, grazing flies (*Paracoenia*, Ephydridae) and water mites (*Partnuniella*, Hydrachnellae). Hydrobiologia 41 (2), 251−271. https://doi.org/10.1007/BF00016450.

Weiss, J.F., Landauer, M.R., 2009. History and development of radiation-protective agents. International Journal of Radiation Biology 85 (7), 539−573. https://doi.org/10.1080/09553000902985144.

Wetzel, R.G., Rich, P.H., 1973. Carbon in freshwater systems. CONF-720510. In: Woodwell, G.M., Pecan, E.V. (Eds.), Carbon and the Biosphere. National Technical Information Office, Springfield, VA, pp. 241−263, 392 pp. https://www.biodiversitylibrary.org/bibliography/4036#/summary/.

Weyant, J., Azar, C., Kainuma, M., Kejun, J., Shukla, P.R., Nakicenovic, N., La Rovere, E., Yohe, G., April 2009. Report of 2.6 versus 2.9 Watts/m2 RCPP Evaluation Panel (PDF). IPCC Secretariat, Geneva, Switzerland. http://www.globalchange.umd.edu/iamc_data/docs/RCPP-Report.pdf.

White, G.C., 1978. Estimation of plant biomass from quadrat data using the lognormal distribution. Journal of Range Management 31 (2), 118−120. https://doi.org/10.2307/3897657.

Whitehouse, S., December 30, 2018. Opinion letter. NY Times.

Whitfield, D.W.A., 1972. Systems analysis. In: Bliss, L.C. (Ed.), High Arctic Ecosystem, Devon Island IBP Project. Dept. of Botany, Univ. of Alberta, pp. 392–409, 413 pp.

Whittaker, R.H., Likens, G.E., 1973. Carbon in the biota. In: Woodwell, G.M., Pecan, E.V. (Eds.), Carbon and the Biosphere. National Technical Information Service, Springfield, Virginia, pp. 281–302, 392 pp. https://www.biodiversitylibrary.org/bibliography/4036#/summary/.

Whittaker, R.H., Marks, P.L., 1975. Methods of Assessing Terrestrial Productivity, pp. 55–118. In: Lieth, H., Wittaker, R.H. (Eds.), Productivity of the Biosphere. Springer Verlag, Berlin-Heidelberg-New York, 339 pp. https://www.springer.com/gp/book/9783642809156.

Whittaker, R.H., 1975. Communities and Ecosystems. Macmillan, New York and London, 385 pp. https://www.biblio.com/communities-and-ecosystems-by-whittaker-robert-h/work/354508.

Wiegert, R.G., 1964. Population energetics of meadow spittlebugs (*Philaenus spumarius, L.*) as affected by migration and habitat. Ecological Monographs 34, 217–241. https://doi.org/10.2307/1948501.

Wiegert, R.G., 1965. Energy dynamics of the grasshopper populations in old field and alfalfa field ecosystems. Oikos 16, 161–176. https://doi.org/10.2307/3564872.

Wiegert, R.G., 1976. In: Ecological Energetics. Dowden, Hutchinson and Ross, Inc., Stroudsburg, PA, 455 pp. https://www.worldcat.org/title/ecological-energetics/oclc/2077851.

Wielgolaski, F.E., 1975. Productivity of tundra ecosystems, pp. 1–12. In: Reichle, D.E., Franklin, J.F., Goodall, D.W. (Eds.), Productivity of World Ecosystems. The National Acadamies Press, Washington, DC, 166 pp. http://nap.edu/20114.

Wigley, T.M.I., Schimel, D.S. (Eds.), 2005. The Carbon Cycle. Cambridge Univ. Press., Cambridge, 292 pp. http://catdir.loc.gov/catdir/samples/cam031/00023735.pdf.

Wigley, T.M.L., 2005. The climate change commitment. Science 307 (5716), 1766–1769. https://doi.org/10.1126/science.1103934.

Wigley, T.M.L., 2005. Stabilization of $CO_2$ concentration levels. In: Wigley, T.M.I., Schimel, D.S. (Eds.), The Carbon Cycle. Cambridge Univ. Press., Cambridge, pp. 258–276, 292 pp. http://catdir.loc.gov/catdir/samples/cam031/00023735.pdf.

Wikimedia Commons contributors, 2019. All forcing agents $CO_2$ equivalent concentration. In: Wikipedia, The Free Encyclopedia. Retrieved from. https://commons.wikimedia.org/wiki/File:All_forcing_agents_CO2_equivalent_concentration.png.

Wikimedia Commons, 2011. Ecological pyramids. Wikipedia, The Free Encyclopedia. Retrieved from. https://commons.wikimedia.org/wiki/File:EcologicalPyramids.jpg.

Wikimedia Commons, 2014. Chloroplast-new.jpg. In: Wikipedia, The Free Encyclopedia. Retrieved from. https://commons.wikimedia.org/w/index.php?title=File:Chloroplast-new.jpg&oldid=142062821.

Wikimedia Commons, 2014. Trophic Level Pyramid. https://commons.wikimedia.org/wiki/File:Trophic_levels.jpg/ /https://commons.wikimedia.org/wiki/File:EcologicalPyramids.jpg.

Wikimedia Commons, 2017. Climate Influence on Terrestrial Biome. In Wikipedia, The Free Encyclopedia. https://commons.wikimedia.org/w/index.php?title=File:Climate_influence_on_terrestrial_biome.svg&oldid=252182132/.

Wikimedia Commons, 2018. Summary of Anaerobic Respiration: The Metabolic Pathway of Glycolysis. Image Credit: ScienceGal4.0 - Own Work. CC BY-SA 4.0, Retrieved 19:14, July 23, 2019, from. https://en.wikipedia.org/w/index.php?title=Glycolysis&oldid=906168029/.

Wikimedia Commons, 2019. The Electromagnetic Spectrum. In Wikipedia, The Free Encyclopedia. Retrieved July 23, 2019, from. https://en.wikipedia.org/wiki/File:EM_spectrum.svg/.

Wikimedia Contributors, 2019. The Krebs Cycle. In Wikipedia, The Free Encyclopedia, Retrieved from. https://simple.wikipedia.org/w/index.php?title=Krebs_cycle&oldid=6475531/.

Wikimedia Contributors, 2018. Cellular Respiration. The Free Encylopedia. Retrieved 7 February 2019 from. https://en.wikipedia.org/w/index.php?title=Cellular_respiration&oldid=901951109/.

Wikipedia Commons, 2019. Food web. In: Wikipedia, the Free Encyclopedia. Retrieved 16:28, July 23, 2019, from. https://en.wikipedia.org/w/index.php?title=Food_web&oldid=899759843/.

Wikipedia Commons, 2019. Radiative Forcing, In Wikipedia, The Free Encyclopedia. Retrieved July 25, 2019, from. https://en.wikipedia.org/w/index.php?title=Radiative_forcing&oldid=906557276/.

Wikipedia Contributors, 2018. Q10 Temperature Coefficient Plot. In Wikipedia, The Free Encyclopedia. Retrieved 7 February 2019 from. https://en.wikipedia.org/wiki/File:Q10Temperature CoefficientPlot.svg/.

Wikipedia Contributors, 2019. Food web. In: Wikipedia, The Free Encyclopedia. Retrieved from 2019. https://en.wikipedia.org/w/index.php?title=Food_web&oldid=899759843.

Wikipedia Contributors, 2019. Citric acid cycle. In: Wikipedia, The Free Encyclopedia. Retrieved from. https://en.wikipedia.org/wiki/File:Citric_acid_cycle_noi.JPG.

Wikipedia Contributors, 2019. Keeling Curve. In: Wikipedia, The Free Encyclopedia. Retrieved from. https://en.wikipedia.org/w/index.php?title=Keeling_Curve&oldid=905651841.

Wikipedia contributors, 2019. Representative Concentration Pathway. In: Wikipedia, The Free Encyclopedia. Retrieved from. https://en.wikipedia.org/w/index.php?title=Representative_ Concentration_Pathway&oldid=905406108.

Wikipedia Contributors, 2019. Life-cycle greenhouse-gas emissions of energy sources. In: Wikipedia, The Free Encyclopedia. Retrieved from. https://en.wikipedia.org/wiki/Life-cycle_ greenhouse-gas_emissions_of_energy_sources.

Wikipedia Contributors, 2015. Photorespiration. In Wikipedia, The Free Encyclopedia. Retrieved February 7, 2019 from. https://en.wikipedia.org/w/index.php?title=Photorespiration &oldid=891785302.

Wikipedia Cintributors, 2018. Photosynthetic Efficiency. In Wikipedia, The Free Encyclopedia. https://en.wikipedia.org/wiki/Photosynthetic_efficiency/.

Wikipedia Contributors, 2018. Carbon Cycle. In Wikipedia, The Free Encyclopedia. Retrieved 3 June 2019 from. https://en.wikipedia.org/wiki/Carbon_cycle/.

Wikipedia Contributors, 2018. Watermellon Snow. In Wikipedia, The Free Encyclopedia. Retrieved 4 February 2019 from. https://en.wikipedia.org/w/index.php?title=Watermelon_snow&oldid= 854721545/.

Wikipedia Contributors, 2018. Calorimeter. In Wikipedia, The Free Encyclopedia. Retrieved 20 February 2019 from. https://en.wikipedia.org/w/index.php?title=Calorimeter&oldid= 860513268/.

Wikipedia Contributors, 2019. Laws of Thermodynamics. In Wikipedia, The Free Encyclopedia. Retrieved 25 July 2019 from. https://en.wikipedia.org/w/index.php?title=Laws_of_ thermodynamics&oldid=888134088/.

Wikipedia Contributors, 2019. Ocean Acidification. In Wikipedia, The Free Encyclopedia. Retrieved 23 July 2019 from. https://en.wikipedia.org/w/index.php?title=Ocean_ acidification&oldid=877963376/.

Wikipedia, 2019. Solar Irradiance. In Wikipedia, The Free Encyclopedia, Retrieved February 7, 2019 from. https://en.wikipedia.org/wiki/Solar_irradiance#Irradiance_on_Earth's_surface/. https://en.wikipedia.org/wiki/File:EM_spectrum.svg/.

Wise, M.A., Calvin, K.V., Thomson, A.M., Clarke, L.E., Bond-Lamberty, B., Sands, R.D., Smith, S.J., Janetos, A.C., Edmonds, J.A., 2009. Implications of limiting $CO_2$ concentrations for land use and energy. Science 324, 1183−1186. https://doi.org/10.1126/science.1168475.

Wolf, L.L., Hainsworth, F.R., 1971. Time and energy budgets of territorial hummingbirds. Ecology 52 (6), 980−988. https://doi.org/10.2307/1933803.

Woodall, P.F., Currie, G.J., 1989. Food consumption, assimilation and rate of food passage in the cape rock elephant shrew, *Elephantulus edwardii* Macroscelidea: Macroscelidinae). Comparative Biochemistry and Physiology Part A: Physiology 92 (Issue 1), 75−79. https://doi.org/10.1016/0300-9629(89)90744-5.

Woodward, F.I., Smith, T.M., 2005. Impacts of climate and $CO_2$ on the terrestrial carbon cycle. In: Wigley, T.M.I., Schimel, D.S. (Eds.), The Carbon Cycle. Cambridge Univ. Press., Cambridge, pp. 248−257, 292 pp. http://catdir.loc.gov/catdir/samples/cam031/00023735.pdf.

Woodwell, G.M., Bodkin, D.B., 1970. Metabolism of terrestrial ecosystems by gas exchange techniques: the Brookhaven approach. In: Reichle, D.E. (Ed.), Analysis of Temperate Forest Ecosystems. Springer-Verlag, New York, pp. 73−85, 304 pp. https://www.worldcat.org/title/analysis-of-temperate-forest-ecosystems/oclc/102449.

Woodwell, G.M., Whittaker, R.H., 1968. Primary production in terrestrial communities. American Zoologist 8, 19−30. https://www.jstor.org/stable/3881529.

Woodwell, G.M., Rich, P.H., Hall, C.A.S., 1973. Carbon in estuaries. In: Woodwell, G.M., Pecan, E.V. (Eds.), Carbon and the Biosphere. National Technical Information Service, Springfield, Virginia, pp. 221−240, 392 pp. https://www.biodiversitylibrary.org/bibliography/4036#/summary/.

World Bank, 2017. Global Solar Atlas. https://globalsolaratlas.info/. Global Map of Global Horizontal Radiation Image by SOLARGIShttps://solargis.com/maps-and-gis-data/download/world/.

World Energy Council, 2019. https://www.worldenergy.org/data/resources/resource/hydropower/.

World Resources Institute, 2005. Worldwide Greenhouse Gas Emissions 2005. https://www.wri.org/resources/charts-graphs/world-greenhouse-gas-emissions-2005/.

World Resources Institute, 2013. Rain Forest Clearing. Brazil's National Space Research Institute. https://www.wri.org/our-work/project/governance-forests-initiative/brazil/.

World Resources Institute, 2018. New Global $CO_2$ Emissions Numbers Are in: They're Not Good. Retrieved. https://www.wri.org/blog/2018/12/new-global-co2-emissions-numbers-are-they-re-not-good/.

Xiang, J., Wenhui, Z., Guobiao, H., Huiqing, G., 1993. Food intake, assimilation efficiency, and growth of juvenile lizards *Takydromus septentrionalis*. Comparative Biochemistry and Physiology Part A: Physiology 105 (2), 283−285. https://doi.org/10.1016/0300-9629(93)90209-M.

Yang, H., 2019. Soybean Growth Simulation Model, Overview Diagram. Institute of Agriculture and Natural Resources, University of Nebraska−Lincoln. https://soysim.unl.edu/soysim_overview_diagram.shtml.

Yousif, M.O., Abdul, M.K., Abdul-Rahman, A., 2014. Growth performance, feed efficiency and carcass composition of African catfish, *Clarias gariepinus* (Pices:Clariidae) fingerlings fed diets composed of agricultural by-products. International Journal of Biosciences 4 (10), 276−284. http://www.innspub.net.

Zaho, C., Liu, B., Piao, S., Wang, X., Lobell, D.B., Huang, Y., Huang, M., Yao, Y., Bassu, S., Ciais, P., Durand, J.-L., Elliott, J., Ewert, F., Janssens, I.A., Li, T., Lin, E., Liu, Q., Martre, P., Müller, C., Peng, S., Peñuelas, J., Ruane, A.C., Wallach, D., Wang, T., Wu, D., Liu, Z., Zhu, Y., Zhu, Z., Asseng, S., 2017. Temperature increase reduces global yields of major crops in four independent estimates. Proceedings of the National Aacademy of Sciences of the USA 114 (35), 9326−9331. https://doi.org/10.1073/pnas.1701762114.

Zheng, J., Thornton, P.E., Painter, S.L., Gu, B., Wollschleger, S.D., Graham, D.E., 2019. Modeling anaerobic soil organic carbon decomposition in Arctic polygon tundra: insights into soil geochemical influences on carbon mineralization. Biogeosciences 16, 663−680. https://doi.org/10.5194/bg-16-663-2019/.

Zinke, P.J., Stangenburger, A.G., Post, W.M., Emanuel, W.R., Olson, J.S., 1984. Worldwide Organic Soil Carbon and Nitrogen Data (ORNL/TM-8857). Oak Ridge National Laboratory, Oak Ridge, TN. https://daac.ornl.gov/SOILS/guides/zinke_soil.html.

# Author Index

# Subject Index